T0073868

Flavorama

Flavorama

**A Guide to Unlocking
the Art and Science of Flavor**

Arielle Johnson

HARVEST
An Imprint of WILLIAM MORROW

FLAVORAMA. Copyright © 2024 by Arielle Johnson. Foreword by René Redzepi. All rights reserved. Printed in Thailand. No part of this book may be used or reproduced in any manner whatsoever without written permission except in the case of brief quotations embodied in critical articles and reviews. For information, address HarperCollins Publishers, 195 Broadway, New York, NY 10007.

HarperCollins books may be purchased for educational, business, or sales promotional use. For information, please email the Special Markets Department at SPsales@harpercollins.com.

FIRST EDITION

Designed by Tai Blanche

Library of Congress Cataloging-in-Publication Data has been applied for.

ISBN 978-0-358-09313-8

24 25 26 27 28   IMG   10 9 8 7 6 5 4 3 2 1

To Tom—for everything

Contents

Foreword

By René Redzepi

For chefs, flavor drives everything we do. If the food doesn't taste good, we've failed at our jobs, and all that hard work is rendered pointless (or is it?).

This is something we have grappled with intimately on a daily basis at our restaurant, Noma, in Copenhagen. We started out twenty years ago with a set of rules: make a cuisine that doesn't just highlight Nordic ingredients—it uses them almost exclusively. Over the years, this idea has changed as I've wanted to put in some of my Albanian/Turkish food upbringing and to work with some of the highlights of the travels that we have done. Now used commonly are kombu from Japan and chiles from Mexico. We still work every day trying to express Nordic seasonality as a general guiding light.

This guidance is easy to follow in the spring and summer, when the days are long and we can get our hands on the most delicious white asparagus, strawberries, and a myriad of wild herbs. In the winter, when the only vegetables available from our farmers are beets, carrots, and potatoes from their root cellars, it's a considerable limitation. Early on, it became obvious that we had to make a choice: treat this limitation like an albatross around our necks, or put it to work as a creative constraint.

Cooking the way I'd been trained—French-style, formal, with ingredients like foie gras and olive oil—was not an option in those early Noma years. So we had to take a step back and consider

new avenues for deliciousness. Slowly but surely, this led us to preserving, extracting, fermenting, foraging, and considering the culinary potential of things we hadn't thought of before as tasty or even edible: weird, finicky, and "lost" varieties of vegetables, seaweeds, wild weeds, flowers, mosses, and even insects.

In those early days, we asked ourselves, Does a dish really need a lemon, or is it simply a certain kind of sourness that's missing? Could it be tangy wood sorrel leaves, plump wood ants, sea buckthorn berries, or plums that we fermented like sauerkraut? Now repeat that, ad absurdum, for every ingredient you *think* is essential in a dish.

It wasn't how any of us had been taught to cook. We had to amass the experience and the knowledge ourselves; a quest for understanding flavor that has led us to do many ridiculous and impractical things. Looking back at it now, it seems so crazy that we spent all those hours harvesting ants or cleaning moss. But we had to. It was our way.

It was 2012, in the middle of all this research, that I first met Arielle Johnson. Now she would be the first to say that she's not a chef. She was in fact a graduate student studying food chemistry, and rather than spending the summer getting her foot in the door for a high-paying job at Nestlé or something, she decided it was a better use of her time to come and do research with a bunch of cooks at a lab run by a restaurant.

Why have a scientist on staff? Here's a story: we'd just started to get interested in edible insects, and Lars Williams, who was then the head of our collaborative project the Nordic Food Lab, was researching how ants communicate using pheromones. It was Arielle, with simply a glance at their chemical structures, who explained to us that those pheromones were also flavor molecules. That they had specific tastes she could predict for us without having to taste them, like grapefruit, lemongrass, and lavender. Flavors that we could now bring into the kitchen without having to use grapefruit, lemongrass, or lavender.

Good cooking is organized chaos, and a chef responds to changing conditions and ingredients every hour. For instance, consider peas in season: Are they harvested and never cooled down—i.e., still warm from the sun? Or have they spent a day or two in the fridge? That will make the same ingredient taste vastly different—one pea might taste starchy and bitter, while the other as tender and sweet as a peeled grape. Why is it that the flavor of horseradish, a spicy root growing underground, can vary so much from day to day? While we know, of course, that farming practices and weather itself have a huge impact on these things, the more we understand that the variability goes down to the molecular level—and how—the better we can put our taste and intuition to work in our cooking.

To understand flavor inside and out, top to bottom, is nothing short of a lofty goal. We were lucky enough to have someone as eager as Arielle to make the connection between what we had discovered through intuition and taste and what she had learned about flavor through science and years of study. After she finished school, the now Dr. Arielle Johnson came to Noma full-time, where she and Lars built our original Fermentation Lab from scratch (literally, from shipping containers) so that we could have a dedicated home for doing this kind of research. Over the years, the fermentation lab has gone through several iterations, becoming the driving force at the heart of everything we do.

Having Arielle around is like a secret sauce that makes everything a bit clearer, and flow easier. She's a legit scientist, but she thinks like a chef, breaking down knowledge into exactly what you need to be more aware of flavor, where it comes from, and all the ways you can create and control it—regardless of whether you're cooking at a restaurant or cooking at home.

Flavorama is a book that every cook, no matter how experienced or science-literate they are, needs in their kitchen, written by someone who knows more about the subject than anyone I know. Arielle changed the way that I think about flavor—if I had this book when we started on our quest twenty years ago at Noma, it would have saved us a lot of time and trouble. But, lucky for you, you're reading it right now, and in these pages, she'll do the same for you.

Acknowledgments

A hearty thanks to:

Kim and Jessica and the rest of the team at Inkwell, the best in the business.

Deb, Rux, Sarah, Jacqueline, and the team at Harvest/HarperCollins, for the expert attention, love, and benefit of the doubt lavished on this long (but rewarding) project as it gradually coalesced into being. This book could not have happened without you. Thank you.

Katie and Kate, angels of mercy (and organization), the best of collaborators.

On the industry side of things, especially those who have been white-knuckling it with me along the whole journey—Nastassia and Dave, Ioanna, Vaughn, Pam, Jorge, Angela, Victoria, Tienlon, Gillian, Alex and Bella, Hiro, Caitlin, Santiago, Ian, Nick, Shannon, Alex, Francis and Bronwen. And particularly to Lars: nobody has had more influence on how I think about flavor—creatively, ontologically, epistemologically. Thank you for that, for seeing something in my offhand remarks in a lab on a houseboat about the smell of insect volatiles, and most of all for your support and encouragement to put these ideas out into the world.

For advice, collaboration, and/or patience back in the day, David, Ferran, Jose, Chris, Mandy, Daniel.

Also: Brooks, who I'm not sure is aware, kicked off this whole thing in 2015 when he casually asked me, "So when is *your* book coming out?" as if that were something I was capable of— plus Sheryl, Darcy, and the rest of the Superiority Burger extended universe; Diego, Fabian and Jeremiah, Danny, Anthony, Karen; Andrew and Adam and the team at Smallhold. For the vibes, ideas, memes, emotional support, and keeping me sane—Jena, Priya, Elizabeth, Helen, John, Tim, Stefani, Ananda, Rita, Carina, Hector, Moth, and Lesley. And thank you, Alton, for all your support.

For support and inspiration from the land of drink, the extended Lyan family, Dave, and Ryan; Dave, Lance, and the rest at St. George Spirits; Don, Kyle, Aaron, Audrey, Ria, Orlando; the teams at Gen Yamamoto and Ben Fiddich; Alice and Hana, Raphael and Honey's, Yesfolk, Gyu+, Endorffeine, and Empirical.

For their peerless help with research: Namae-san, Yukari, Nancy, Hye Joon Michelle, Mina and Kwang, the teams at Higuchi Moyashi, Iio Jozo, Jook Jang Yeon, Murakami Jyu Honten, and Onjium.

My academic mentors, colleagues, friends—Kent, Amy, Anne, Fabio, Krishnendu, Marion, and everyone from the NYU ECC days; from UC Davis, Hildegarde, Sue, Roger, Andrew, Alyson, Doug, Wender, Michael, Line, Anna, Maya, Ellie, Helene; Claudia, Hildreth, Gordon, Andreas, Diego, Lynda and Becky, Bridget, and Celine at/around MIT; Mike and Rosemary at Drexel; David at UC Riverside; Christina and Nick, Jason, Ben, Pia, Ben, Rachel, Lane, Rob, Josh, Mark, Maria, Dana, Paul, John, Austin, Andreas, and Sanjay. And especially to Harold, for being a constant inspiration, taking me seriously when I was nobody, and always knowing just the right thing to say when I needed it.

My friends and role models in media: Helen, Nicola, Cynthia, Christine, Danielle, Mike, Lisa, Evelyn, Nadia, Devita, Paul, Matt, Lisa, Hillary, Maya, Samin, Nik, Paula, Sarah, Tamar, Darra, Christopher, Corby, Mitchell, Greg, Dana, Paz, Ness, and June.

My Noma extended family, thank you for taking a chance on me, and creating a place where these ideas could incubate and thrive: René and Nadine, Peter, Thomas, Rosio, Annika, Devin, Nate, Traci, Melina, Arve, Jack, Gabe, Ben, David, Jason, Jan, and Kristian.

And of course, my literal family: Muggie, the initial enabler of my flavor obsession; Adriana, Margot, Julian, and Sam; Eban; Mom and Blair; Dad and Miren; Alice, Pete, Grace, Jack, Joe; Alison and Andy, and most of all, to Tom—it would take a whole book just to scratch the surface on all the ways you support and encourage me. I'm so lucky to have you.

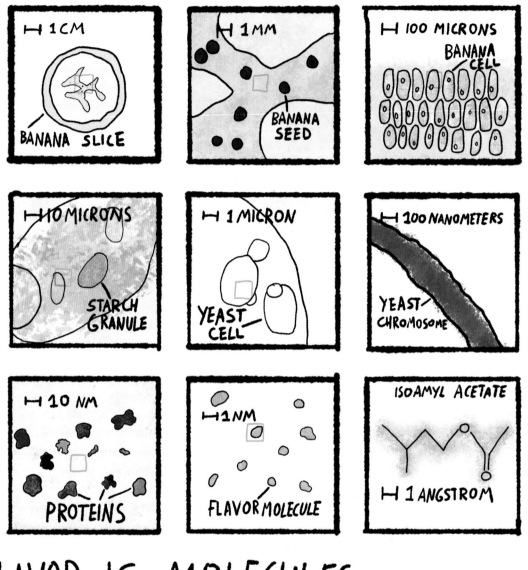

FLAVOR IS MOLECULES

tiny amounts of matter make up everything you can taste and smell

The Unbridled Science of Flavor

And How to Get It to Work for You

This is a book about flavor.

You know, the thing that drives us to drop serious money on heirloom tomatoes. The reason we don't just subsist on Soylent. The town where Guy Fieri lives.

When flavor is really good—an absolutely perfect, sweet-sour-perfumed nectarine, an intensely minerally and beefy grass-fed steak, some overwhelmingly aromatic, freshly ground spices—it's enough to give me chills. If you're reading this book, you probably feel the same way. It's become a truth (nearly) universally acknowledged that cooking is science—but I've always hungered for more focus, more knowledge about the most important part: what food actually tastes like, and the *how* and *why* behind it. I don't mean what the objectively "perfect" chocolate chip cookie is, or fifteen hacks for making February asparagus taste less like February, or the engineering behind Cool Ranch Doritos. I want to know everything that goes into making real, delicious food real and delicious—and then, how to use that knowledge to make it even better.

For me, it's never been enough to just have great flavor experiences—I have this need to dig into how they work, the deep mechanisms that create them. I'm so obsessed with understanding flavor that I got a Ph.D. in the subject, studying it with analytical chemistry (I'm technically Dr. Johnson, in academic settings—and to my mother). In the process, I learned that there's a whole cosmos of flavor knowledge out there, in chemistry, microbiology, psychology, ecology, neuroscience, ethnobotany, even economics and history. And there's leagues more fascinating flavor science happening in any great croissant, traditional barbacoa, or organic blueberry than in the most hyperengineered, flavor-optimized packaged food.

Some cooks hear "science" and they think "straitjacket": oh, great, someone calling themselves an objective authority, here to squeeze the life out of creativity and soul with a bunch of rules about what's "correct."

I find that as tedious as you do. I'd also be lying if I told you that there was some scientific way to bypass ever running into problems while cooking. (Sorry.)

Fortunately, science is remarkably useful for something much more fun: if it's exciting to know what you're doing in the kitchen, it's positively elating to know *why* you're doing what you're doing—and how you can do it smarter. Not just being certain that a certain recipe, followed correctly, will get you where you want to go, but understanding the underlying mechanisms well enough that you can predict how whatever technique you're using or action you're taking will affect the flavors of your ingredients, and adjust on the fly. It's why I've spent so much of my time working in restaurants, using the science I know to chase deliciousness, rather than uniformity. Flavor science as a liberating tool for improvising.

The place that I like to start? **Molecules**. Just trust me—

Flavor Is Molecules

Each of our senses specializes in one kind of information to paint a picture of some important aspect of the world around us. Vision detects the physical details of our environment through reflected light. Hearing lets us monitor our surroundings for things that we can't see, but that *do* create mechanical vibrations in the air, which we perceive as sounds. Flavor is not its own distinct sense, but a composite of our senses of taste and smell, the two senses devoted to detecting molecules directly. Your sense of taste specializes in molecules that touch your tongue; your sense of smell, on molecules that can travel through the air.

Flavor is a personal, emotional thing (that's why there's not really such a thing as a "perfect" chocolate chip cookie; it's subjective)—but all that starts from sensing real, physical molecules. Everything else about it cascades from the mechanistic things that molecules do.

Flavor is molecules, which means we can use chemistry to answer some of your most common kitchen questions. Why does lemon juice taste so much less sour if I don't wait until the end of cooking to add it? What makes a sprinkle of sea salt so much different than table salt? Why is cooked ginger so much less spicy than fresh? What's really happening that makes caramelizing onions so much slower than caramelizing sugar? Why are the spiciness of chile oil, hot sauce, and fresh chile so different? What gives an amazing peach so much more mouth-filling creamy flavor than a just-okay one?

The mechanisms these follow aren't particularly mystical or secret. There's a huge body of scientific work out there covering every conceivable angle of flavor, but it's hardly ever brought together in one place, let alone put into the context of cooking, or even shared with cooks at all.

So I've specialized in trying to fix this, helping chefs and cooks understand what's going on beneath the flavor hood so they can work even more intuitively and deliciously. I get asked a lot of flavor questions, and to decipher them, I rummage around in the deep cuts of the scientific literature, then figure out how to translate what I find into terms a cook can actually use. Around the thousandth time doing this, I finally noticed that I kept saying, "Wow, I really wish there was some kind of guidebook I could give you that sums this up in one place." Which is what this book is: a condensed version of all the amazing stuff we know about how flavor works and how we can maximize it.

While I can't pretend I've managed to cover absolutely *all* of flavor here (that would take about a million pages), I can promise three things: a strong foundation in flavor's scientific mechanisms, nuggets of scientific wisdom I've dug up from the archives that tend to get glossed over elsewhere, and a focus on how all that science is being used, right now, in the real world—to notice patterns, enhance cooking intuition, and inform creativity. Everything here, from the overall structure to the individual recipes, is shaped by how I've actually used the science of flavor in cooking—and how really great cooks use it, too. Hence, the title. Think of this as a panoramic look on all things flavor.

Flavorama is organized around five core laws. These are predictable features that hold true over and over, the basic rules of the road for flavor.

0. **Flavor is molecules.**
1. **Flavor is taste and smell.**
2. **Flavor follows predictable patterns.**
3. **Flavor can be concentrated, extracted, and infused.**
4. **Flavor can be created and transformed.**

As with other phenomena that follow basic laws (thermodynamics, black holes), there is one law that is so fundamental that it shapes and informs how all the other laws work. This law is numbered "number 0," as in, it's even more important than number 1. In the case of flavor, we've already touched on the 0th law: flavor is molecules.

Flavor isn't just ineffable vibes, or an abstract quality that imbues a lemon with lemoniness. Absolutely everything required to set off the wonderful perceptual experience of flavor in your brain is contained within the food you are eating, in the form of molecules. Like you and everything around you, food is made from molecules. Some give it strength and texture, some give it moisture and bounce, and some give it flavor.

Chemistry is the grand project of hunting down and describing all these molecules. We know how fast they evaporate (useful if you're

trying to get them to stick around in a dish), whether they break apart under heat or acid (useful if you're trying to make a broth, sauce, tea, or cocktail or add some oomph to butter), what happens when they are fermented (useful if you want to make kimchi, sauerkraut, or miso). People who are really literate in chemistry can do this just by looking at how a molecule's atoms are arranged, like a fortune-teller reading your heart line and predicting when you'll get married—except, actually real and accurate.

The good news is that all these things follow patterns and trends. Sour flavors all come from acids, which do predictable things. Aromas are all volatile, which means they will float away when you heat them. Flavor infusions will always be different, in predictable ways, when using fat vs. using water. You don't actually need to memorize long tables of precise values to make use of the science of flavor—you just need to pay attention to spotting patterns. It's an enhancement, not a replacement, for using your senses and intuition while eating and cooking.

The fact that flavor is molecules underpins this whole book: where those molecules come from, how they were made, and how to get them to do what we want. Those molecules are what we sense when we sense flavor; they're what lets us infuse and concentrate and create flavors. Separately, chemistry and your sensory experience of flavor are a mysteriously tangled piece of rubber and an insubstantial puff of air. Together, they're an intricate balloon animal. There is no flavor without molecules, *and* the tastes and smells created by flavor molecules are what make knowing anything about those molecules interesting to a cook in the first place. The two are inescapably important to each other. But let's not get ahead of ourselves. If we're going to understand flavor, let's start with the first law: understanding taste and smell.

PART 1

The First Law of Flavor

Flavor Is Taste and Smell

When you're eyeing your friend's plate at a restaurant and want to know if it's any good, what do you ask them?

"How does it taste?" We talk *all* the time about how things taste—do they taste good, do they taste bad, do they taste like vanilla or dirt or raspberries or cardboard?

The thing is, most of the times we talk about "taste," what we're actually talking about is **flavor**. From a scientific point of view, and a culinary one, too, taste is *necessary* for flavor, but it's not *sufficient* for flavor on its own. That's because taste is just a part of flavor.

You only get flavor when you combine taste *and smell*. Smell is just as important, sometimes more so, in this equation. That's not the way it feels, certainly. Your smell sensations are so seamlessly integrated with taste while you're eating that they don't feel like smell at all. But even if you don't feel like you have the nose of a master perfumer or sommelier, anytime you eat, you're using smell to sense flavor.

So You Want to Tell Apart Taste and Smell?

If you've ever had congestion from a bad cold—or experienced the smell-obscuring symptoms of COVID-19—you know that nothing tastes good when you have a stuffy nose. It's blah, it's one-note, it's boring. Usually, though, your sense of taste is working fine. The "bland taste" you're experiencing is just taste—flavor without smell. Smell molecules can't reach your smell receptors through your blocked nose, your brain gets no smell data, and the flavor you experience, devoid of smell, feels turned down and flat.

Taste is salty, sour, sweet, umami, and bitter. Just about everything else is smell: the clove-floral flavors of basil, the fruity or caramel notes in varietal coffee, the creamy greenness of a cucumber or honeydew melon, the jammy-winey quality of a great raspberry, the roasted flavor of a chicken, the piney-woody-cozy flavor of cardamom. I could go on and on (and I kind of do, starting on page 123).

Understanding flavor means understanding taste and smell, and I think the best way to understand taste and smell is a two-pronged approach: paying attention to what your senses tell you, what tastes and smells you experience; and knowing what's going on under the hood—what it is about how taste or smell function that creates those sensations. I think of it as a robust give-and-take between theory and practice.

Taste and Smell: Complementary Counterparts

I talk a lot about patterns being important, so here's a useful one: there's a lot of yin and yang to the functions of taste and smell. If something is true of taste, the opposite is probably true of smell:

- Tastes are single sensations, and only a few types of molecules can cause each one. Smells are multidimensional sensations, created by many different molecules.
- There are a very limited number of tastes, but it's fundamental for the flavor of your food to get them right. There are a near-infinite number of smells, but there are many different ways they can be used in a dish and still taste good.
- Taste has simple overall mechanisms we can understand quickly. Smell has much more complicated mechanisms that take time to understand.

- Taste is for getting very clear information about a few things. Smell is for getting fuzzier information, but about lots and lots and lots of things.
- Taste is close to hardwired into us. Smell is mostly shaped by experience and memory.
- Individual tastes each have unique and specific uses that we need to go deep on. We get much more from looking at broad patterns in smell, rather than trying to catalogue each one individually.
- Changing or substituting tastes—sweet for salty, sour for bitter—makes huge, even catastrophic differences in flavor. Changing or substituting smells makes noticeable, but more impressionistic, differences in flavor.
- Equally true of taste and smell: they respond to molecules.

Taste and Smell: Sensation Created from Molecules

Flavor is taste and smell—but no food actually contains taste, or smell, or flavor in the literal sense; foods contain molecules. Your nose and tongue function as detectors for these molecules, catching them and sending signals to your brain. Only then does your brain cook up your perception of taste, smell, and flavor from those signals. In other words, technically flavor is entirely a figment of your perception, built in your brain from signals sent to it by molecules from the natural world.

Chapter 1

Taste

We'll start with taste: stalwart, tidy, bold strokes, straightforward to categorize, and easy to understand. There are five options: salty, sour, sweet, umami, and bitter.

Let's Taste Some Molecules

Let me set the scene: It's July, you're sweaty but not unpleasantly so, and you just bought (or, if you're lucky, picked) some good tomato specimens, maybe some big Brandywines, and they're still warm from the sun. You cut off a fat tomato slice and eat it with a sprinkle of salt.

What happens next? You taste salt, obviously. Next, you might pick up on a pleasant tang of sourness, just enough sweetness to be noticeable, maybe a thread of bitterness if you bit down on some seeds, and, as you're swallowing, the mouth-filling richness of umami. Start to finish, it's pretty great, and you remember why you look forward to this every year.

All those tastes from the tomato—salty, sour, sweet, bitter, and umami—don't just come from the ineffable and intrinsic tomatoey-ness of the tomato. No tomato, or anything else, contains "sweetness" or "sourness." What it does have is taste molecules, molecules with the power to coax your brain into producing the sensation of

sweetness or sourness. Any foods that have a taste have *that* taste because they've got the right molecules.

Let's take a closer look at that tomato. What is a tomato, really? Looking at it like a chemist, we'd say, well, it's made of molecules, and the majority of those are dietary fiber and water. They're combined in some parts like a thick gel, and in

others like a sponge soaked in juice, all encased in a balloonlike skin. The tomato's water-based juice has a cocktail of other molecules dissolved in it. Some are acids or sugars; there are also amino acids and minerals.

When you bite into the tomato, its juice (and these other molecules dissolved in the juice) spread all over your tongue and mix with your saliva. And all over your tongue, you have taste receptors, which are like catcher's mitts for taste molecules. Each taste receptor is uniquely molded to grab on to a specific molecular shape and, once it grabs one, send off a signal that gets relayed to your brain. Your taste receptors grab on to the sugars, acids, amino acids, minerals, and tannins from the tomato and send off messages to make your brain create the sensation of sweetness, sourness, umami, etc. just for you.

OLD (WRONG) TASTE BUDS MAP

NEW + IMPROVED TASTE BUDS MAP

MSG'd Zucchini Carpaccio

MSG is pure umami, which can be hard to spot if you're not used to looking for it. It will make the silky crunch of zucchini feel superrich and intense-tasting, and heightens saltiness and sweetness while cutting down bitterness.

Cut **1 medium to small tender zucchini** into thin rounds on a slight diagonal, between ⅛ inch and 1/16 inch thick. Sprinkle very lightly with **a small pinch (perhaps ⅛ teaspoon/0.5 g max) of Aji-no-moto Monosodium Glutamate crystals or other powdered MSG.** Let it soak in for a few minutes, then immediately eat with your hands.

Makes 1 snack

What Molecules Have Tastes, and Why?

Like people, molecules have personalities. Taste molecules are all pretty solid and steady. They don't do anything wild like create vapors or gases: if you boil down a pan of seawater, the salt molecules stick around in the pan, rather than escaping with the steam. They all dissolve well in water, which makes sense, given the watery environment your tongue is constantly bathed in.

On the tongue, you have millions of finely tuned, grabby catcher's-mitt taste receptors that only come in a few different varieties. Think capsule wardrobe, not a hundred-look haute couture show. The way each receptor functions is the result of the long endless process of evolution, adaptation, and ruthless editing.

Taste or otherwise, a receptor's whole purpose is to find and grab a particular type of molecule and send off a useful message about it to the brain. Taste receptors are primed to grab common molecules from food, the thing you most regularly put inside your mouth. And because taste is the last piece of information you can get about something before you choose to swallow it or spit it out, the messages that taste carries are strong and simple.

Sweet taste receptors grab sugars, telling you if a food has easy sources of energy. **Salty** notes the presence of the essential mineral sodium, which the body needs for generating nerve impulses and maintaining blood pressure and fluid-electrolyte balance. **Umami** signals free amino acids, the basic building block for proteins. **Sour** responds to acids, which lets you know if a fruit is underripe, has a lot of useful vitamin C, or is safely fermented. And **bitter** alerts you to potential toxins, many of them made by plants. The list of molecules you can accidentally poison yourself with is pretty long and pretty diverse, so bitter receptors come in something like twenty varieties to cover the whole field, like interchangeable socket wrench heads. Each bitter receptor sends its information along the same bitter channel, making most bitter molecules taste more or less the same. The other tastes stick to a simpler, specialized receptor.

Researchers are finding that there are probably some other things we can sense with taste, like carbonation and fats; and spiciness is a major component of flavor that feels like taste but is actually sensed with touch, as pain (more on that on page 108). Those minor details aside, there are only five tastes to learn to recognize and to look out for when you're tasting food, five clear ways to categorize ingredients by taste, and five elements of flavor you can add to your cooking through taste.

Too Salty?

Okay, there is one other cool exception: your tongue actually generates two different signals for sodium, a "regular salty" one and another "too salty" one that only gets activated at really high salt concentrations. Your sense of saltiness tells you "oh good, salt" or "whoa, too much of a good thing," like the near-acidic burn of undiluted soy sauce or licking the outside of a pretzel.

Tasty Delights

- Taste detects five distinct things: sweet, sour, salty, bitter, and umami.
- Each taste evolved to detect one type or genre of molecule: sugars, acids, sodium, toxins, glutamate.
- Taste molecules usually have just one primary taste quality—salts are salty, sugars are sweet, and there aren't salty-tasting sugars.
- Each taste is a separate signal, with some tastes amplifying or suppressing one another's signals, like umami increasing the signal for salty, or salty decreasing the signal for bitter.

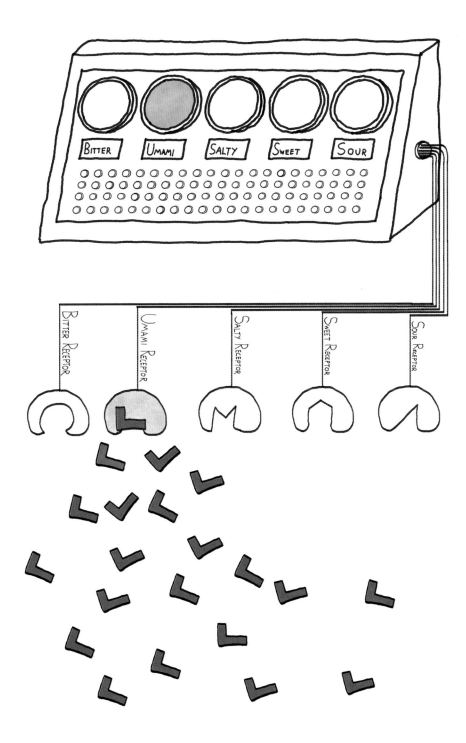

Chapter 2

Smell

Smell, odor, scent, and aroma are all interchangeable words for the same thing—and I'm using all four as shorthand for "a flavor element that we sense with smell or aroma," the same way we use "taste" for "a flavor element we sense with taste."

With that out of the way, it's worth taking a minute to consider smell on its own terms before getting into what it does for flavor.

Our sense of smell, like our sense of taste, detects molecules. When you hold a ripe peach in front of your face and inhale, floaty, airborne volatile molecules diffuse out of the peach and enter your nostrils. They fill your nasal cavity, where, way up at the top, there's a small, sticky carpet, a kind of welcome mat for aroma, full of densely packed smell receptors. This is called the *olfactory epithelium*, and it does for smells what the taste buds do for tastes: funnels those molecules you sniffed toward the grasping and catching grip of receptors, pairing off the compatible ones, and sending out signals for your brain to create the sensation of smell.

The Overlooked Sense

Smell, which goes by the technical name "olfaction," is our most underappreciated sense. Ancient Greek philosophers like Aristotle believed that smell was unimportant, primitive, and a bit frivolous, a reputation that unfortunately stuck. More recently, we've learned new things about smell that make clear how much we've been underestimating it.

In 2004, the Nobel Prize for Physiology or Medicine went to the biologists Linda Buck and Richard Axel for their work uncovering the receptors we use for smell, the genes that code for them, and the ways they function. They found that humans have about four hundred different types of olfactory receptors, each with its own gene. Depending on how you estimate, that means that smell-detecting genes account for 1 to 2 percent of all our genes. That might not seem like a big deal . . . until you consider that smell receptors make up the largest group of genes in our genome—and that humans are only a few percent genetically distinct from chimpanzees, roughly the same magnitude of genetic information that we dedicate to smell. It's kind of a big deal for us!

Smell on the Brain

Our molecule-grabbing receptor equipment for generating smell signals is unique and special. Our setup for carrying those signals to the brain and processing them is special, too.

Most of the information that enters your brain is collected by nerves and receptors far out in your body, passed up to the spinal cord, and then funneled through the simplest part of the brain and its primary entry gate, the brain stem. Signals filter from there through mostly subconscious areas of the brain for processing before lining up at the thalamus, the clearinghouse and gatekeeper for your cerebral cortex, the pink and wrinkly outer layer of your brain where conscious perception happens.

Smell breaks basically all these rules.

Your smell-sensing receptor cells don't just pass information to be relayed on to brain cells, they *are* brain cells themselves. They're long and skinny, one end embedded in the *olfactory bulb* at the underside of the brain, the other passing through tiny holes in the bottom of your brain case and into your nasal cavity. This end has the receptors for catching smell molecules. Right now, as you read this, you have brain cells dangling out of the bottom of your skull, exposed to the air inside your nose at all times. And we all walk around like this is totally normal.

The signals your smell receptors generate are shot directly into your olfactory bulb, bypassing the brain stem entirely. The physical connections between the olfactory bulb and the rest of

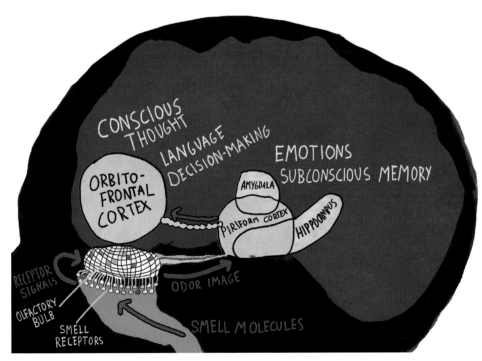

SENSING SMELL

CONSCIOUS THOUGHT
LANGUAGE
DECISION-MAKING
EMOTIONS
SUBCONSCIOUS MEMORY
ORBITO-FRONTAL CORTEX
AMYGDALA
PIRIFORM CORTEX
HIPPOCAMPUS
RECEPTOR SIGNALS
OLFACTORY BULB
SMELL RECEPTORS
ODOR IMAGE
SMELL MOLECULES

the brain make smell more enmeshed with both emotion and intellect than the other senses. The olfactory bulb connects directly to the *piriform cortex*, a bunch of daisy-chained bits of the limbic system—the subconscious, mammalian part of our brain where our feelings and long-term, impressionistic memories live. Most of our sensations are eventually analyzed here, but only after we process them in other areas of the brain. Smell signals enter the limbic system directly—preconsciously, unfiltered, and straight into our emotional memories, which are particularly intense, immediate, and vivid when triggered by smell (and, by extension, flavor). And while the other senses have to wait to pass through the thalamus to enter the prefrontal cortex, the highest cognitive center of the brain (the part we use for conscious thought, language, and rational decision-making), smell signals get VIP entry via direct connection between the piriform cortex and part of the prefrontal cortex called the *orbitofrontal cortex*.

What does all that mean for you?

RETRONASAL OLFACTION

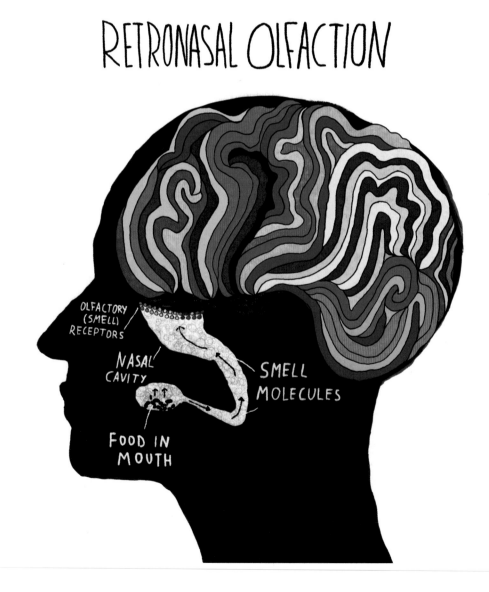

OLFACTORY (SMELL) RECEPTORS

NASAL CAVITY

SMELL MOLECULES

FOOD IN MOUTH

Well, we're deeply social animals, and lots of our fundamental brain processes are geared toward interpersonal relationships. Smell is no exception to this: get a whiff of the cologne (or even the laundry detergent) that a friend or a lover whom you haven't seen in years used to use, and the flood of feelings and memories it triggers can stop you in your tracks. Even if you never paid much attention to that smell, it was there when you were with them and subconsciously knitted itself into those memories.

And because smell + taste = flavor, they amplify each other's brain signals when you sense them together, making eating (and flavor) an even more intense emotional experience than smell on its own. When you bite into an on-the-mark rendition of your grandma's beef stew, you can get sucked back in time, emotionally, to eating it with her and all the feelings of security and home and identity that went along with that. The well-worn examples of this in storytelling (Proust's madeleines or, y'know, *Ratatouille*) are, neurobiologically speaking, right on the money.

Secret Smelling: Retronasal Olfaction and Flavor

It can be easy to forget that smell is half (or more) of flavor. After all, it's not like you're inhaling big huffs of the food on your plate while you're chewing, right?

Let's return to that tomato we were considering earlier. While its taste molecules were busy flooding your tongue and creating sweetness, sourness, and the other tastes, the tomato was also releasing volatile, airborne molecules into your mouth—the same molecules you'd pick up on if you sniffed it.

Your throat and nasal cavity are connected to your mouth. That's how, when you laughed too hard at your middle school lunch table, milk could shoot out your nose. It got pushed from inside your mouth into the nasal cavity. The same thing can happen, and constantly does, with the smell-molecule-saturated air that's released into your mouth when you're chewing—it fills all the space it can, meaning it wafts up the back of your throat and into your nose, where you can smell it. Every time you chew or swallow, the movement pulses those molecules up into your nasal cavity. What you sense is smell, but smell that sneaks in the back door. In technical talk, we call this *retronasal olfaction* (and yes, I do have dibs on that as a band

name). "Olfaction," because you're smelling, and "retronasal," because it goes into your nose backward.

And what do we get out of retronasal olfaction? Most of the flavors that make it a tomato, rather than just a collection of sweet, sour, salty, umami, and bitter: rich, malty-fruity, rosy, violety flavors; vegetal tomato-leaf and grassy ones; subtle hints of cocoa, honey, and wintergreen. The fruity, vegetably tomatoeyness of it—the stuff that taste just doesn't cover. Really, the distinctive flavoriness of most flavors (the chocolatiness of chocolate, the raspberryness of raspberries, the thymeyness of thyme) doesn't have much of anything to do with tastes—it all comes from retronasal smell.

If you feel a little skeptical or like you're missing something, that the tomatoeyness is more than taste but doesn't *feel* like smell, that's reasonable. The parts of flavor that come from smell feel like they're happening inside your mouth—because your brain tricks you into thinking that what you're smelling is happening on your tongue. So you don't actually feel it in your nose or as a smell; it gets fully integrated into flavor and perceived as happening inside your mouth—i.e., where the food is.

Perhaps you've been told, when seasoning something with lemon, to wait until the very end to add it, in order to prevent its losing some of its sour punch and tasting flat. And, indeed, if you were to add lemon juice to a gently simmering sauce and leave it cooking for a little while, then taste it, it would seem a bit disjointed, flat, and muted. But this isn't because any of the tastes have changed! When you heated it, the smell molecules boiled off. That flatness you perceive is just the tastes of lemon juice alone, flavor minus smell.

Citrus Peel Dashi

Dashi is a quick kelp-and-smoked-tuna broth that's foundational in Japanese cuisine. I sometimes like to drink it as a broth, kind of like you would a bone broth or miso soup. Dashi has an extremely clean umami depth to it, and you can add some flavor complexity in the form of aroma, without messing with this already intense, but very pure mix of tastes.

Making dashi has two steps: steeping dried kelp (kombu) in not quite boiling water, then

removing it and simmering katsuobushi flakes (dried, smoked tuna loin) in the kelp water. Kombu and katsuobushi flakes can be found online, at some higher-end conventional grocery stores, and at Japanese specialty stores. You can substitute a *dashi packet* (essentially a teabag full of kombu and katsuobushi) for the raw ingredients; in this case, simply bring the water to a boil, then turn the heat down and simmer for a few minutes before removing the packet.

Take **a small piece of kombu kelp**, about 2 by 4 inches or 5 grams, and combine in a small pot with **2 cups (450 ml) good-quality tap or filtered water**. Heat slowly over very low heat until it is steaming but not quite simmering, 15 to 20 minutes, then remove the kelp before it actually boils (boiling the kelp extracts slimy molecules you don't want in your dashi).

Add **⅔ cup (6 to 10 g) katsuobushi flakes** to the pot and bring to a simmer over medium-high heat. Simmer for about 1 minute, then remove from the stove, put a lid on the pot, and steep for 15 minutes. Skim out the katsuobushi or strain the dashi through a fine-mesh sieve.

Put the hot dashi in an individual bowl, or two small mugs, and cut a **fat swath of yuzu peel** (if you can't find fresh yuzu, use lemon peel or Meyer lemon peel) with a peeler or sharp knife. Point the skin surface toward the hot dashi, squeeze the peel to push out the flavorful oils, and drop it in the dashi. Then serve immediately, and drink up.

Makes 1 big mug

There's an easy experiment you can do on yourself to see the effect of smell on flavor (you may remember this from grade school). All you need is a jellybean or gummy bear, preferably in a distinctive fruity flavor like raspberry, pineapple, pear, or peach. (My personal go-to is a Jelly Belly juicy pear or a Haribo peach candy.) Take the candy in one hand and pinch your nose shut with the other. Put the candy in your mouth and start chewing it, with your nose still pinched. What

FLAVOR MINUS SMELL

FRESH LEMON JUICE

COOKED LEMON JUICE

SOUR + BRIGHT

"FLAT - TASTING"

(because you boiled off the smell molecules)

flavors can you notice? It's probably difficult, because blocking your nose makes it hard to discern any flavor and is muting the candy's tastes. You'll probably sense some sweetness and sourness, but not much else. Before you swallow it, unpinch your nose and breathe normally. The distinctive layers of flavor, the peariness or the peachiness, will feel like they slowly fill your mouth, turning up the tastes, and popping the full flavor out into 3-D. This is your sense of smell, returning in full retronasal-olfactive force to make bigger, better flavor than taste can do on its own.

The Smell of a Molecule

Taste and smell are both sensed with receptors, like all the senses. But their similarities end there. Smell takes a much more complex route to get from molecule-in-your-nose to feeling-in-your-brain than taste. Our taste receptors attach to five different signaling lines—one for each taste, about twenty-five kinds of receptors total, counting all the extra ones for bitter. Separating tastes into clear categories (sweet, salty, bitter, etc.) is easy, since its mechanisms work to create clear categories. With smell, we have to get creative.

When it comes to smell, we have *four hundred types of receptors*, each with its own distinct line of signaling to the brain; this is already eighty times more types of signals than taste has. And four hundred receptors doesn't mean that there are only four hundred aromas, one for each type of receptor and signaling line. We can sense far more than that, because somewhere along the line evolution decided it was okay for smell to get chaotic, if it meant we could smell more things.

Taste molecules bind to taste receptors pretty monogamously: a lactic acid molecule activates the sour receptor but doesn't really fraternize with the four other tastes. Or when a glutamate molecule gets onto your tongue, it looks around until it bumps up against a receptor for umami. They bind together, with a secret handshake, and the umami-specific cell attached to the umami-specific receptor sends an umami-specific signal to your brain.

Smell molecules and receptors are essentially polyamorous—there's no single "right" pairing between a specific smell molecule and one kind of receptor. Each smell molecule has an affinity to several different varieties of smell receptors, getting them all to send off their signals to the brain at the same time. And each type of receptor has the ability to latch on to several different smell molecules, even *lots* of different smell molecules, especially ones that share some type of molecular architecture. When a molecule of the primary aroma of cloves, eugenol, enters your nose, there's no one receptor poised to grab on to eugenol and eugenol only. There are five or ten (or even more!) receptors that can bind eugenol, but each can also bind to other, different molecules. And when eugenol finds this bevy of receptors, it fires off each one to send their unique signal up their own, individual signal-carrying lines.

The receptors don't organize themselves to send the brain an obvious, unified "CLOVE SMELL" signal at this point either. In between the crackling signal lines coming out of your smell receptors and the parts of your brain that create the perception of smell, there's a switchboard whose job it is to collect the raw signals into a packet of information to send up the phone tree.

This switchboard is the olfactory bulb, a little nodule of brain tissue that sits on the very bottom of your brain case, connected directly to the olfactory neurons that extend into your nasal cavity. It takes the Morse code–like barrage of signals from the receptors activated by eugenol and marks them all down into a single *odor image*, as a neuroscientist would describe it. This bundled odor

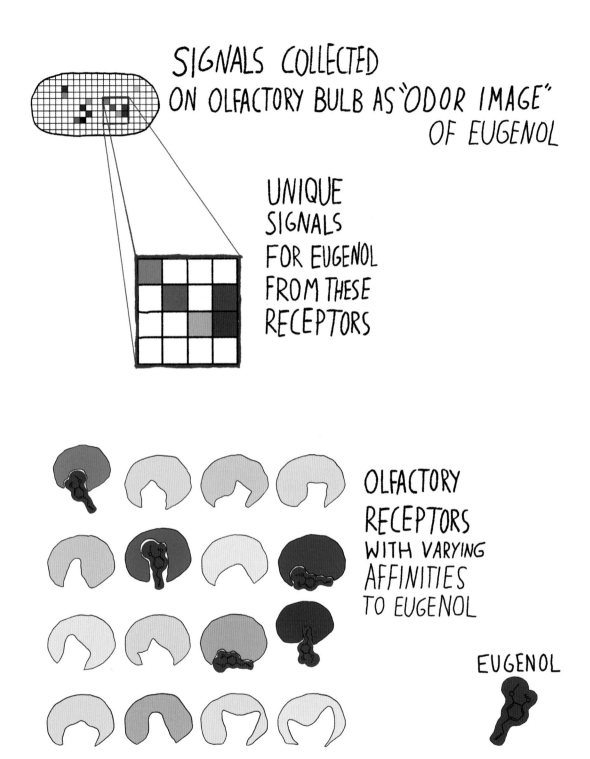

SIGNALS COLLECTED ON OLFACTORY BULB AS "ODOR IMAGE" OF EUGENOL

UNIQUE SIGNALS FOR EUGENOL FROM THESE RECEPTORS

OLFACTORY RECEPTORS WITH VARYING AFFINITIES TO EUGENOL

EUGENOL

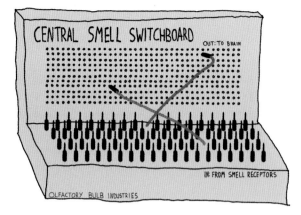

image doesn't smell like cloves yet; the olfactory bulb sends it off to other areas of the brain, where *finally* it's decoded and translated to the feeling of a clovey smell.

The signal the brain gets from a taste molecule is like hitting one key on a piano: it activates its own distinct indicator (one note or taste) that's easy to tell from the others. The signal that a smell molecule activates is more like a QR code: a two-dimensional pattern of many unique indicators.

The brain's QR code of the smell of a molecule is made up of multiple elements—and the way most molecules smell to us has multiple elements as well. Eugenol smells like cloves, but if you sit with its smell for a minute (and have some practice naming your perceptions—check out page 19 for more on that), you'll pick up on nuances: clove, but also woody and a little bit like cinnamon; a "sweet" vibe that's also just a bit savory and smoky and medicinal. Some smell molecules lack one primary message and have more evenly weighted elements. Furaneol is both toasty-caramel-burnt-sugar and jammy-fruity, and is equally important in the aromas of strawberries, pineapple, and roasted coffee, for instance.

It's impossible to pin down single smell elements in isolation, the way you can with taste elements. At the most basic, molecular level, smells have multiple sensory qualities compounded together. The way we perceive them is more like seeing a face than tasting a taste—a highly organized mix of many features, which is impossible to break down into one descriptor but immediately recognizable as a (complex) whole. This is frustrating for simple categorization but, in its limitless variety, very fun and delicious for flavor.

Mixing Molecules

Let's imagine that every possible pattern you could encode on your olfactory bulb by switching smell receptors signals on and off corresponded to a distinct smell you could smell. If you ran through them all, you'd count 2^{400} of them—2 times 2 times 2 times 2 times 2, until you've multiplied 400 times. This is a really, really big number—about 2,580,000,000,000,000,000,000,000,000, 000,000,000,000,000,000,000,000,000,000, 000,000,000,000,000,000,000,000,000,000, 000,000,000,000,000,000,000,000,000,000 (or 2.58×10^{120}). That's quite a bit larger than physics estimates for the number of atoms in the observable universe (10^{80}). Larger, as in roughly the size of the difference between the smallest distance we can measure with an electron microscope (many times smaller than the size of an actual atom) and the distance between Earth and Earendel, the most distant star we've ever detected, 28 billion light-years from us.

In short, our biology sets us up with a lot of flexibility for smelling different smells—maybe to the point of overkill. Scientists have used other techniques to estimate what this translates to in reality.

Older estimates pegged the number of smellable molecules at 10,000. More recent research based on a really big database of known molecules calculated that there could be 40 billion that have the right chemical properties to be smellable. What we know for certain, though, is that there's a hell of a lot more molecules we can smell than molecules we can taste, each with its own set of sensations.

Although it's useful to understand what flavor elements come from individual molecules, it's a somewhat academic exercise: you never smell just one type of molecule, unmixed with any others, outside of a laboratory. The smell, as well as the flavor, of any ingredient comes from a blend of multiple smell molecules. For very simple things, like herbs, spices, or fruits, that might be a few dozen. For something more complex like wine,

coffee, or chocolate, we're talking hundreds or even thousands.

When you smell any of these mixtures, each of these many smell molecules bind to its own set of olfactory receptors at the same time, and they pass their signals on to the olfactory bulb simultaneously with one another.

The olfactory bulb is a diligent record keeper, and generates its QR-code smell image from those signals, whether they're from pure eugenol molecules or a gingerbread cookie, with cloves and other spices and all their retinue of flavor molecules. Since individual molecules fire off multiple

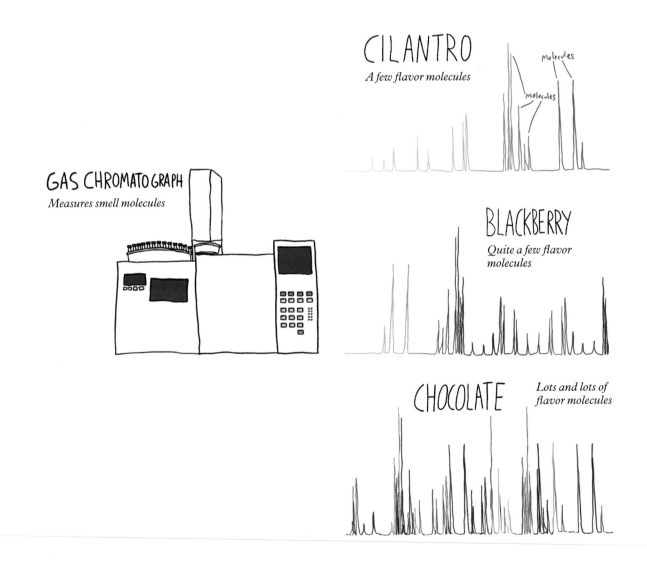

GAS CHROMATOGRAPH
Measures smell molecules

CILANTRO
A few flavor molecules

molecules

molecules

BLACKBERRY
Quite a few flavor molecules

CHOCOLATE
Lots and lots of flavor molecules

receptors anyway, the olfactory bulb doesn't really distinguish whether the signals it gets are from just one kind of molecule hanging out in your nose or a mix of molecules. (There might not even be a way for it to tell at all.) It generates its patterned, indexable, QR code regardless.

Even the very chemically complex foods just mentioned—wine, coffee, chocolate—get recorded and encoded in a single snapshot. As a result, they're all experienced as distinctive, memorable aromas rather than confused jumbles of a thousand different signals all talking over one another. These aromas are greater than the sum of their parts. Not only do they combine flavor notes you can recognize from their individual component molecules, but the mixture will also have new and unique ones that, a bit spookily, only exist when its components are brought together.

Most foods mix so many components into their flavors that very few of the molecules we can smell are unique to any one ingredient—more often, molecules pop up in the ensembles of quite a few, like cameos from a well-loved character actor. The pure, rich, deep note that adds weight to the "raspberry!" in raspberry? It's from a molecule called *beta-damascenone*, and it's also in apples, roses, wine, tobacco, black currant, and aged rum. Herbs and spices produce so many molecules in common that their flavors have more to do with the proportions of those molecules than the presence of any single one. Mixtures of aroma molecules can have significant overlap and still be unique and distinguishable. Basil, ginger, and cumin have notes in common, but each tastes like itself and isn't easily mistakable for one of the others.

Even with all this overlap, we're really good at assigning distinct odor images ("QR codes") to mixtures. I've seen estimates that there are at least one trillion mixtures-as-smells that we can sense and tell apart.

If you look at cooking as creating combinations of ingredients, and thus mixtures of aroma molecules, you can see you're set up to create flavor synergy like this every time you step into the kitchen. Your olfactory bulb doesn't really care what you throw at it—it's set up to make a complete pattern from whatever you're smelling, and your brain is set up to read these as holistic impressions of flavor.

Smells on Smells on Smells

- You can smell way more things than you can taste.
- Individual aroma molecules have multiple aroma qualities.
- No aroma molecules are totally unique to one ingredient.
- Ingredients contain many different aroma molecules.
- Mixtures of aroma molecules (which describes all ingredients you cook with) reflect some of the aroma qualities of their components, as well as some qualities that only exist in the mixture.

Chapter 3

Flavor First

Learning to Taste and Smell

The most important thing you can learn to make your experience with food better—cooking, eating, professional, amateur, whatever—is learning how to notice and name what your senses of taste and smell are telling you.

The ubiquitous phrase "salt to taste" reveals a lot. If you're not tasting what you're cooking, and not noticing the flavors that are there, how can you possibly know if you like the amount of salt in it, and how much more you need to add if you don't? No one else can tell you what you like. That's up to you.

And if I've piqued your interest in understanding more about how flavors work—and using that understanding to become a more creative and improvisational cook—know that theoretical knowledge will only get you so far. There's no way to build your ability to notice patterns in flavor, and develop your own internal flavor compass, that doesn't involve paying close attention to those flavors as you're experiencing them. I think of it as training yourself to perceive reality as directly as possible—not what we've been taught to believe something is "supposed" to taste like, but what it actually tastes like right here, right now.

If that sounds intimidating, take a breath. When you hear that someone has a "good palate," it can sound as if being able to taste something and know that it "needs acidity," or coming up with illustrative language about a wine you've just sampled, is a sixth sense that you either were born with or made an unsavory supernatural deal to get.

Fortunately, this is mostly bullshit. We're built to taste and smell; most of us are just in need of practice. I know, because I've trained dozens of people to become precise analytical tasters for real, published research—and you don't need to be nearly that precise for tasting and smelling to be supremely useful.

The gold standard for rigorous, scientific tasting—the kind used to painstakingly generate reproducible data for research—goes by the rather bland name of "descriptive analysis." Descriptive analysis has a few steps that work together to get tasters really good at tasting and identifying flavors, and we are going to steal them so we can improve as tasters, too.

If you were one of my research subjects (we'd call you a "judge" or a "trained assessor") and made it through an initial screening to make sure there wasn't something medically strange about your sense of taste and smell, we'd start off with an experiment on simple tasting and talking. You and the other judges would taste examples of whatever it is we were going to study—sparkling wines with different kinds of fizz, gins, artisanal vinegars—and make notes about everything you could taste. Orange peel? Leather? Dirt? Raspberries? All on the table. We'd taste several examples, because articulating a difference in flavor by comparing two things is often easier than pulling words out of thin air.

The next time you came in for the experiment, we would taste the examples again—but I would also have collected tangible, smellable references for all the flavors you said you thought were important, and then I'd make you smell all of them. For "orange," you'd smell orange peel, orange candy, and orange marmalade. For "raspberry," fresh raspberries, frozen raspberries, raspberry jam. For "dirt," potting soils. For "leather," suede, chamois, and garment leather, which all smell a bit different. Then you'd have to tell me which one matched what you were smelling in the actual samples. This step is crucial: it ties the sensations inside your head to concrete, real-world examples that everyone else is smelling, too.

Then, we taste, and taste, and taste and taste and taste. Every time you came for us to collect data—and there would be many sessions—I would make you smell all the references blind, and quiz you on them. If you missed any, you'd have to go back and try again. With those flavors fresh in your mind, you'd taste the samples we were actually doing research on—tasting one sample, going through each of those reference flavors one by one and noting how much of each you could detect, and then repeating it with the next sample. We'd do this multiple times for practice until you were consistent, then, *finally*, taste for real.

All of these steps boil down to a few rules to follow.

Pay Attention

Rule #1 of tasting and smelling: Palate is 99 percent paying attention plus practice. Scientifically speaking, most people are actually pretty excellent at distinguishing between flavors, which is the foundation of palate. Give an average person two fairly similar foods—two types of orange juice, or tea brewed by different methods—and they'll usually be able to tell their flavors apart, even if they can't name exactly why. This ability is part of being human, like learning speech. You already have more experience with it than you think.

At your next meal, smell, taste, and sit with your perceptions. You might start by tasting a dish you've made from a recipe you love. What do you notice about it? Do some flavors show up on first bite? Do others layer on top of them? Are some cutting through others, maybe providing some lightness? Is salty or sour or spicy sticking out? Is there anything dark or heavy?

You should also try tasting raw ingredients. (Maybe not raw meat, but things that don't pose a major safety issue.) Are they very sour or sweet or pungent? Do you notice umami? Are there fruity or herbal qualities? Are those more berry-fruity or apple-fruity? Are they sagey and a little heavy, or more fragrant and delicate? Do any of the flavors in the raw ingredients show up in the finished dish? Different versions of those flavors?

These questions will have you referencing your internal catalogue of flavors. I think of it as the memories of flavors you've experienced before, where you experienced them, and what they went together with, all arranged like paint chips at the hardware store or in a thick deck like a Pantone guide. When you're trying to put a name to a flavor, you flip through them like a graphic designer flipping through their Pantone to find just the right color. But no one is born with this catalogue—you build it for yourself by paying attention and carefully considering flavors and smells.

Compare

You might find yourself sniffing and tasting a cup of coffee really carefully and, frankly, it still smells like . . . just coffee? It smells good, sure, but you're not getting any of the florid notes described on the bags of beans at an artisanal roaster. Naming what you're tasting gets a lot easier when you add in a little of what humans are really good at: telling flavors apart.

Next time you're able to, compare two versions of the same kind of food side by side. When you're at a coffee shop with a friend, each get a cup of pour-over made with different beans—maybe one from Kenya, one from Honduras. Or if you're at a bar, order a flight—three or four scotches, amari, or even tastings of beer. If you're shopping for fruit, get plums, peaches, and apricots, or three or four different kinds of citrus.

Taste one, then the other. Sniff them deeply one by one. Go back and forth from one to the other, and pay attention to what you get from each. Also pay attention to what jumps out when you smell and taste the two samples next to each other. You'll often get a "tip of the nose" feeling—you know they're different, you can *feel* why, you just can't name it. But if you can notice a difference in flavors, you have a starting point for learning how to describe it. You have a starting point for naming what's similar about flavors, if anything, despite that difference. And you have started the process of remembering and being able to recall those specific flavors, just by paying more attention to them.

Now that you have experience paying attention, start to articulate differences. You can probably pick out which one tastes more acidic, bitter, sweet, or salty after a few tries. Maybe one smells brighter, the other more muted. Maybe one is spicier and the other is fruitier. Maybe one reminds you of an old book, and the other smells more like toasted bread. Comparing like this gives your mind something to latch on to, to tease out associations to the differences you know you are detecting, rather than smelling just one thing and casting about in a sea of possible associations to make an identification.

Use References

I've definitely been in tastings where there's that one guy (or girl, etc.) who just has to take it beyond useful shoptalk—noticing and describing five or six key flavor notes in a coffee, and what those say about where it was grown and how it was processed and roasted—and show off how special their palate is. They insist that they taste not mere caramel, but specifically salted caramel. Not mere banana, but specifically bananas foster, with an unstated edge that anyone who *doesn't* detect it is an idiot.

It is not a problem if your immediate association to a flavor is something like salted caramel or bananas foster. Your sense of flavor, especially smell, works by making associations to things you've tried and noticed before. If it works for you, it doesn't really matter if what you are calling "rose" isn't perfectly identical to what you'd smell while smelling a rose, or if your brain goes to "blueberry cake" in something that only partially smells like blueberry cake.

But, remember, you also don't need to make up romantic word poetry just for extra points. Flavor is its own poetry! More language is useful if it helps *you* clarify, but for its own sake, it can muddy the definition of what you were intending to explain better.

Maybe you taste something and make an association to salted caramel. Is that because it is salty and caramel-smelling? Caramel is generally just caramelized sugar and water, whereas salted caramel typically has cream added. Are you detecting creamy dairy aromas, or perhaps browned-milk-solid qualities? Each of these associations is slightly different and conveys different things. Likewise, bananas foster. Do you detect banana in a particular stage of ripeness? Is it cooked banana, rather than raw banana? Is "bananas foster" conveying two different flavors simultaneously, banana-y and roasted? A recent tasting note I made says, "Pine nut flavor (mildly nutty, but not distinctively sweet-feeling like almond/hazelnut/pecan)." Probing a little deeper will help you reflect sensory reality with your words.

I highly encourage you to steal the technique we use in flavor research: using references to ground your flavor connotations in reality.

As an example, do you find yourself often noticing "fruity" flavors in wine but have difficulty probing further? Try buying, smelling, and tasting a couple of different kinds of apple, pear, and berries. Fresh raspberries, blueberries, and strawberries next to raspberry, blueberry, and strawberry jams.

For other flavors you're curious about, green tea, oolong, and black tea. Smell caramel and butterscotch. A greenish banana and a ripe banana. Cloves, allspice, and cardamom. Dry and fresh ginger. Rose petals and rosewater.

As you smell your references: Is there one flavor or aroma note that sticks out? More than one? Are they sharp or soft? How does one evolve into another?

All this hard work isn't just coming from a misplaced sense of perfectionism. If you deeply pay attention to flavor and compare flavors enough, you'll get to the point where you can sniff and taste something new and be able to summon up some word associations on the spot, without a lot of angst. Flavor molecules all came from somewhere, so if you can distinguish a flavor more accurately, you can make more inferences about it. "Herbal" is one thing, but "parsley" or "cilantro" is much more specific. And the more specific you can be, the more carefully you can fine-tune your cooking.

Using references can be especially useful for tastes. People can generally figure out sweet and sour without much difficulty. Sometimes they call something bitter when it might just be astringent or tannic, like a black tea. Salty, umami, and spicy, though, can be surprisingly difficult, even for really experienced tasters. There is a tendency to latch on to "salty"/"saline," "umami," or "spicy"/"peppery" when describing something that's actually an aroma or aroma-plus-taste that's difficult to tease apart. It's fine if you can't tease it apart! But "salty" or "saline" wines and other alcohol very rarely have any noticeable sodium chloride in them—perhaps you are detecting sourness without sweetness, or a particularly marinelike or unsweet aroma? Umami can be hard to describe as well as to consistently notice—it helps if you actually taste some MSG. You might find that something you had been calling "umami" was actually not a taste at all, but a meaty, funky, or browned aroma. "Too sour" and "too bitter" can be easy to confuse, and you'd deal with both of those things differently when deciding how to adjust a dish.

Practice Makes Perfect

Tasting and naming what you're tasting rapidly gets easier with practice. The more you pay attention, the more you compare, the more you think about a fixed reference, the faster you'll be able to access your flavor memories and to describe what you're tasting or smelling.

When I started graduate school, I volunteered as a subject for a sensory analysis of wine and chocolate (yeah, it was a real tough one). It was a pretty neat experiment—everyone talks about pairing wine and chocolate, but it turns out that no one had done a controlled, precise experiment on what actually happens, sensorially, when you taste different wines and chocolates together.

I was about twenty-two at the time, had just recently begun drinking wine, and knew nothing about wine tasting. I was intimidated when we got together to taste and free-associate flavor words, intimidated as well when these got made into references we could train on, and slightly lost during the first few proper tastings.

One term I had particular difficulty with was "oak." We had a reference for oak, of course, and I smelled it a lot, but I had a lot of difficulty picking it out in the actual wines. I'd smell, hoping to latch on to something, and come up short. When I eventually got it, it just snapped into place. I could smell oak! It was a little woody, a tiny bit resiny-pungent, and a little spicy and vanilla-y. Training with the reference helped, of course, and I wouldn't have gotten there if I wasn't deeply paying attention; but the thing that really made the difference, on top of those things, was simply doing it a lot. You could devise drills to replicate this, if you feel so inclined, but the takeaway is: your ability to learn flavors will develop over time. You can help it along by creating an environment of careful smelling, and lots of it, but you will need some patience. It takes repetition, and that's fine!

Be Nice to Yourself

Unless you are literally participating in a sommelier exam, it's not a competition. Experiencing flavor is something you're going to do multiple times a day for the rest of your life. If you feel like chilling out for a while—or stop getting joy from paying deep attention and making references—feel free to take a step back! Or feel free to keep tasting and being okay with slow progression. If you don't get oak now, you probably will get it a hundred smells, or fewer, from now. It's all fine. It's a long game. Nobody is born with an incredible palate—we all have to earn it.

PART 2

The Second Law of Flavor

Flavor Follows Predictable Patterns

Flavor follows predictable patterns

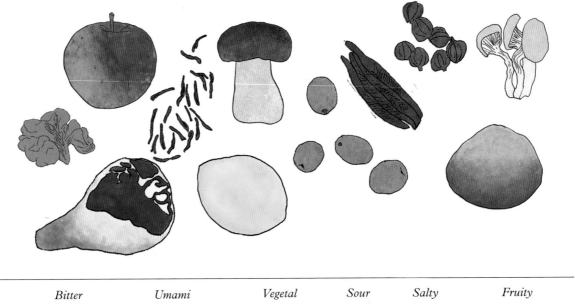

| Bitter | Umami | Vegetal | Sour | Salty | Fruity |

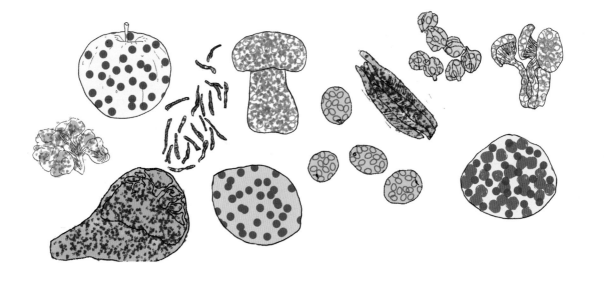

Understanding taste and smell, and *how* to taste and smell, is a great way to appreciate all the nuances of delicious things that are working the way they're supposed to—for instance, an excellent olive oil, a brilliant artisan sourdough. But it's a superpower when things go off the rails.

When a recipe goes *just right*, it's a beautiful thing.

Your butcher had short ribs that are plump and well marbled and not leaking any juices into their tray. The carrots, onions, and celery you add to the pan are dense and sweet, the wine you deglaze with has a great backbone of acidity, and the tomatoes, really good umami. You make a gremolata to go with the short ribs: the parsley is fresh, mild, and fragrant; the garlic is hot but not aggressive; and the lemon you zest for it has a delicate, oily rind. You've followed every step to the letter, and your dedication paid off.

Good recipes are a start. But even if you don't think of yourself as an improvisational cook, paying attention and responding to taste and smell is a huge part of keeping things on track. You smell for the well-integrated, rich beefy-oniony flavor that marks a well-maintained braise. You taste for salt close to the end of cooking to see if more is needed. You know how much garlic or tanginess you prefer, so, like the penciled-in annotations in my mother's *Joy of Cooking*, you know when to make your own alterations.

And in the real world, ingredients are variable. They go out of season. Or the market offers up only the saddest, most tasteless version of what you're shopping for. What if your butcher is out of short ribs? Maybe all the parsley is watery, bland, and bitter. Maybe you forgot to buy the bottle of wine. At times like these, I am reminded of the immortal wisdom of Mike Tyson:

> *Everybody has a plan until they get punched in the mouth.*

Good cooking isn't really about perfect predictability and control—that's for tubes of cookie dough—it's organized chaos. It's about responding in the moment to subtly changing conditions. Taste and smell molecules contain all kinds of information about those conditions, even before you see any visual cues. The better you get at tasting and smelling, the better off you are, even with a recipe.

I'd rather learn to adapt to chaos, to expect it from the get-go. To be able to pivot around a lack of lemons, different herbs, or the wrong kind of protein, rather than accept mediocrity—or defeat. Most of the really great cooks I know share an uncanny ability to reconfigure on the fly, pulling improvisations or entire dishes seemingly out of thin air. Of course, like a "great palate," as much as this can look like inscrutable mysticism, or being so experienced that you literally just know everything, that's not actually how it works. It's more like thinking in patterns.

When I encounter a problem I don't immediately know how to deal with, I could take the approach of throwing myself at it until I've solved it with brute force. I could wait for divine inspiration. But I'm pretty lazy and impatient and prefer to take a third approach: figuring out how to turn the problem into a different, easier problem I already know how to solve.

If I need parsley for gremolata, and all the parsley sucks, I could work really hard to somehow rehabilitate the parsley. Or I can think about what flavors the parsley contributes to the gremolata—a fresh and green, herbal backbone to the heady prickle of garlic and the sweet, light, citrusy aromas of lemon peel—and go looking for those qualities somewhere else. Maybe I emphasize "green," and go for cilantro—or emphasize "herbal," and use basil instead. It's unreasonable to expect that whatever results is going to be exactly the same as the original gremolata—but that's fine, because I'm flexible, and I'm confident that it's going to taste pretty good, in a similar way, because it follows the same pattern.

This is about getting into a mindset: you shift your point of view on something until you can see the ways it actually looks like something else. That's pattern recognition at its most basic, and the engine driving most strokes of inspiration: having the flexibility to observe something from more than one angle, seeing what you can generalize, and making connections to other things with features in common.

Seeing in Patterns

Let's consider a lemon.

Even better, go get a lemon.

So: What is a lemon?

It's a round yellow citrus; it's the fruit of the *Citrus limon* tree; it's originally a hybrid between a sour orange and a citron. While these three things are all true, they're not necessarily the most useful way of looking at a lemon.

Let's shift our point of view a little. Instead of reiterating what you've been told about lemons, focus on what your senses tell you about the lemon. It's not just a uniform object containing the cosmic quality of lemon-ness. If we go from the outside in, we find three distinct layers: the zest, the pith, and the juice-infused flesh.

While you taste and pay attention to each one, you'll notice each has distinct flavors it contributes to the whole concept of "lemon." The zest is powerfully aromatic, with citrusy flavors both soft and sharp, as well as herbal and floral ones. The pith is spongy and bitter. And the juice is bracingly sour, with a lighter version of the perfume of the rind, and some subtle sweetness. Read like this, a lemon is an ingredient with a lot of smells to give, and a strong sour taste as well. It's a bracing, sour ingredient (with fresh and bright aromas). It's also a citrusy-aromatic ingredient (with a tangy undercurrent). Both are equally true, and both are important roles for a lemon to play.

If you have a recipe that calls for lemon, which role is the lemon playing in that recipe? Is it there mostly for citrusy aroma or sour taste? Are both equally important? I'm not asking this as a rhetorical question—you need to taste and think to answer it, and that answer is going to be unique for every recipe.

Substitute a different citrus, and you'll see different iterations of the same general qualities of aroma, juiciness, and sourness. A lime more or less matches the sourness of a lemon, but its aroma is much more herbal and resinous. An orange has the same citrusy undercurrent in its zest, but a couple of its distinct notes can really only be described as orangey, and it also has a sour but much sweeter juice. A grapefruit trends further in the same direction: really intense, really distinctive grapefruity aromas on top of the citrusy base, and sweet-sour juice with a definite element of bitterness. Each of these fruits could conceivably play understudy to a lemon where citrusy flavor was important, but they're all different shades of citrusy, and all relate to different qualities of that lemon.

"Citrus" is one way to define ingredients and group them together, but, changing our perspective a little, we could just as easily decide on "sour" as a uniting category. (That would mean, for whatever purpose we have in mind, we care about the lemon mostly as a sour ingredient, and its citrusy aroma is more incidental.) The lime could work here, but the less acidic orange and grapefruit might need some help.

But if we're looking from the angle of "sour," we don't have to stick to citrus—let's rope other strongly sour ingredients into our search. Is strong sourness important, and will it be okay to have a little pungency? In that case, I'd look for a vinegar that fits the bill. Does it need to be sour *and* fruity in some way? Then think about pomegranate molasses or tamarind as an option. Do you just need something a little sour, now, and don't have any of those on hand? That's fine, do you have any yogurt? Maybe it's the "excess" brine in the pickle jar's time to shine.

There's nothing stopping you from bringing this line of thinking to every ingredient—or whole dishes. You could even try to build a mental flavor model when you read a recipe.

Let's go back to the short rib recipe, the one we wanted to make, but this time, the butcher didn't have short ribs and you forgot the wine: beef short ribs braised with onions, carrots, celery, tomatoes, and red wine, served with a parsley–garlic–lemon peel gremolata. What are the basic flavor impressions of a dish like this? Not just the ingredient list—the tastes and smells you'd experience? Browned, robust, and meaty, from a well-exercised cut that can cook for a long time. Sweet, vegetal, and slightly sulfury flavors from the vegetables. An acidic liquid with depth and a savory-sweet-fruity element from the tomatoes and wine. Along with a green, fragrant, and punchy herb sauce.

Approaching it from this angle, you've just identified a pattern of flavors, layered together. And if you have none of the ingredients in the original recipe, you can look at this pattern you've analyzed, and treat those layers as targets to hit. Maybe after registering the distinct unavailability of short ribs, you notice that pork shoulder looks good at the meat counter instead. You can improvise along the pattern from there, and you might come up with something like . . .

Pork Shoulder Braised with Sherry Vinegar, with a Gremolata of Cilantro, Oregano, and Orange

Preheat the oven to 300°F. Put **2 to 3 pounds (1 to 1.5 kg) of boned pork shoulder, cut into large cubes** into a large, heavy Dutch oven.

Around it, add **2 roughly sliced medium-sized red onions, 1 medium celery root, cut into large cubes**, and **2 cloves of smashed garlic**. In a blender, puree **1½ cups (360 ml) jarred roasted red peppers with their juices** and **½ cup (120 ml) sherry vinegar,** and pour into the Dutch oven. Sprinkle lightly with **salt.** Cover with the lid, and cook in the oven until the meat is tender, about 3 hours. Check it every 40 minutes or so and add a little water if the pan starts going dry. If desired, turn the heat up to 450°F, remove the lid, and cook for another 10 minutes for greater browning.

While the pork is cooking, make the **cilantro-oregano-orange gremolata**: in a small bowl, combine **¾ cup (30 g) chopped cilantro leaves, grated zest from 1 medium or ½ large orange,** and **the minced leaves from 1 to 2 sprigs oregano.**

Serve the pork, celery root, and juices in bowls, with the gremolata sprinkled over to taste.

Serves 3 to 4

Or, did you see lamb neck but weren't sure of the best way to tackle it? Try this on for size:

Lamb Neck Braised with Leeks, Black Garlic, Dried Cherries, and Apricots with Tarragon–Preserved Lemon Sauce

Black garlic, if you haven't had the pleasure yet, is fresh garlic, hot-aged until it browns to nearly black. It's sweet and funky and just a bit garlicky, and I love to buy and use it in powdered form. Versions of it can be found from small producers to Trader Joe's, but if you're having trouble sourcing it locally, you can order it online—I like Kalustyan's myself.

Preheat the oven to 300°F. Put **2 to 3 pounds (1 to 1.5 kg) of boned lamb neck, or 3 to 4 pounds (1.5 to 2 kg) of bone-in lamb neck** into a large, heavy Dutch oven. Around it, add **2 medium bulbs fennel, thickly sliced, 4 leeks, cleaned and sliced crosswise, ½ cup (80 g) sliced dried apricots,** and **¼ cup (40 g) dried sour cherries.** Combine **1 cup (240 ml) yogurt whey** with **1 cup (240 ml) water** and **1 teaspoon (3 to 4 g) black garlic powder,** and pour into the Dutch oven. Sprinkle lightly with **salt.** Cover with the lid, and cook until the meat is tender, 3 to 4½ hours, depending on connective tissue. Check it every 40 minutes or so and add a little water if the pan starts going dry. If desired, turn the heat up to 450°F, remove the lid, and cook for another 10 minutes for greater browning.

While the lamb is cooking, make the **tarragon–preserved lemon sauce**: in a medium bowl, combine **¼ cup (60 ml) chopped preserved lemon, ½ cup (20 g) roughly chopped tarragon leaves, thyme leaves from 1 to 2 branches of thyme,** and enough **good quality extra-virgin olive oil** to thin into a loose paste. (Makes about ¾ cup/180 ml.)

Pick and cube the lamb meat, then serve in bowls with the vegetables and some of the juices. Garnish with the sauce to taste.

Serves 3 to 4

Compared with the less experimental pork, the lamb might sound a little more off-the-wall. Maybe like something you read about in a restaurant review and think, "But how the *hell* did they come up with that combination?" They most likely went through a flavor journey similar to this one, tasting a lot, recognizing patterns, and successively iterating on them to fit the ingredients they had.

Rules Suggestions for Approaching Flavor as Patterns

Flavor is your ultimate guide to putting ingredients together.

Try: thinking of flavors themselves as categories of ingredients (like sour or umami).

Notes and shades of the same flavors appear in many ingredients, in different combinations.

Try: tasting to distinguish flavor layers from taste and flavor layers from smell.

Try: tasting, noticing, and naming the layers of flavor in an ingredient: Sour? Salty? Fruity? Floral?

Try: tasting the similarities and different "shades" of a flavor in similar ingredients.

Ingredients that share flavors can perform well in each other's roles.

Try: noticing and naming the layers of flavor when you're eating, cooking, or even reading about a recipe.

Try: substituting ingredients with similar flavors for each other, and noticing how the substitute changes the flavors of the dish.

There's no wrong way to taste.

Experimentation is your friend, and practice makes perfect!

Chapter 4

Putting Patterns to Work

A Field Guide to Tastes and Smells

Do you want to make flavor work for you? When it comes to putting flavors together, it helps a lot to have flavor experience—in other words, a good working knowledge of your materials.

While producing his most dynamic drip paintings, Jackson Pollock didn't slap on paint at random and call it a day. He had to start by stepping away from mentally categorizing paint as the thing you apply with a brush in a particular way to make a figurative painting, and framing it as a kind of sculptural material, with its texture and flow behavior dictating how to work with it. Methodically experimenting with different viscosities of paint, nonstandard application tools like syringes and sticks, and the physical movements of his body, Pollock created completely new techniques based not on what paint was supposed to be, but on the properties it shows you when you look at it on its own terms.

For cooking, a working knowledge of material is something you can get by tasting lots of ingredients, making and reading lots of recipes, and eating with an eye (tongue?) for patterns. As the saying goes, the best time to start building this experience is ten years ago. The second-best time is right now. And all the eating you get to do? It's hardly a terrible burden. It starts by paying attention and thinking while you taste—but keep in mind that there is no "wrong" way to do this.

Your goal is to build your own personal mental map and directory of flavors—like those beautifully arranged hardware store paint chips, or an irresistibly flippable Pantone color formulation guide, which is what mine looks like. It could be like an artist's palette, a Rolodex, an Audubon guide to wildflowers, or a card catalogue—or an entire library or mind palace; whatever speaks to you.

And in this endeavor, it's not like you have to go out there with no preparation or structure. The most basic patterns in flavor to understand are flavors themselves—like salty, bitter, or meaty. Understanding those, and the subcategories and gradations within them, is the perfect place to start. The next several chapters are my attempt, using what I've learned about chemistry, perception, cuisines, history, even botany, to organize flavor into a 101-level field guide to start you out. It may not be a surprise given the form of my mental flavor map: we're going to use color to help us understand how to think about it.

Color and Flavor

I'm obviously going steady with taste and smell, but our sense of sight has a lot of cool things going for it, too—especially color, which, kind of like smell does, blends the responses of different types of receptors into a single perception. Let's say you want to paint your bedroom blue. Now you need to communicate what blue you want. How do you distinguish between pastels, deep jewel tones,

green-tinted tealy blues and red-tinted purply blues, or darker and drabber navies? Even if you're specific about wanting, say, a cobalt blue, there are plenty of blues that express cobalt in much

lighter or darker versions, or shades that are more or less saturated or desaturated.

Color theory has all kinds of systematic models for describing individual shades, as well as their similarities and differences from one another—not unlike what we've been doing with ingredients and flavors. Color models can define any shade pretty precisely as a blend of a few variables—my favorite is called *HSV*, for reasons that will be immediately obvious. It starts with defining the base *hue* first—its exact spot on the red-to-green-to-purple spectrum of visible wavelengths—then dials in the exact shade of that hue as a combination of *saturation* (washed-out pastel to throbbingly intense) and *value* (clearest and brightest to darkest and most shadowy). There are also models like *CMYK*: colors as combinations of cyan, magenta, yellow, and black inks (which is the color model this book is printed with). Or *RGB*: red, green, and blue light (the one used if you're reading it on a computer screen).

This approach of breaking down colors and describing their relationships to one another has a lot of potential for thinking about flavors, too. How great would it be to look at a dish, or survey the ingredients available to you, with the same eye for the systematic as a color theorist?

Like the lemon we analyzed earlier (page 26), individual ingredients have several layers of different flavors, just as colors come in different shades and hues. Knowing these layers, let's group these ingredients with others that have the same flavor layer in common. This could be as specific as grouping brown sugars (page 74), or more broadly thinking of all ingredients with a significant layer of sweetness as a group (page 70). A grouping of herbs based on a shared layer of vegetal greenness (page 170). Cured ingredients that are all notably salty (page 44).

Within a group, different ingredients can play similar flavor roles in dishes. They can often be interesting substitutes for each other—think of using dill instead of parsley, caraway instead of cumin, maple syrup instead of brown sugar, umami-rich miso instead of umami-rich bacon—with the knowledge that, while they're grouped together, they'll each bring something unique of their own to the flavor party.

Let's say you need something to fill in a layer of green-herbal flavor—maybe for a garnish, treated like a salad green, or stirred into a sauce. All of the several options you have on hand have different flavors and different moods: do you want the almost minty savory greenness of dill, the slightly bitter-herbal greenness of parsley, or the ultragreen, citrusy, slightly perfumed green of cilantro?

It would be pretty overwhelming to keep a totally unstructured mental list of absolutely every possible flavor and ingredient you might ever use. I'd feel adrift in decision paralysis. Instead, I like to build mental groups or categories of ingredients that share a pattern of flavor. That way, I can feel my way around from a zoomed-out, pattern-recognition point of view, and then think about nuances and specifics.

They're technically not "mental groups" anymore, since I've written them down here so they can be flipped through and perused like that graphic designer's Pantone guide, which shows thousands of colors as a spectrum, and describes them both in relation to one another and broken down into individual combinations of inks. So start with those broad categories of flavors, and then get into shades and variation within them. What is salty, and where can you get it? What about warm-spicy, or citrusy? And how do their other flavor layers shade them, changing their mood: palm sugar and orange marmalade are both sweet; sumac and rice vinegar are both sour; basil and tarragon are both fragrantly fresh-rich-herbal. Rather than being perfectly interchangeable (because that would be boring), they speak the same flavor language but in different accents.

Flavors from Taste vs. Flavors from Smell

The biggest distinction, the most obvious and fundamental pattern, is the huge difference between taste and smell. Flavors that come from taste work in a far different way than flavors that come from smell—whether it's the kinds of molecules, the way our brain forms signals into perceptions, or the roles each plays in your cooking. You need to take different approaches with flavors that come from the sense of taste and flavors that come from the sense of smell.

Tastes are the foundation and primary structural elements of a stone building, like a cathedral. Nothing stands up, flavor-wise, without them.

Smells are the stained glass—detailed, complex, varied, even narrative. They transform a barely tolerable stone cave to an airy kaleidoscope to marvel at.

If smell is baroque counterpoint, Vivaldi at his most ambitious and unhinged, Glenn Gould playing the *Goldberg Variations* at the height of his talent, distinct lines with their own rhythms and moods, thrilling and shimmering in their juxtaposition—taste is a single-melody Gregorian chant. Beautiful in its simplicity, but leaving nowhere to hide errors or missteps.

You've got to get tastes—salty, sour, sweet, umami, and bitter—right. None of them can substitute for another, and while they can modulate one another to a certain degree (sour reining in sweet, sweet taming bitter, umami lifting up salty), if you drastically change the sourness or sweetness of a dish, you drastically change what that dish is. On the flip side: there are near-endless ingredients that can get you the taste you're looking for.

If you're making a dressing for a salad, and you're working from a frame of "vinaigrette," you can think laterally from using red wine vinegar to using other vinegars: white wine, sherry, rice, apple. And yes, that does open up options. But if you think of it as "a sour sauce for leaves," rather than "a vinegar," suddenly you have many more options, because lots of things are sour: vinegars, but also citrus, fruit molasses, even sumac powder, yogurt, or pickle brine.

Same story with sweet: white sugar, sure, but how about honey, palm sugar, or sorghum syrup? Salty? Sea salt, yes, but consider herb salt, anchovies, or salty cheese. Depending on how far afield you reach, you may need to experiment just a little to get the sourness intense enough, as well as a balance in the other flavors that ride along—but fundamentally, if you can check that box on "sour" (or "sweet," or "salty"), you're in business.

You may be thinking, "But what about those other flavors that ride along? Isn't that something I need to deal with?" Yes—however, framed another way, they are also *potential*: shadings and opportunities. Put off by too much vinegary pungency? You don't have to think "vinegar." You can just think "sour." Find lemon too fruity to finish earthy vegetables? "Sours" like kimchi juice or strained yogurt will set you free. Your sister is pregnant and can't stand the smell of fish sauce, but you both agree the sauce needs some umami oomph? Try a Parmesan rind, nutritional yeast, or dried shiitake, as you see fit. The trick is learning to cut through the noise to accurately spot tastes in ingredients, and spot the roles they play in dishes. Tastes themselves are fairly inflexible, but the package they come in is very flexible.

Smell has the most material for expressing cultural or personal style—an "eggplant and tomato dish" can be from almost anywhere, but the aromas of the thyme, or oregano, or coriander, cumin, cloves, and other spices blended into the garam masala used to season it give it a sense of place rooted in Provence, Greece, or Punjab.

The space to layer, contrast, and juxtapose smell flavors in a dish or recipe has an elasticity that's very different from the clearly defined borders, and inherently limited (five) options for tastes. "Umami"—not the funk or mushroomy flavors that can come with umami, but the taste-bound umami itself—differs from ingredient to ingredient mostly in intensity, not in the qualities of how it feels sour. There's endless variation and shading to what "herbal" can mean, and even to the balance and impression that the flavor of a specific herb like basil can have—more clovey, more floral, more licoricey—and still recognizably be the flavor of basil.

If you're improvising on the flavor patterns within a recipe, you can make safer or bolder stylistic choices by selecting (smell-based) flavors closer to your target or further afield. Using pear in a salad or dessert where you're usually accustomed to using apple is one thing, but if you think instead of the need for a fruity element, not necessarily a botanically similar one, suddenly you can put guava or honeydew melon to use. It's using another warm and sweet-feeling spice like clove or star anise in a cookie where you usually use sweet and warm cinnamon, versus giving that "spiced" role to black pepper or even a little cumin.

With taste, our task is teaching ourselves to spot and accurately categorize a flavor as one of five. With smell, it's to develop a sense for the landscape, for what to look out for when you're surveying what's in season, for flexible organizing principles and tools that let us gauge how much a flavor embodies something like "basil," "herbal and fragrant," or simply "herbal"; just as important is developing a sense of the *degree* to which two flavors in the endless space of distinguishable flavors are similar or different.

I'd be a bad scientist if I told you there were obvious or highly objective rules or categories for these patterns in smell. We don't even have simple rules and categories for smells that all scientists agree on, let alone cooks. So, in this book, the nice overlapping area between science and cooking, we're even less in the realm of what's perfectly "correct"—we care about using the science to make things work. And, as professors of statistics say, "All models are wrong—but some are useful." Where taste has a narrow range of categories, each with a lot of depth, working with smell is more about making connections and detecting patterns over lots of breadth.

So, I've taken a long look into flavors and molecules in order to develop what I think is the best model for navigating them. We start with higher-level genres of flavor like "herbal," "spiced," and "fruity," and divide them into subgroups based on ingredients that have the greatest relative similarity. Sometimes the clearest path toward understanding a flavor like "meaty" is mostly about how and why certain smell molecules got there.

In each case, none of these genres or models cover every single possible facet, ingredient, or situation definitively—they're all "wrong" in plenty of ways—but they're all useful.

CHAPTER 5

The Five(ish) Tastes

Salty

Chemically speaking, a salt is a simple, not very exciting, ionically bonded type of molecule. Culinarily, salty is the lift-everything-up, balanced-with-an-edge taste of sodium chloride. Getting saltiness right is so fundamental that, in professional kitchens, it's referred to simply as "seasoning," as in "Season that before you send it out." Salty, more than any other flavor, is just about essential to incorporate into anything and everything you cook. Get the salt level tasting right, and everything else will be basically okay. Miss the mark, and the other flavors will have trouble bringing the dish back into balance.

Salty: The Only Taste That Senses Rocks

As a taste, salty's outlier status extends from the culinary to the biological. It's the only taste perception that doesn't come from biomolecules: while creatures from bacteria to mammals can create taste molecules like sugars, acids, or amino acids, no organism can make sodium or chloride from scratch. There's more or less a finite amount of salt on Earth, most of it dissolved in the oceans or deposited in mineral formations (where it goes by the geological name "halite").

Salt is a loosely bound duo of sodium and chloride, held together by their opposite ionic charges (sodium is the positively charged ion, chloride, the

Rules of Salty

Salty is the taste of the elemental mineral sodium, which we usually encounter in the form of sodium chloride (salt).

Saltiness makes other flavors taste more intense, focused, and balanced while suppressing bitterness.

Nearly all salts come from either evaporating seawater or mining mineral deposits.

Most of the flavor differences among salts are because of their texture, and sometimes trace (nonsodium) impurities.

Salt is an excellent preservative, making brined, cured, and preserved foods excellent ways to add saltiness to a recipe.

"Salting to taste" is the number one flavor technique to master for better cooking. Professional kitchens consider it so essential, they just call it "seasoning."

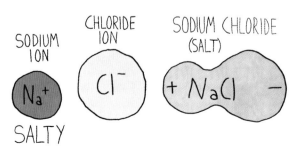

negative). They split apart in water, and it's these newly single sodium ions that actually create salty taste. Our main taste receptor for salty is shaped like an exceptionally tall donut or a hollow éclair, sized specifically for sodium ions to squeeze through into the taste cell.

Besides making things taste right, saltiness helps keep important biological systems going. Sodium ions are used in the nervous system as neurotransmitters, enabling us to pass information from our senses to our brain and have thoughts and feelings. Sodium also helps regulate muscle function, ensuring that the protein bundles and fibers contract and relax on schedule. And it's important for our blood pressure and cell operations that the sodium in our body fluids maintains a concentration of around 0.3 percent—if we have significantly more or significantly less sodium than that, we can get sick and die. Sensing and seeking out salty is a vital, but tricky game. When you get bloated and thirsty after you eat a lot of salty food, that's your body seeking out and retaining water to keep you in the proper range. We also have a second, reserve taste receptor for salty to help avoid oversaturation (or hypernatremia). It gets tripped only at very high concentrations of salt, and then sends that much less pleasant "too salty" signal to our brains.

Salt Loves Water

A lot of salt's versatility in roles culinary and biological comes from the powerfully magnetic friendship it has with water. Anywhere you put salt, water nearby will try to flow out to meet it. Salt's affinity for water makes it *hygroscopic*—it pulls moisture right out of the air. That's why salt left out in humid air gets palpably damp overnight and why people put rice in their saltshakers to absorb the water and prevent clumping.

Salt's ability to pull and push water makes it essential for maintaining fluid balances in our cells and tissues. We call this flow of water in response

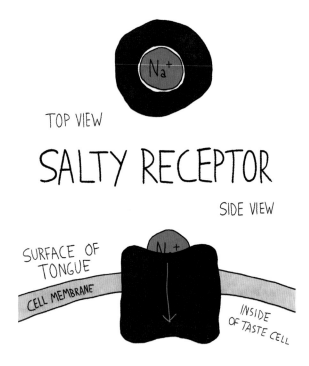

to salt *osmosis* (the source of the term "learning by osmosis," the idea that if you immerse yourself in information, some of it will flow into your brain to correct the imbalance, without any concerted studying by you involved).

Sodium's thrall over water happens for three reasons. First, it dissolves very easily in water, creating a salt brine wherever it does. Second, nature hates an imbalance, like a patch of briny water surrounded by saltless water, and will shake things up to correct one. Third, water is easier to move around than sodium. You can see this for yourself anytime you slice up cabbage for coleslaw or sauerkraut and then salt it. You create an imbalance: a concentrated brine solution where the salt has dissolved on the wet surface, and a barely salty inside. To correct the imbalance, either sodium ions have to go into the cabbage and make it saltier, or water has to come out of the cabbage to dilute the briny surface. Because water is easier to push through cell membranes than sodium, a little bit of sodium will slowly diffuse in, but a lot more water will flow out until there are equal concentrations of salt to water inside and outside.

The result? The sliced cabbage wilts in a puddle of exuded moisture.

As formerly living things, most raw ingredients are collections of cells, so osmosis happens when you salt most ingredients. If you salt onions when you start to cook them, osmosis will cause their water to start weeping out, and they'll soften faster. Salt the outside of a steak and osmosis will pull some of the meat juices to the surface, dissolving the salt. And because salt can (slowly) diffuse through watery ingredients and membranes, that salty meat juice will eventually diffuse back into the steak, seasoning it throughout.

Salty Flavor Patterns

Salty is a make-or-break flavor: everything tastes better when it's right and everything tastes flatter and less exciting when there's not enough of it. In between "salt's important, have at it" and absolutely every nuance there is to know about salt, we're working in the realm of common trends and patterns that salt and saltiness tend to follow. Which is lucky for us, because patterns are something we can recognize and use. (That's thinking like a scientist, by the way.) Keep these patterns in mind, and you can spot and improvise on them, or navigate the flavor landscape by feel rather than strict directions. Pick your metaphor—either way, you'll radically improve your cooking.

Salt Goes with Everything

When a recipe tells you "salt to taste," you should follow that advice. In fact, you should salt to your taste even when you haven't been told to. And, unless you're very sure otherwise, assume that you need to add salt to pretty much everything you cook: to a piece of meat before you sear it, a pot of potatoes on the boil, and to your mapo tofu.

And to desserts, where salt does its regular thing of deepening and intensifying flavors while also suppressing bitterness. Salt even selectively enhances sweetness, boosting it at lower levels of sugar and backing off at higher ones, all in all making everything immediately taste more dynamic, without that dull, I'm-getting-tired-of-eating-this feeling from too much sugar. Salt plays particularly well with browned flavors, especially in brown-buttery salted caramel and a well-salted chocolate chip cookie. In sweet dairy desserts like homemade vanilla ice cream and vanilla custard, going just a little higher than you think you should on the salt will make you feel like you're seeing them with multidimensional eyes. Salt even refreshes the taste of chocolate, enhancing its acidity and balancing its bitterness.

Try This

A good trick: if you're serving chocolate ice cream or chocolate mousse, drizzle a little of the best olive oil you can find over it, and then sprinkle some flaky salt on top. Fancy!

When to Salt?

The well-informed salter doesn't wait until the end of cooking to season. Like nineteenth-century political-machine-style voting, salting while cooking is best done early and often. Using frequent, gentle doses of salt throughout cooking works to slowly diffuse it into foods, giving it time to work its way through. If this sounds like an affectation, consider making some pasta two different ways: one boiled in well-salted water and the other boiled in plain unsalted water, then salted after draining. The first one will be close to perfect. The second one is both too salty and not salty enough, unbalanced to the point of unpleasantness. On the other hand, if you only add salt at the beginning of cooking something and never again, you'll actually increase the salty flavor, because water evaporation concentrates the salt. It's best to season with just a little salt at the start and more as you go, then taste and add a bit more if the dish tastes flat. For grains, pasta, root vegetables, and other vegetables just salting the water well at the beginning is sufficient, since you'll drain most of it off at the end.

Finishing a dish with salt—that is, sprinkling salt over it right before you serve it—wakes up and excites the palate. Think of the big white salt flecks on an otherwise bland soft pretzel or the salted rim on a margarita—each bite or sip starts with a salty jolt, which fades into all the other flavors.

Sea salts of different sizes and shapes and flavored salts work well for finishing, since they don't dissolve much and their crystal structure and shadings of flavor come through.

Which Salt to Choose?

For an ingredient whose taste comes from just a single molecule, salt has a surprising amount of variety in shape, color, and flavor. The very different-looking salts available—including chunky kosher salts, mined rock salts, sea salts evaporated in ponds or pans next to the actual sea, recrystallized sea salts with big flakes—are all based on sodium chloride. And sodium always tastes the same kind of salty, just in different intensities. The perceptibly different flavors in these salts are mostly based on crystal size, shape, or the presence of impurities, which are just molecules that are not sodium chloride.

Are You a Well-Informed Salter?

- Salt early and often.
- Consider using salty ingredients instead of plain salt.
- Finish with special salts.

For general savory cooking, the type and shape of salt you use for seasoning is mostly a matter of personal preference. Sea salt, flaky salt, kosher salt—all good. The only salt I don't recommend using if you can avoid it is ordinary table salt. (I make a special exception for fast food: it's just right for sticking to slightly greasy popcorn, fried fish, or French fries.) And in recipes for baking or fermentation, volume measurements of salt are calibrated to the variety of salt the recipe specifies, so you shouldn't mess around with it.

In my kitchen, you can usually find kosher salt, flaky Maldon, and fine sea salt from Trapani in Sicily, which tastes good, has a nice feel in the fingers for seasoning while cooking, and, even better, runs about four dollars or less for a two-pound bag.

The goal of seasoning while you're cooking is to get simple, even saltiness, evenly distributed. (Remember: early and often.) Salts with distinctive flavor qualities don't add much of note here, because they'll just get muted and diluted. Think of the saltiness added by this kind of seasoning as a straightforward primary color, rather than complexly patterned layers.

When I'm doing regular cooking, I never measure salt when I season. I go by taste and sight, sprinkling the salt in with my fingers. The size and feel of the granules is important for this to work: a bit of heft and roughness is helpful. You can get a feel for how much salt is in different-sized pinches and easily direct it where you want it to go. "Regular" table salt is essentially useless for this: the grains are so small it's impossible to get a decent grip on them, and they tend to slip and spill everywhere if you pour any into your palm. For most restaurants, fancy or casual, seasoning means kosher salt. It's inexpensive and easy to get, and the grain size is small enough to dissolve quickly but large enough to pick up in a pinch and control how much you're adding. I have worked at restaurants where the ingredient budgets were large enough to use flaky sea salt as a regular seasoning salt. More than a conspicuous affectation, the large, thin flakes are even more tactile to use than kosher salt, giving that much more precision to sensing with the fingers how much salt goes into things.

Finishing a dish with salt gives you a chance for final calibration: you can add more or back off to bring saltiness to just the right level. This is also a great opportunity to show off salt with different qualities—whether that's different crystal shapes or different flavors, either from small amounts of natural impurities or from other components that have been mixed together with the salt.

Salty and Crystalline: Sea Salts

Fleur de sel (or fiore de sale) is a solar-evaporated sea salt. It is often a little bit sticky, because it hasn't been dried quite so intensely as refined salts like table or kosher salt. It usually has smallish, irregular crystals, and its light crunch makes it a good finishing salt for lighter proteins like fish and lighter roasted vegetables. These salts are made by corralling seawater into shallow ponds called *salt pans*, where crystals begin forming on the surface of the sea as the sun starts to evaporate it and are skimmed or raked off to harvest them. Many well-known fleur de sel–style salts are named for where they are made: Trapani, Camargue, Guérande, Île de Ré.

Sel gris (or gray salt) is usually produced from the same solar-evaporated salt pans that make fleur de sel, but instead of being skimmed off the top, sel gris is collected by scraping and raking off the crystals that form on the bottom surface of the pan itself. They're gray from their higher clay content, and they're much larger, coarser, and crunchier than fleur de sel. Sel gris works great for finishing more robust dishes, like meat and root vegetables. Sel gris is also much higher yielding than fleur de sel, so it is often more economical to buy if you want a handmade sea salt but are on a budget. It's also less likely to be labeled with a distinct origin or producer than fleur de sel, even though it is made in many of the same places.

Flaky sea salts have a distinctly different shape than fleur de sel and sel gris: larger, crushable, hollow pyramids. Their salt pans are often heated to evaporate faster, and salt crystals left in the brine longer to grow larger and lacier. I like flaky salts especially because their texture is great as is, sprinkled onto a leafy salad, where you crunch through it as you bite; it's substantial, but shatters. The crystals are also easy to crumble in your fingers if you want to finish the dish with smaller flakes. Jacobsen and Maldon both make excellent flaky salt.

Try This

Salts are soluble in water but poorly soluble in alcohol and oil, so it's much easier to make a salty broth or boiling liquid than a salty oil or a salty liqueur. But you can also use this rule to your advantage: if you oil something and then sprinkle it with salt, the oil film prevents the salt from mixing with water and dissolving, creating crunchy pops of salt for a little excitement. Try drizzling leaves like arugula or tender radicchio with olive oil before sprinkling sea salt on them. They'll stay plumper and less wilted, and you'll taste a lively saltiness.

Salty and Craggy: Mined and Rock Salts

Besides the salt harvested from the ocean, there are also large mineral deposits of salt called *halite* that can be mined out of the ground. Mined salts are usually much drier (less sticky) than sea salts, and they come in blunt, chunky crystals that can be the size of gravel or as large as a baseball or even larger. We usually grind mined salts before we use them, either in a salt grinder (like a pepper grinder, but for salt) or with a Microplane or rasp.

Trapani Fiore de Sale: light, crunchy

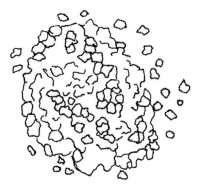

Fleur de Sel: light, crunchy

Sel Gris (gray salt): crunchy, slight minerality

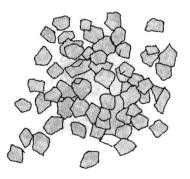

Flaky Sea Salt: crushable, briny

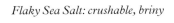

Pink Himalayan Salt: slightly earthy, mineral

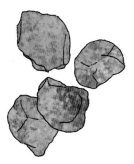

Kala Namak: funky, sulfury

Salty Cumin-Fennel Limeade

When you hear "salty and citrusy drink," you might think, ". . . like the salt rim on a margarita?" and yeah, kind of. Making this, I was thinking about a few drinks that hit a particular salty-sweet-sour bull's-eye: Vietnamese chanh muoi and xi muoi sodas, made with salt-preserved limes or plums (respectively), sugar, and seltzer; and the aforementioned jaljeera, the Indian summer tonic with sour limes, dried mango amchoor, and/or tamarind; cumin and other herbs and spices; and slightly sulfurous-funky black salt. In all of them, saltiness holds the sweetness back from the edge of cloying, but also gives it what I can only describe as a kind of neon, hyperreal edge.

Roughly crush ½ **teaspoon (1.25 g) cumin seeds** and ½ **teaspoon (1.25 g) fennel seeds** and combine them in a heatproof cup with a **scant ¼ teaspoon (0.5 g) ground cinnamon** and ¼ **cup (50 g) demerara sugar**. Pour over ½ **cup (125 g) boiling water**, stir to dissolve the sugar, and steep for 5 minutes.

Pour the mixture into a blender with **1 lime cut into quarters, ¼ teaspoon (1.5 g) fine sea salt, ¼ teaspoon (1.5 g) Himalayan black salt (kala namak**), **2½ cups (600 ml) of cold water**, and **1 cup (125 g) ice cubes**. Pulse to blend for 15 to 20 seconds, stopping before the lime gets completely pulverized. Strain the solids out and taste the limeade to adjust the dilution or sourness to your liking (it should taste quite salty, but not unpleasantly so). Drink right away, served over fresh ice.

Makes 1 very large or 2 smaller servings

Pink Himalayan salt gets its pink color from traces of iron oxides (i.e., rust). It is not actually mined in the Himalayas, but rather from mines in the Salt Range foothills in northern Pakistan, a few hundred miles from the western edge of the Himalayas. You can readily find plate-sized slabs of pink salt for twenty to thirty dollars at restaurant or kitchen supply stores. Since minerals like salt can absorb lots of heat and hold on to it for a long time, one neat trick is to gently heat up a slab on the grill or on a burner, and then use it as a flat-top surface for searing foods. Shrimp, fish, scallops, or thin pieces of meat cook and brown on the slab and pick up a nice salty crust from it.

Kala namak is a special heat-treated salt mined in northern India and Pakistan. It has small amounts of sulfur-based minerals in it, which make it appear black-brown in chunk form and medium pink when ground. These sulfury minerals also translate to pleasantly funky flavors, and the salt is prized in Indian cuisines for seasoning chutneys and salads, as well as the salty cumin-lemonade drink jaljeera.

Salt and Complexity: Salts Infused with Other Ingredients

Spices, herbs like thyme, flowers, chiles, citrus, and lots of other flavorful ingredients blend beautifully with salt and get a much longer shelf life than their dried forms in the deal. When you use these infused and compounded salts for seasoning and finishing, you create a subtle veil of complex flavor above and beyond pure saltiness. One sea-infused salt hybrid worth seeking out is **moshio,** or seaweed salt. Shio, or Japanese salt, is often heated rather than solar-evaporated to crystallize it, since much of coastal Japan has a cool and humid climate that slows dehydration. To make moshio, seawater is boiled with seaweed before evaporation, giving it umami depth and enhanced oceanity on the palate.

Oaxacan **sal de gusano**, or worm salt, mixes dried chiles, sea salt, and toasted *chinicuil*, the agave-eating larvae of the moth *Comadia redtenbacheri*. A traditional accompaniment to mezcal (especially sprinkled on an orange slice), sal de gusano delivers both fragrant heat and warm and toasty umami in addition to saltiness. Not only a partner to distilled agave, it also gives salty depth when sprinkled to finish agua frescas, stews, tomato salads, corn and squash dishes, salsas, tacos, or even simple cut fruit like pineapple, mango, or watermelon.

Making Flavored Salts

Compounding salt and flavorful ingredients yourself is pretty straightforward. It gives you control over how intense you want it to be, you'll know how fresh it actually is—no worries about its sitting indefinitely on a warehouse shelf—and it's an order of magnitude cheaper than and of equal quality to all but the most artisanal flavored salts.

To make a flavored salt, blend together kosher or sea salt (I use an inexpensive Trapani sea salt or Trader Joe's sea salt) with your chosen flavorful ingredient according to the ratios listed below, until the aromatic component breaks down into tiny pieces. Unless otherwise specified, pulse in a blender with a decent motor or a spice grinder. When you use fresh ingredients like shiso or chile, the resulting salt will be fairly damp. You can dry ingredients before blending them if this bothers you. I spread them out on a baking sheet and dry them in a turned-off oven for a day or so, which keeps a bit more of their flavor intensity.

Rosemary Salt

Sprinkle on pork or beef, roasted mushrooms, or on vanilla ice cream with a drizzle of olive oil or pistachio oil. Other good herb options: thyme, winter savory, sage.

The ratio by weight is 1 part rosemary to 12 parts salt.

Combine **2 tablespoons fresh rosemary leaves** with **3 tablespoons sea salt or kosher salt**. Pulse the mixture in a powerful blender or food processor a few times, until the rosemary leaves have been reduced to small particles, around the size of coarse-ground black pepper. Store tightly covered and away from light at room temperature, for up to 4 months.

Makes 4 to 5 tablespoons

Vanilla Salt

Since you're shelling out for whole vanilla pods, you may as well use a nice flaky sea salt for this recipe. The salt is splendid sprinkled on lobster, crab, sea urchin, raw scallops, or on a cookie right before you bake it.

The ratio is 1 bean (which yields 1 to 2 grams seeds) for every 50 grams salt.

Split **1 fresh vanilla pod** in half lengthwise along the natural seam, then scrape the blunt side of the tip of a knife down the length of the pod, gathering the sticky seeds on the blade.

Combine the seeds with **3 tablespoons flaky sea salt** (heaping tablespoons for very large flakes, shallow tablespoons for smaller or denser flakes). Mix the seeds into the salt with a fork, pushing with the tines to break up the mass of seeds. Store tightly covered and away from light at room temperature, for up to 4 months.

Makes 3 tablespoons

Habañero Salt

This salt brings out both the floral and spicy sides of the habañero. Look for aromatic, plump pepper specimens. Sprinkle sparingly, preferably with a lime squeeze, on avocado, grilled fish, corn, or any shellfish. The salt is also good as a milder version with the same amount of aji dulce, or with dried chiles like pasillas, smoky chile moritas, or chipotles. The ratio is 1 part habañero to 7 parts salt, which works out to about 2 habañeros for every 125 grams salt.

Using gloves, remove the stems, halve, and seed **2 fresh habañeros**. Combine with **7 tablespoons sea salt or kosher salt**, then pulse in a blender or food processor until the pepper is in small flakes. The salt will be fairly wet; you can use it immediately as is or spread it out on a plate in an out-of-the-way dry spot for 2 days. Store tightly covered and away from light at room temperature, for up to 4 months.

Makes about ½ cup

Celery Seed Salt

Celery salt is classic on a pickle-piled Chicago hot dog. I also like a sprinkling over olives, cheese cubes, salami, and other preserved pork. It's also good with caraway, cumin, or fennel seeds instead of celery seeds. The ratio is 1 part celery seed to 4 parts salt.

Mix **1 tablespoon celery seeds** with **1½ tablespoons salt**. Pulse in a blender or food processor until the seeds are coarsely ground. Store tightly covered and away from light at room temperature, for up to 4 months.

Makes 2 tablespoons

Lemon Salt

Sprinkle this salt over salads, sliced tomatoes, cooked leafy vegetables, sliced roast chicken, or sorbets or ice creams, especially when topped with olive oil. All citrus peels—lime, mandarin, grapefruit, Meyer lemon, and even yuzu or calamansi, if you can get your hands on fresh ones—make amazing salt. The ratio is 1 part zest to 10 parts salt, or about 2 zested lemons per 100 grams salt.

Mix **1 tablespoon lemon zest (from 1 lemon)** with **3 tablespoons kosher or sea salt** with a fork. Store tightly covered and away from light at room temperature, for up to 4 months.

Makes about ¼ cup

Pine Salt

Yes, this is made with ordinary conifer needles. It's resinous, fresh-citrusy, and lively. Sprinkle it on grilled asparagus or on braised meats, along with a little crème fraîche. The ratio is 1 part pine needles to 10 parts salt.

Find a nearby member of the pine family, which are all edible (see Note), including spruce and firs. If you're spoiled for choice, you can crush a few needles to release the aromas and see what you prefer. I especially like grapefruity Douglas fir. In the early spring, you might find lime-green new-growth spruce tips, which are tangy and tender enough to eat whole. This salt uses mature needles, but pick out ones on the less woody side if you can. Pluck some twigs, then pull off the needles.

Combine **3 tablespoons pine needles** with **3 tablespoons sea salt or kosher salt**. Pulse in a blender or food processor until the pine needles are reduced to small particles. Store tightly covered and away from light at room temperature, for up to 4 months.

Makes about ⅓ cup

Note: Make *very* sure you don't harvest anything from a yew tree, a non-pine conifer with bright red berries that's extremely poisonous. Some pine species like the ponderosa pine and the lodgepole pine have been known to induce labor in cattle, so be cautious or perhaps sit this one out if you're pregnant. Don't trespass to forage, and don't forage from trees that are likely to have been sprayed with insecticides or other treatments.

Shiso Salt

Shiso, a cultigen of *Perilla* that's a relative to mint and basil, has fragrant, slightly resinous, and fresh herbal notes, combined with slightly spicy, cumin-y flavors. *Perilla* varieties are widely used in Korean, Japanese, and Chinese cuisines as herbs or vegetables, so shiso or closely related sister herbs like deulkkae or Korean perilla are usually available in grocery and produce stores serving those communities, and reasonably often in farmers' markets. Sprinkle this salt on sliced plums, rice bowls, or rare salmon or mackerel. You can use red or green shiso leaves—however, note that because of the mineral ions in the salt, red ones will turn blue-green in salt form (but will usually change back to red when you sprinkle them on something). Other delicious options in the fragrant, delicate herb family for salts are spearmint, Thai basil, or lemon verbena. The ratio is 1 part shiso leaves to 5 parts salt.

Combine **2½ tablespoons chopped fresh shiso leaves** with **3 tablespoons sea salt or kosher salt.** Pulse in a blender or food processor until the leaves are reduced to small particles. Spread the damp salt in a thin layer on a plate in an out-of-the-way dry place for 2 days. Store tightly covered and away from light at room temperature, for up to 4 months.

Makes about ¼ cup

Salty and Funky: Cured and Fermented Ingredients

Crystalline salt may be the most precise way to season food, but using salty cured and fermented ingredients imparts complexity and depth along with salinity.

Easy Ways to Add Flavor Complexity with Salty Ingredients

- Chop up anchovies and sauté them with garlic or onions when you start a tomato sauce or braised dish.
- Balance slow-cooked meats and tagines with chopped preserved lemon.
- Add your favorite soy sauce to dressings or marinades for meats or sprinkle over just-cooked broccoli or green beans.
- Grate hard salty cheeses like Parmesan and pecorino over pastas, tomatoey or oniony soups and stews, or nearly any vegetable (start with brussels sprouts, roasted sweet potatoes, or cauliflower).
- Shred prosciutto or other dry-cured ham (jamón American Country, Chinese jinhua) onto fruit or sandwiches or stir them into potatoes, brussels sprouts, celery root, or squash about 10 minutes before you're finished roasting them. Toss little cubes of them or ham ends into soups and broths.
- Make like the avant-garde Basque restaurant Arzak: Gently dry and crisp thin ham/jamón slices on a rack in a 200°F oven for 1 hour, then blitz them into a powder in a blender and sprinkle this "ham salt" over cooked rice, tuna salad, or cantaloupe cubes.
- Cook up some high-quality dried spaghetti, season with butter and pepper, then grate bottarga (salted mullet roe, available online or in Italian specialty shops) all over it for a briny, slightly fishy, salty-umami delight.

Anchovies: cleanly fishy, silky, savory

Cured Pork: deep umami, porky, nutty

Soy Sauce: deep, malty, umami

Miso: deep umami, malty, fruity-floral

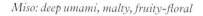

Capers: pickley, herbal

Olives: pickley, vegetal, slightly bitter

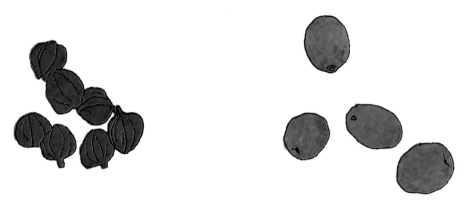

Salted fish like anchovies; salted porky meats like bacon and prosciutto; brined plants like capers, olives, and preserved lemons; fermented plants like kimchi and sauerkraut; and fermented sauces like soy sauce, miso, and doenjang (Korean soybean paste) all contribute savory, funky, and pickley flavors.

Caper-Preserved Lemon Dressing

Use this salty dressing on flavorful leaves like arugula, shaved fennel, or chicories. It also works well with crisp romaine hearts. Dressings like this, which have plenty of oil, make fantastic sauces for roasted carrots, brussels sprouts, or sweet potatoes; freshly steamed rice or other grains; and grilled fish or chicken.

Mix **2 tablespoons (30 g) minced preserved lemon (from about ½ preserved lemon)** and **1 tablespoon (10 g) capers.** Stir in **2 tablespoons (30 ml) fresh-squeezed lemon juice**, **1 tablespoon (15 ml) Dijon mustard,** and **1 grated garlic clove**. Whisk in **½ cup (120 ml) olive oil.** Serve within a few hours.

Makes about ¾ cup

Salt Loves Water, Part Two: Osmosis, Juiciness, and Preservation

Let's revisit that unassuming little scrap of salted cabbage from earlier. It's a powerful demonstration of salt's irrepressible talent for osmosis: it was rendered a pliant, crumpled heap lying in its own leaking juices, just from a brief encounter with salt. While roughing up cut vegetables has its uses, salt's other tendencies in the osmotic direction have applications that make it a sophisticated culinary operator in lots of arenas beyond finishing a dish.

Brining

Gastronomic chatter around brining tends to peak around Thanksgiving, when you'll see exhortations and procedures for submerging your turkey in salty water for a few days before you plan to roast it. But a cold soak in a 5 to 8 percent salt brine is also touted for chicken, pork chops, and other more everyday cuts of meat. The endgame? Thoroughly and evenly salty flavor, for one. The other benefits have less to do with tastes and rely on the significant changes salt can make to the texture of proteins.

Meat is a muscle, and a muscle is basically a striated sponge, made from incrementally layered and twisted fibrous proteins, and soaked in juices. Most muscle proteins just sit next to water, rather than dissolving in it (like mixing pieces of boiled egg with water, instead of a raw liquid egg). Facilitated by salt, a few of these proteins can dissolve into the juices of the muscle like melting gelatin, leaving a loosened solid matrix behind, and creating a juicy, binding effect. (This is easiest to see in sausages, which, when well-salted, have a bouncy, cohesive texture rather than a crumbly one.)

Proteins are all long strands of amino acids, each of which has unique stick-outty bits that act kind of like different types of fasteners: magnets, hooks, Velcro. A protein achieves, and holds, its particular globular or fibrous shape when it crumples up, and these different genres of fasteners catch on their matching mates. Speaking more literally, many of these amino acids carry ionic charges as part of their shape-holding mechanism, and when salt ions slowly diffuse into the muscle, they mess around with those charges a bit. Now, the protein strands repel and push away from one another, like strands of your hair ionically charged with static electricity. Since those proteins are interconnected in the muscle, it's kind of like your diaphragm pushing down and away from your chest cavity when you breathe. Instead of pulling in air, the proteins pull in the watery juices they're bathed in, and hold on to them. In real terms, it means a noticeably juicier piece of meat.

Brine Without the Brine: Dry-Salting

With its big containers and overnight soaking in quarts of liquid, brining can feel like a production. But you can still get some of the benefits of salt and osmosis by generously salting meaty ingredients and keeping them in the fridge for a day or so before you cook them. Whole chickens or skin-on chicken parts, steaks, pork loins or chops, and turkeys all become juicier, from the same protein-juice push-pull as brining, when given a salty rubdown and rest. (And, of course, seasoned into their bulk, rather than just on the surface.) Salt these generously—think confident, seasoning sprinkle rather than a whole-handed toss of salt—sit them on a plate in the fridge, and let the salt diffuse in for 4 to 6 hours or for up to 24 hours, when the process may begin to dry out the meat and make it disagreeably salty.

Preservation and Curing

In the same way that one cup of coffee will perk you up but ten cups of coffee will make you nauseated and twitchy, salting has a dose-dependent effect. At those lower and tamer levels of salt for brining or short-term dry-salting, a little bit of water flows out to balance the salt, then much of it diffuses back into the muscle along with the salt it just dissolved, and the texture is juicy.

Preparing pork legs and other cuts of meat for charcuterie and hams, or anchovies for curing, uses significantly larger amounts of salt, sometimes literally burying ingredients in salt for a time. Compare a hunk of prosciutto to a raw cut of fresh pork: the cured meat has almost leathery density in comparison. The extreme dose of salt involved creates a much harder osmotic pull on the water inside, and much more of it diffuses out of the muscle, compressing and firming its texture.

Why bother with this? That much salt has a preservative effect; its mechanism is basically the same thing that happens to the muscle. Picture this: you're a microbe, perhaps a *Listeria monocytogenes* bacterium or a *Cladosporium herbarum* mold spore. You're floating through the air and you happen upon a huge, uncooked pork leg just hanging around out in the open. You ready yourself for the trashed-hotel-room party you're about to inflict on this choice specimen, but as you land you feel an odd burning and rushing feeling just on the edges of your awareness. Before you can fully register what's happening, you're yeeted out of existence as the heavy-duty salt around you pulls all of the liquid out of your unicellular body. High-salt osmosis: no cell is safe.

Heavily salting meats and other ingredients likely caught on as a technique because of this preservative ability: got more hog than you can eat? Salt it, and save it for later. Aging over months and years also intensifies and funkified flavors in salt-cured foods, creating those highly salty complex-flavored ingredients we love so much: the porky, salty rainbow of cured meats from bacon to prosciutto, anchovies, capers, olives, miso, and soy sauce. ("Funkification" includes creating savory flavors, which you can read about in chapter 10.)

Sour

Sour is the bracing, refreshing, mouthwatering taste of acids. ("Tanginess," "tartness," and "acidity" are all different ways of saying "sour.") Next to salty, sour is the most important flavor to master to make your cooking good. Sourness lifts and brightens, saving dishes from tasting flat; it balances and reins in sweetness and bitterness when they threaten to overwhelm. It cuts beautifully through rich and fatty flavors, energizing your palate through contrast and banishing any sense of flabbiness. And it can infuse almost anything with lip-smacking liveliness.

What would a taco al pastor be without its little chunk of pineapple, a hot dog without sauerkraut or mustard, a BLT without rich and tangy mayonnaise? A little too much. Same for a daiquiri without lime or dark chocolate without its fruity-acid backbone. Acids make us involuntarily salivate, so sour quite literally cleanses the palate, cutting through and clearing any lingering fatty, starchy, or proteiny heaviness. We also know from sensory experiments that sour tends to suppress our perception of bitterness and sweetness, while enhancing subtle amounts of saltiness and even suppressing it when there's too much. Sour is the perfect balancer, the ultimate contrasting condiment of the tastes.

Sour has many more forms to choose from than salty: pure liquids like vinegar; squeezable fruits like lemons; dry powders like sumac; and thick goos like pomegranate molasses or yogurt. Likewise, there's an impressive range of other flavors to be found in sour ingredients: bright or dark fruitiness, vinegary pungency, salty-funky flavors, or creamy ones. This depth of field comes from the many different types of acid molecules that make sour, and the many ways to make them. Acids can be made by lots of different plants (a fruit like lemon, a stem like rhubarb, a leaf like sorrel), or created in a non-sour ingredient by fermentation (like vinegar, yogurt, or sauerkraut).

The Rules of Sour

Sour is the taste of acids, which are mostly made by plants or by fermentation.

Sourness balances sweetness, bitterness, and fattiness, and makes flavors seem brighter and livelier.

Sour enhances gentle saltiness and tames excessive saltiness.

There's a sour for every occasion—textures from flaky powders to liquids thick and thin, and flavor shades from bright to creamy to pungent.

The evolution of our sense of taste is a bit like a wardrobe, where we've acquired garments as we've attended events with specific needs. We have a ball gown, winter-weight wool sweaters (sweet, for getting easy energy from sugars), summer linen, a wet suit, or cycling cleats (salty, for making sure we get the sodium we need to function). Going by the current thinking in evolutionary biology, sour is more like the little black dress of tastes: not specialized for any one situation, but perfect and appropriate for lots of very different occasions.

Going back a few million years, a taste for seeking out sour foods would mean our ancestors were more likely to get sour-tasting vitamin C (also called ascorbic acid) in their diets, which is essential for connective tissue production and maintenance and is nice to have if you'd prefer not to develop scurvy and lose all your teeth. It's also sour-tasting and in lots of fruits.

Beyond vitamin C, fermenting microorganisms produce a totally different set of acids (namely, lactic and acetic acids) by totally different mechanisms, and the resulting acids also taste sour. If you put yourself in a prerefrigeration, even precooking mindset, a lot of the food that you'd forage or scavenge would be colonized by microorganisms. In other words, it would be naturally fermented or rotten. Producing acids is a fairly good rule of thumb for identifying a relatively harmless, even beneficial microbial process. On the other hand, spoilage microbes don't make much acid as they spoil things. Being able to taste sourness while you are weighing your food options tends to lead you away from things that are nonacidic and potentially harmful, and toward things that are acidic and generally less dangerous.

Humans also need a lot of calories to grow and develop our big brains, and foods that go through fermentation (and, as we've just established, often become acidic in the process) tend to be easier to chew and digest and have more caloric availability than their nonfermented counterparts. Chewing may seem like a minor point, but think of how much time animals like sheep or cows have to spend grazing and chewing in a day. It's not insignificant!

Combined with the human penchant for developing tools, all this may have even been self-reinforcing: you eat fermented foods, which you've been able to identify by their sourness, and start developing a liking for the sourness you can taste in them. So to keep getting those sour foods you like, you learn to cultivate and control fermentation, and start fermenting your foods. As yet another added benefit, acidic fermented foods not only haven't been spoiled by spoilage microbes, but their very acidity protects them from further contamination by those spoilage microbes. Now, you and your family and your buddies can keep the food you fermented around longer without it going bad—the fermentation literally preserves it. And with a store of food, you're all less vulnerable to the ebb and flow of food availability. You all stick around longer, passing on those fermentation skills (and sour-sensing genes) to your descendants. Thanks, evolution!

The Science of Sour

From a chemical point of view, acids are an easy-to-spot family. They're extroverted, reactive little molecules, ranging from gregarious social butterflies to outright pugnacious bullies, depending on their innate strength and their concentration. Like most food molecules, they have a backbone of carbon atoms studded with hydrogens, often with a few oxygens thrown in. Their principal trait is that they tend to break off and lose some of their hydrogens, trailing them in their wake like loose sequins from a vintage dress. Acidity is, quite simply, loose hydrogen ions. The more loose hydrogen ions something has, the more acidic it is.

Hydrogen ions get up to some wacky stuff in food, and they especially like to mess with the internal forces that keep protein molecules folded into their particular functional shapes. A little bit of hydrogen jostles milk proteins with a gentle force that coaxes them to form soft gels like yogurt or crème fraîche. In marinades, they latch onto muscle-fiber proteins in meat and push them to repel one another like two wrong-ended magnets, which draws in the fluid around them and makes the meat especially juicy and tender. When too

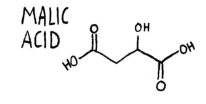

MALIC ACID

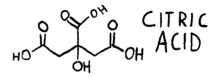

CITRIC ACID

ACETIC ACID

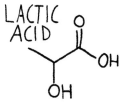

LACTIC ACID

many hydrogens dogpile on, the proteins reverse direction and start to tighten and coagulate into a firm, almost "cooked" texture (think ceviche).

When you eat something sour, it (and all its hydrogen ions) covers the surface of your tongue. Some of your taste buds come equipped with tiny, donut-shaped pores of a size just right for hydrogen ions—and nothing else—to fit through. (They are similar to the donut-shaped pores for sensing salty but much smaller.) As they file through, your taste bud cells diligently count them up, sending a signal of appropriate intensity to your brain to create the corresponding feeling of "sourness."

Sour doesn't just keep to itself either: it balances and dampens our perceptions of sweet, bitter, and excessive saltiness, by interfering with the joining-up of those taste buds and their taste molecules or by turning the volume down on their competing signals in the brain. This is why high acidity in (sweet) dessert wines and a squeeze of lime in your (bitter) gin and tonic are just about obligatory, and why I find half-sour pickles too salty and full-sour pickles just right, even if they have the same amount of salt in them.

A Guide to Our Acid Friends

Even though sour mostly comes down to hydrogen ions, and all hydrogen ions are identical, they come from a bunch of different acid molecules that have subtly (and sometimes, not so subtly) different vibes to their sour flavors: sharper or smoother, or even (pleasantly) pungent.

Just about every acid that matters to cooking comes from one of two places, from a plant or from fermentation. The plant versions are citric, malic, and tartaric acid; the fermentation ones, lactic and acetic acid.

Fruits are by far the most common plant parts we use as sour ingredients, and the list includes all citruses as well as apples, pomegranates, tamarind, and sumac. Pretty much all fruits are sour when they're underripe (as a kind of don't-eat-me defense mechanism) and accumulate sugars as they ripen. Some lose their acidity in this process, but the fruits that hold on to it stay appreciably sour. Besides fruits, some plants have acidic leaves or stems, which is what gives us sour ingredients like rhubarb, sorrel leaves, hibiscus, oxalis, and spruce tips.

Most fruits contain a mixture of citric acid and malic acid, which are both pretty clean-tasting acids, with citric acid suggesting a kind of neutral lemoniness and malic acid slightly sourer, lingering a little longer on the tongue.

Malic and Citric in Name Only

While malic acid is named after apples (*malum* in Latin) and citric acid is named after citrus, apples usually contain plenty of citric acid and citrus usually has a decent amount of malic acid.

The third major plant acid, tartaric acid, is found mostly in grapes and tamarind. (Archaeologists sometimes check amphorae and other ancient storage containers for lingering traces of tartaric acid, which is a good sign that they once contained wine.) Molecule for molecule, it's sourer than citric acid and about as sour as malic acid. Its naming story is a weird one: if you store wine for a while, a weak form of tartaric acid will fall out of the wine as solid crystals, which settle to the bottom of the cask or tank it's in, or form little encrustations on the walls. "Tartar" in the general sense refers to a scale or deposit on something (exactly the same as tartar on your teeth—gross, right?). At some point people went from saying "you know, that acid that forms a crusty scale or deposit" to the synonymous but slightly more elegant-sounding "tartaric acid." These lightly acidic crystal deposits are basically a waste product of wine making, but ground into a powder, they become **cream of tartar**, which has been used for ages as a handy cooking ingredient, helping stabilize whipped egg whites and the like.

There are two other acids worth knowing about, both of which come from fermentation. Fermentation is a bit like shepherding: if you herd sheep onto a plot of grass, they'll eat the grass and use that energy to, among other things, grow wool, something we want and can't get without the sheep. In a roundabout way, they're a living machine for transforming grass into wool. In the case of sour flavors, we shepherd a couple of special bacteria into ingredients, and they transform components of those ingredients into new, tasty acid molecules.

Minor Acid Players

Malic, citric, and tartaric cover most sour flavor in most sour things, but there are a whole bunch of minor acids in the foods we eat, too. Oxalic acid is a big one for rhubarb, sorrel, and oxalis leaves, the source of its name. It has an especially bracing, lightly astringent sour character, but it's slightly toxic in large quantities (this is why we don't eat rhubarb leaves, which have way more oxalic acid than the stalks). Ascorbic acid boosts up the sour flavor of citruses and other fruits. Limes contain succinic acid, which is astringent and a little bitter, and coffee has a complex blend of quinic, chlorogenic, and caffeic acids, which add a bit to sourness but mostly create bitter and astringent flavors. There's at least one acid we use in food that doesn't come from plants (or fermentation) at all: Coca-Cola and some other sodas get their sour flavor from phosphoric acid, created by heating phosphate-rich minerals. In excess, unfortunately, it's linked to dental erosion and even kidney stones.

Lactic acid bacteria do this by consuming sugars and creating lactic acid. They were named after their propensity to sour milk, which we've harnessed to make yogurt, cheese, crème fraîche, and other cultured dairy. After naming them, microbiologists figured out that lactic acid bacteria are tough and versatile powerhouses for creating sour flavors. They chew through all kinds of sugar: lactose in milk as well as the glucose, fructose, and maltose in different plants and grains. And they do it in all kinds of conditions, from salty, brined kimchi and sauerkraut to bubbly, gooey sourdough starters and bread levains. Lactic acid has a pleasant, almost rich-tasting, mouth-filling kind of sourness that's a bit stronger than malic or tartaric acid.

Acetic bacteria are little weirdos—the black sheep of the acid-producing bacteria, if you will—picky eaters that need a bit of coddling, consuming just the alcohol in things like wine or beer, ignoring the sugar, and demanding a warm environment with plenty of oxygen to do their work. You probably know of the acid they produce, acetic acid, as vinegar. Like them, vinegar is pretty weird. Most types of acid molecules have a sour taste and no aroma. They physically *can't* be smelled, because their hydrogen-popping, ionic ways mean they stick hard to water and don't ever float up to the olfactory receptors at the top of our noses. Acetic acid is the rare molecule that has a taste *and* a smell: it creates both the sour taste of vinegar and its vinegary, pungent aroma. It's both an acid, creating tastable hydrogen ions, and a volatile, able to take on a gaseous form and thus be smelled. This volatile feature gives it a special culinary edge, too. Other acids don't evaporate, so if you heat up something like apple cider or wine, they'll get more sour as the water boils off and the acid molecules concentrate. If you boil something vinegary, the acetic acid molecules will boil off as the water does, making it taste softer, not more sour.

Sour Flavor Patterns

Sourness is a bit like a highly skilled session musician who occasionally plays virtuosic solo shows. She can slip right into almost any genre, intuitively supporting the other players and adding an inscrutable sparkle that keeps her in high demand. But she has the chops, and the personality, to give a spotlight performance that leaves you wanting more. Sauerkraut and kimchi make a delicious base for deeply tangy stews, but each is also a lively accompaniment in small doses to fatty grilled beef or pork. Yogurt can sauce up vegetables as a contrasting finish and marinate meats to thoroughly suffuse them with acidity, while it's satisfying and complex enough as well to simply eat by the bowlful.

Easy Sour Additions to Make Your Life More Delicious

- Buy tinned anchovies. Open the tin and squeeze lemon juice on the anchovies. Wait five minutes for the lemon to soak in a little, then eat the lemony anchovies on buttered bread.
- Add a big spoonful of lemon juice, lime juice, sherry vinegar, rice vinegar, reduced wine, or pomegranate molasses to soup, to brighten it up.
- Take a leaf from vinegared sushi rice and splash a little vinegar, lemon or lime juice, or pickle brine (from cucumber, sauerkraut, kimchi) over hot cooked starches: rice, potatoes, or pasta.
- Sprinkle a little sumac or black lime powder on roasted or poached foods (cauliflower, sweet potatoes, eggs, chicken, white fish, salmon, carrots, broccoli) right before you serve them.
- Plop a spoonful of plain, whole-milk yogurt on root vegetables, roasted chicken, green beans, rice, and grilled lamb, as a sauce.
- Make a tangy pan sauce for cooked meat. To the same pan you browned the meat in, add a scant cup of slightly acidic liquid (red or white wine; sherry; vermouth; sour cherry juice for duck; apple cider for pork; or pineapple juice for pork belly or shoulder; crème fraîche for almost any meat, sherry vinegar diluted 1:1 with water or stock) and cook over medium heat, scraping up the browned bits. Simmer until it reduces and gets more syrupy. Stir in 1 to 2 teaspoons Dijon or other mustard you favor. Add fresh herbs if you like at the last minute. Pour over the meat you cooked.

You undoubtedly incorporate some sour flavors into your cooking already—maybe by drizzling something fresh and acidic on spinach after it's cooked, slipping pickled red onions into sandwiches or tacos, or keeping a favorite vinegar on hand to dress your salads.

Sour's range and versatility set it a bit apart from other tastes. You might spot it as an intense drizzle-able juice (lemon and vinegar), a powder to sprinkle (black lime and sumac), leaves or stems to chop or steep (rhubarb, hibiscus, and sorrel), viscous goos or pastes (pomegranate molasses, tamarind, yogurt), aromatic fruits (tart cherries, apples, green mango, sea buckthorn berries), or stolid, solid vegetation (sauerkraut, kosher pickles, kimchi, pickled mustard greens).

To keep the possibilities from becoming overwhelming, start like this: augmenting dishes with sour ingredients, using them as a finish or tasting and adjusting for acidity as you go, will make just about everything taste brighter, more balanced, and more finessed. Sour is a nearly all-purpose balancer. It makes sweet things less cloying, salty things less fatiguing, and takes bitter things from shuddering and wincing to intriguing and sophisticated. It lightens the mouth-load of rich, heavy, and fatty foods, physically cleansing the palate by making your mouth water.

Develop a more liberal hand with sour and you'll get all these benefits, even if you stick to safe mainstays like lemon juice and vinegar. After that, it's a matter of tasting around, getting comfortable, and figuring out which of the options out there for sour you like the most. There might be just a couple of players you want to add to your bench, or you may find yourself going on an acquisition spree. Either way, expanding your acidic intuition is one of the simplest ways to add interest and variety to your food.

Sour and Sharp: Vinegars

Vinegars, as you may recall from earlier, get their acidity from fermentation, which creates sour and pleasantly pungent-piercing-smelling acetic acid. Since vinegars can be fermented from almost anything alcoholic, they can have a huge range of flavor variation, but they all share this layer of aromatic, vinegary tang. (By definition, since acetic acid is one of the only food acids you can smell, that's how it smells, and you can't have vinegar without it.) The aromatic synergy of vinegar pushes that contrasting, palate-clearing power of sourness even further.

Vinegars range from 2 to 4 percent acidity for rice vinegars and 6 percent or more acidity for wine vinegars, and the extra verve of acetic acid makes them punch above their weight class, tasting stronger than sour fruits or pickles with similar percentages of acidity.

Wine vinegar, coming in red and white varieties, is the most classic and basic vinegar for cooking with a French or Italian inflection. Both varieties are lightly winey, strong on acidity, and definitely vinegar-pungent. **Red wine vinegar** is a little brasher and has a classic use as the vinegar in vinaigrette. **White wine vinegar** is a little bit softer and subtler, melting into the landscape in sauces like béarnaise and beurre blanc.

Beurre Blanc

Beurre blanc is a rich and sour butter sauce, made with white wine and white wine vinegar. Serve it with fish, scallops, or chicken, or with earthy, crisply roasted or sautéed mushrooms, brussels sprouts, or cabbage. Many cooks strain out the shallot before serving, but I like the texture. Right before serving, you can flavor beurre blanc with a tablespoon of fresh chopped herbs (chives, parsley, dill, tarragon) or a teaspoon of thyme leaves, grated orange zest, or smoked paprika.

Be sure your butter is cold when you add it; also, don't overheat it, or it will break and become oily instead of maintaining the emulsion you need.

Heat **1 tablespoon (15 g) unsalted butter** in a medium saucepan over medium-low heat. Add

Wine Vinegar: softly fruity   *Sherry Vinegar: nutty sherry, apple, musty*

Balsamic Vinegar: sweet, resinous, dried fruit, caramel   *Rice Vinegar: mildly malty to deeply caramelized*

1 tablespoon (15 ml) finely minced shallot (half a shallot) and cook until soft and translucent. Add **2 tablespoons (30 ml) dry white wine** and **2 tablespoons (30 ml) white wine vinegar**. Bring to a simmer and reduce until syrupy. Remove from the heat and let cool until very warm (baby bath temperature, not hot). While it's cooling, cut **2 sticks (225 g) unsalted butter** into small cubes. Whisk the butter into the warm shallots and wine, a few at a time. The first few will melt, but gradually the added cubes will start to slump into a thick fluid. This is the texture you're aiming for. Keep whisking in butter until you've used all of it. Season with **salt** and **pepper** to taste.

If the sauce gets too cold, rewarm over very low heat for a few seconds. Serve immediately.

Makes 1 generous cup

Variations

Once you've made beurre blanc the textbook way and understand how its flavor profile works, it can be an excellent canvas for your own experiments with sour. For instance, you could go bold and completely swap out the wine and the wine vinegar for completely different, non-wine-based ingredients, or play it a little safer and start by replacing a tablespoon of one of them with a sour ingredient you want to get to know better.

Some ideas to get you started thinking about experimentation:

Substituting a tangy, high-acid sake for the wine and rice vinegar for the wine vinegar is awesome with delicate fish and shellfish; dry sherry and sherry vinegar, subbed similarly, make for tanginess with a little nuttiness.

Maybe a strongly acidic ingredient taking on all or part of white wine vinegar's role: perhaps lemon or lime juice, finished with a couple of gratings of corresponding zest, or pomegranate molasses.

Or a more softly sour ingredient augmenting white wine: kimchi brine, sauerkraut brine, or umesu (Japanese red plum vinegar) for pork, sausages, or vegetables; or a not very hoppy sour beer, like a lambic.

Taste at the syrupy reduction stage of the recipe, and at the end, to see if you like where it's going. Augment with a little dash of white wine vinegar before serving if it seems a little flat or unbalanced.

Left to its own devices for long enough with any contact with air, any alcoholic drink will eventually sour into vinegar. That's why the classic vinegar of any locale tends to be made from the alcoholic beverage of that place. In England, which produces very little wine, the alcohol is beer, and the vinegar is malt. The aromatic pungency of malt vinegar is part of the quintessential fish and chips experience sprinkled over both. **Malt vinegars** are usually straightforward and uncomplicated, with a little bit of a caramelized character and a pleasantly British fustiness, reminiscent of a picked onion. I wouldn't necessarily use them in every instance I'd use a wine vinegar, but they go well with mustard in dressings or as a final seasoning to anything fried.

Sherry vinegar is also, technically, a wine vinegar, since sherry is technically a wine. But sherry's complex, sometimes nutty, sometimes sweet, sometimes green-apple-inflected flavor layers make for a more complex wine and, as a result, a more complex vinegar. The most basic and inexpensive sherry vinegars have a nice little dose of this complexity; more artisanal varieties, made in limited quantities, have a lot more. Sherry vinegar is a classic addition to brighten up gazpacho, and the fruity-nutty-caramel hints in sherry vinegar can pair with almost anything ripe-tomato-based. Sherry vinegar also slips into vinaigrettes and sauces for meat (try deglazing the drippings from chicken with sherry vinegar!) with great ease.

Complexity is the name of the game with **balsamic vinegar**.

To be clear, I don't mean the caramel-color-added four-dollar bottle of supermarket balsamic. In Italy, traditional balsamic vinegar is the rich, syrupy, dark, dark brown condiment that has earned the right to carry the initials "D.O.P." on its label (the abbreviation in Italian for "Protected Designation of Origin"). TBV, as we might flippantly call it, is a case study in earning flavor by letting it compound and accumulate over time. I've visited one vinegar "factory" in Modena that is more like a manor house that happens to have floors dedicated to long-term vinegar storage. Grapes are unloaded in the basement, pressed to make must (grape juice), reduced down into a light syrup, yeast-fermented for about a year, and carried up to the warm attic. There, it's poured into the largest barrel in a decreasing-sized set called a *batteria*, lined up from biggest to smallest like unpacked matryoshka dolls. Every year, a portion of semifinished vinegar is carefully taken out of each barrel, and moved down the line of the batteria to fill up the next, slightly smaller barrel. You can tell the age of the vinegar in any barrel in a batteria by counting down the line: the biggest barrel has the youngest vinegar. The fifth barrel down the line is full of vinegar that has gone through four annual barrel transfers, and is nearly five years old. The smallest barrel (the twelfth in the smallest *batteria* sets, the twentieth or even higher in larger sets) holds the oldest vinegar, and in the yearly transfer down the *batteria*, it has no smaller barrel to go to. Instead, this is the barrel that gets bottled as finished vinegar. All this bother results in a sweet, oh-so-lightly pungent, richly resinous, dark-caramel-fig-molasses ichor, sealed up in tiny glass bottles like reliquaries holding the blood of a saint, and is worth every euro of the dozens to hundreds you'll part with to get it.

Try This

If you get your hands on real, traditional balsamic vinegar, make a judicious but luxe drizzle over buratta or sliced mozzarella at the beginning of a meal. Or, do so over the best strawberries you can find, sliced, at the end of the meal. Really, just a few drops will do.

Try This

Like salt, many of the standout uses for sour involve using it to finish things at the last minute: a squeeze of lemon right before you serve something, or lime wedges so whoever's eating can do it themselves, for maximum freshness. There's also a lot to be said for really working sour into things early in the cooking process, not unlike brining or dry-salting.

Marinating vinegar all the way through ingredients makes them excellent high-acid condiments for contrast and balance in their own right, to add to sandwiches, tacos, salads, etc. Vinegar pickling preserves a lot of raw and fresh flavors; it also tends to make textures softer and limper, but not "cooked" soft.

Here's a basic, adaptable method: heat up 1 cup vinegar (white for the most neutral, rice makes a nice softer pickle, sherry is robust and a little sweet) until it's almost boiling, then pour over about 1 cup sliced red onion, and cover with a plate or lid. Marinate for 1 to 3 hours, longer for thicker or tougher things.

Further delicious options: ¼-inch-thick sliced cucumber, fennel, jalapeño slices, carrot sticks or coins no more than ½ inch thick, small mushrooms, sturdy cabbage leaves, mustard seeds.

Rice vinegar, at least in the style most accessible in the United States, tends to have a lower amount of acid than other vinegars (about 2 to 4 percent, as opposed to 6 percent or more), and I think its mild, tart-but-almost-malty quality makes it incredibly versatile. I often use a fifty-fifty blend of rice vinegar and lemon to season cooked vegetables, raw cucumber, and salads. In addition, straight rice vinegar makes an excellent quick pickle out of almost any vegetables, a style of pickling called *suzuke* in Japan. Besides these mellow white-to-yellow rice vinegars, many cities in East Asia produce their own unique styles of black vinegars, another example of time and patience buying color and complexity. Two of the best are toasty-flavored Kurosu, from Kagoshima Prefecture, made by aging fermented rice outdoors in ceramic jars, and Chinkiang or Zhenjiang vinegar, which involves making a pei, a pastelike brick, from alcohol-fermented rice, wheat bran, and rice hulls, slowly fermenting the pei into vinegar over the course of months, and then leaching the vinegar itself off the brick. Another good bet is Shangxi aged vinegar, made with similar techniques, with sorghum, barley, and peas in the mix; its deep, malty, caramelized, slightly smoky flavor is a go-to for seasoning noodles.

Black Garlic–Black Vinegar Mayo

This sauce is roasty, malty, sweet, funky, and creamy all at the same time. Some suggestions: use as a dipping sauce for grilled shishito, Padrón, or even bell peppers; as a sauce for boiled or roasted potatoes; in place of mayo in tuna salad, with a little bit of finely minced white onion; or go oeufs mayonnaise (plop it on hard-boiled eggs).

Chinkiang/Zhenjiang black vinegar, especially the Koon Chun brand, is usually available at well-stocked Western grocery stores and specialty markets serving Chinese diaspora communities. Some good higher-end brands are Soeos and Yuho. Shangxi black vinegar or Japanese Kurosu are both good, albeit stronger-tasting substitutes. You can also dilute regular rice vinegar 1:1 with water and add ¼ teaspoon dark brown sugar or molasses per tablespoon of diluted vinegar.

Black garlic is fresh garlic that's been hot-aged until it blackened, sweet, a little funky, and leathery to dry in texture. I recommend buying it ground rather than grinding it yourself. You can find versions of it from small producers or even from Trader Joe's, but if you have trouble sourcing it locally, order it online—I like Kalustyan's myself. There's no equivalent substitute for it, so if you can't find it and need to leave it out, you'll get a tangier, less complex (but still tasty) version dominated more by the aromas of the black vinegar.

Whisk **1 large egg yolk**, **1 teaspoon (5 g) Dijon mustard**, **a pinch of salt**, **1 tablespoon (15 ml) black vinegar**, and **½ teaspoon (2 g) powdered black garlic** in a medium bowl until well combined. Starting with one drop at a time, whisk in **about ¾ cup (180 ml) grapeseed or other neutral vegetable oil**, building up to a thin stream, and continue whisking until the mayonnaise is stiff. Alternatively, you can use an immersion blender to speed the process up. Store tightly covered in the fridge for up to 2 days.

Makes 1 scant cup

I've heard some so-called foodies say that they only use **distilled white vinegar** to clean the floor. This snobbery means that they miss out on the brilliant applications of white vinegar. Its basic appeal is like good use of negative space; it adds acidity and vinegary pungency—and no other flavors—to whatever you mix it with. In other words, there's no way for it to muddy the flavor of things like perfect vinegar-pickled cucumbers, watermelon rind, or peaches, or mayo and black pepper Alabama white barbecue sauce.

Sour and Bright: Citrus

Lemons have a reputation of being the most basic, go-to sour seasoner, which does a disservice to all the other excellent sour ingredients out there, as well as to the lemon itself, which is powerfully and complexly aromatic and not at all basic. Lemons have a lot of acid (5 to 8 percent acidity) and just a little sugar, making their sourness taste pretty uncomplicated. They have way more aroma in the peel than in the flesh, though a tangible population of lemon-flavored smell molecules end up in the juice. Lemons have a strong, general-citrus aromatic backbone overlaid with intense and heady peely-lemongrass flavors. There's a bit of sweet orangey-ness, and floral character there, too. The lemon's aromatic bouquet is a deceptively important part of its success as a seasoner— add lemon juice to a sauce or dish while you're still cooking it, and it tends to feel oddly flat. That flatness is the feeling of absent aroma. Squeezed over greens, in a chicken pan sauce, or as part of a vinaigrette, lemons boast a bright and fruity-sourness that shines. They do a good job lifting up European herbs (parsley, thyme, rosemary), lighter meats, and bitter greens. Lemon doesn't always work as well with more assertive or deeply complex flavors, where its relative delicacy can get overwhelmed and feel a little too sweet-smelling and simple. In many of these cases, you'll want to reach for a lime.

With **limes**, expect a more muscular, bright piney herbal green-resinousness that makes itself known, standing up perfectly in gin and tonics, guacamoles and salsas, and many funkier Southeast Asian broth, noodle, and meat dishes— without overwhelming them. Limes' acidity is comparable with lemons', if occasionally a bit higher (6 to 9 percent), which makes them reasonably interchangeable from the perspective of pure sourness. (It's the aromas of limes and lemons that account for their differences in flavor when you swap them.) And even within the category of limes, different varieties have different flavor shadings—"regular" supermarket limes are technically Persian limes, an offshoot of the longer-established **Key lime**, which is quite a bit smaller and noticeably more sour, but worth getting if you can find them. Some cocktail historians have pointed out that many classic lime-containing cocktails were originally based on Key lime juice, not Persian lime juice, and Key limes definitely do something delicious to a daiquiri or a margarita.

The further you delve into citrus on the search for just the right shade of sour, the more intriguing, intense, and even haunting the flavors you'll find. Extra-aromatic citrus varieties are often not quite as sour as lemon and limes but have flavor shadings that would otherwise be impossible to find. The **Meyer lemon** is actually not a lemon at all but a different hybrid of the same ancestral varieties; it has the lemon background flavor as you'd expect from its namesake, but also herbal, woody, and extra-floral qualities and sometimes flower-petal notes. Its sourness is between that of an orange and a lemon.

Grapefruits are less sour still (1 to 2.5 percent acidity) and sweeter but with an extra complexity from their bitterness. Their grapefruitiness has some dimension to it—there are more penetrating, almost pungent aspects; superfresh, top-notey peely qualities; and a complex base note complex with sweet-candied flavors and some herbaciousness. Grapefruit's reduced sourness often means that you need to use more of it than you would lemon or augment it with another sour ingredient.

Chilled Soba Noodles with Grapefruit

I like an acidic component in chilled noodle things. A more typical version of this recipe might have a chilled broth with a bit of rice vinegar. Here the bitter-sour-sweet-aromatic grapefruit is a nice counterbalance to the earthy buckwheat.

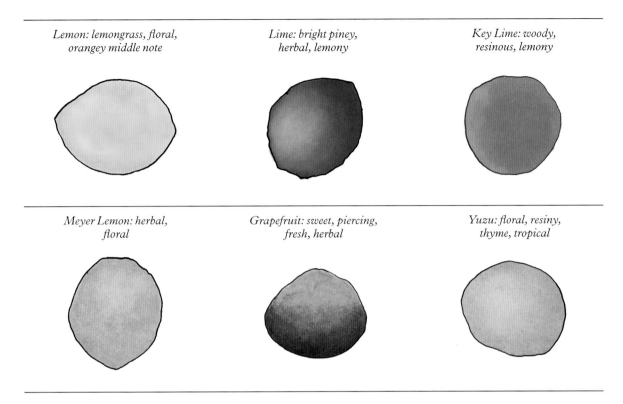

Lemon: lemongrass, floral, orangey middle note

Lime: bright piney, herbal, lemony

Key Lime: woody, resinous, lemony

Meyer Lemon: herbal, floral

Grapefruit: sweet, piercing, fresh, herbal

Yuzu: floral, resiny, thyme, tropical

Bring a large pot of **salted water** to a boil. Add **6 ounces (170 g) dried soba noodles**. Boil for 4 to 5 minutes, until just cooked through (or follow the package instructions if the timing is different). Drain and toss the cooked noodles with ice cubes in a colander to rapidly chill them.

Cut **1 medium grapefruit** (red or oroblanco) in half. Cut supremes from one half (slice off the end, then stand upright and cut down on all sides to remove the peel and pith, then cut between the membranes to release the segments) or just peel and separate segments. Set aside. Squeeze the other half, reserving the juice.

Whisk together **5 tablespoons (75 ml) excellent grassy extra-virgin olive oil** with **4 tablespoons (60 ml) of the reserved grapefruit juice, 3 grinds of black pepper**, and **a pinch of sea salt**.

Toss the supremes or segments in the dressing, remove, and set aside. Dress the chilled noodles with the dressing until they are well coated;

a little pool of dressing may collect at the bottom, but the noodles should not be dripping. Gently toss the dressed grapefruit with the dressed noodles. Garnish with **2 tablespoons (30 ml) minced chives** and more freshly ground black pepper. Serve immediately.

Serves 2 to 4 as a substantial starter or side

You should taste every single one of the vast family of delicious, sour, powerfully aromatic citrus that you can get your hands on. But I want to especially call out **calamansi** and **yuzu**. Calamansi is a beloved Filipino lime variety with a super-sweet-smelling, floral-citronella aroma, and sour juice. Yuzu is possibly my favorite citrus. It produces slightly less juice than a lemon and is almost as sour. The aroma profile shading yuzu's sourness takes the concept of citrusy-lemony and blows up several of its dimensions. These fruits are extra-floral, and extra-fruity, with hints of resiny thyme, coriander, and tropical fruit that I find hauntingly addictive. Calamansi and yuzu have become easier to find in the United States in the last few years. Both are worth seeking out to broaden your sour palette, and they make amazing additions to dressings and sauces, and drinks, or as the final squeeze over a dish.

Sour and Darkly Fruity: Other Sour Fruits (and Flowers!)

Every sour ingredient you might use in the kitchen has dimensions of flavor beyond acidity (unless you're working with powdered lactic acid or something). While citrus runs a whole gamut of these, there are delicious possibilities of "sour and fruity" beyond those boundaries.

Most fruits can't match the sourness of citrus on their own—they're more like "plenty sour" instead of "really sour."

You can look at this as an opportunity: use more of them, and get more of their extra flavors.

To increase storability and to get a little more sour oomph, a lot of sour fruits are reduced, semidried, or dried all the way into a powder. The effect on their aromatic flavors is fewer heady top notes, and instead a kind of dark-toned depth that can range from notes of berries and fruit leather to bass-deep and caramelized.

Pomegranate molasses is a darkly fruity, syrupy sour condiment par excellence. It's a simmered reduction of pomegranate juice, a process that removes water and concentrates its acids fourfold or more. The viscosity of pomegranate molasses means you can drizzle it on dishes and it will stay put rather than waterishly seeping everywhere. Other heat-reduced fruit juices in the category of sour condiments include grape saba and vincotto from Italy, Turkish pekmez

Pomegranate Molasses: deeply fruity, sweet

Sumac: clean, fruit leather

Tamarind: roasty, spicy-citrusy, leathery

Hibiscus: sweet red fruit, plum, cherry

Black Lime: resinous, leathery, musky

As a finishing drop or drizzle, pomegranate mo-lasses is similar in acidity to vinegar or lemon juice. However, it comes on a little slower and less sharply until it fills your mouth with deep, lush sourness. I really like it as a counterpoint to sweet and browned flavors from sautéing, roasting, or grilling.

Drizzle grilled corn kernels with a tablespoon of pomegranate molasses and a few tablespoons of chopped mint, adding more molasses if you need a bigger acid fix.

Here are more places to try it: Drip it over yo-gurt, grilled or roasted chicken, or lamb. Stir a spoonful into soups or drinks (iced tea, lemonade) to impart a noncitrus tang.

made from grapes or mulberries, and American boiled cider (essentially, apple molasses); all of these express the flavor base notes of their in-gredients, and share a similar thickened, textural versatility.

The **sumac** shrub sets fruit that looks like dozens of fuzzy, spherical red flowers, but they're technically drupes, structural analogues to a grape or a cherry, but much smaller. They're also, short of citrus, one of the sourest fruits you can find, with a clean, berrylike fruity flavor. Their texture is best expressed ground into a fine, dark red, slightly damp powder. It works well as an ad-dition to a wet sauce or marinade, swelling up a bit and creating a kind of tangy coating, as well as for sprinkling: it sticks, and stays put, without adding extra moisture. In many Middle Eastern cuisines, it is sprinkled into cooked rice, on ke-babs and other meats, over salads and meze, and mixed with sesame seeds and a wild thyme called za'atar for the all-purpose dry condiment of the same name.

Dry-Sautéed Green Beans with Sumac

I love a tangy, brightly sour green bean, but I sometimes dislike how much a juicy sauce will simply slip off of one. Sumac keeps a solid, some-what sticky texture throughout cooking, and its red-fruitiness is a nice dimension with the vege-tal beans. I also find that while lemon juice can taste quite flat when heated, sumac retains its liveliness.

Combine ¾ **pound (350 g) green beans or haricots verts** (washed and trimmed, and still wet), **1 tablespoon (15 ml) olive oil**, **1 table-spoon (10 g) sumac**, and **2 pinches of flaky sea salt** (about ¾ teaspoon) in a large skillet, and heat over medium-high heat. Cover and let them half-sauté, half-steam, removing the cover and turning the pile over frequently, re-covering the pan in between stirrings and adjusting the heat as needed so the sumac doesn't burn. When the beans are thoroughly cooked, 8 to 12 minutes, they will be coated in sumac and oil with not much liquid left, but crank the heat to boil off extra liquid if necessary. Sprinkle ½ **tablespoon (5 g) more sumac** and **1 more pinch sea salt** over the beans before serving. Eat immediately or cooled to room temperature.

Serves 2 to 3 as a substantial side

While limes are already plenty sour, drying them to make **Omani black limes**, also known as loomi, for long-term storage concentrates their acids even further. Once dry, they keep many of the deep and resinous citrusy tones of a fresh lime, layered with a dark, fruity, leathery muskiness. They're especially associated with the Gulf states around Oman but are at home in many dishes throughout the Middle East. Like sumac, black limes work splendidly ground into a flaky powder, and they work in many of the same places: sprin-kled over rice, vegetables, fish, beef short ribs or kebabs, yogurt. Whole black limes feel as light and hollow as a snail's shell, and in this form, they're especially good for infusing into things while they're cooking, then fishing them out at the end (right into the rice cooking water, for example, for more flavor penetration than a sprinkle).

Sour Fruits (and Near-Fruits) That Steep

Tamarind, in its fresh state, is a brittle brown seed pod full of aged balsamic, spicy-citrusy, lip-smackingly tart pulp. It needs a little processing to show its best side, often dehusking, soaking in boiling water until softened, then pureed and strained. An amazing agua fresca, called *de tamarindo*, employs just such a technique, with about ¼ cup water and an added tablespoon of sugar per pod, then mixed with water to taste over ice.

Tamarind gives a roasty, tangy edge to chutneys and pad thai; I usually keep jarred tamarind concentrate in the fridge and add a teaspoon to chickpeas or chicken, especially when braised or otherwise saucy.

Hibiscus is actually not a fruit at all, but the fleshy leaves that protect the blossom of a hibiscus flower. Like tamarind, it benefits from a steep in water, infusing into a deep red, sour liquid with notes of plum and black cherry. It makes a fantastic drink that's drunk all over—as jamaica in Mexico, bissap in and around Senegal, and sorrel in the Caribbean. If you're a Celestial Seasonings fan, you already drink steeped hibiscus, as the main ingredient in Red Zinger tea.

Black Lime and Cilantro Broth

Really good broth is a perfect canvas for flavor: you can chuck lots of things into it to infuse, and it will happily soak up their essences and deliver them together in one tight package, with no textures or solidity to distract. Here, black limes bring mouthwatering acidity and deep, funky citrus notes, and the added cilantro brings heady fresh aromas, hitting flavor layers all the way down to the meaty core of the broth.

Bring **1 quart (1 liter) of good, preferably homemade chicken broth** to a simmer in a medium-sized pot or Dutch oven over medium heat. When it's simmering, add **half of a dried black lime** (saw it apart with a serrated bread knife) and about **½ cup (35 g) cilantro stems** (from a medium-sized bunch). Cover and continue to heat over low for 20 minutes. Remove the stems and the dried lime and add **½ teaspoon (3 g) salt**, or to taste.

Right before you serve the broth, stir in **½ cup (20 g) finely chopped cilantro leaves**, allowing them to wilt a bit. Serve hot, and immediately.

Serves 2 to 4

Sour and Creamy: Cultured Dairy

Until relatively recently in human history, tangy and fermented dairy was more the rule than the exception. Dairy is such a hospitable environment for beneficial bacteria to do their fermenting thing that unpasteurized and unrefrigerated milk begins to go pleasantly sour within a few hours of milking. This relationship is immortalized in the naming of these bacteria, and the acid they produce: lactic acid bacteria, and lactic acid.

Tangy, fermented dairy products come in all kinds of variations: yogurt, sour cream, labneh, crème fraîche, buttermilk, and kefir are some of the big ones. They all share a thickened texture, from the unwinding and remeshing effect of acid on their proteins, and certain "dairy" flavors that only really appear after fermentation—yogurty, for one, but also aromatic notes of creamy and buttery.

Crème fraîche and **yogurt**, which are fermented into thick semisolids, in particular make really good one-ingredient sauces. Crème fraîche has a higher fat content and softens and thins out as you heat it. Plop it over hot steamed peas, stir into simply cooked white beans or lentils, or use it to deglaze all the browned meat bits on the bottom of the pan when you cook meat (then, pour the

Yogurt: gelled, runny, mild to very tangy

Labne: thick, pastey, slightly creamy, tart

Crème Fraîche: fatty, silky, buttery, slightly cheesy

Buttermilk: buttery, creamy, slightly funky

Yogurt Whey: clean, green-apple, creamy-smelling

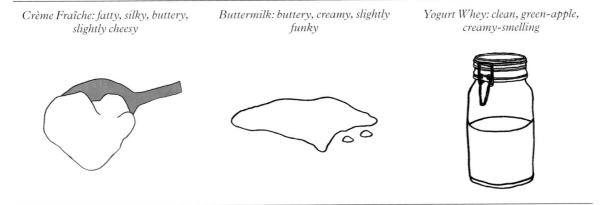

creamy deglazings over the meat and the other foods on your plate). Yogurt is more gel-like, with less fat and more protein. It's delicious on so many things with just a little bit of salt and black pepper stirred in: roasted potatoes or sweet potatoes, chunks of squash, cucumbers, or slow-roasted meat. When you open a container of yogurt, you'll often see that the gel has broken slightly and liquid has seeped out. For most of my life I considered this to be "gross yogurt water," something to be poured off and discarded. It's actually **yogurt whey** (also called acid whey), a delicious sour ingredient in its own right, with an incredibly clean flavor that only hints at dairy creaminess.

Yogurt Whey and Strained Yogurt

Straining yogurt allows some of its whey to drip out, leaving a thicker, more luxurious yogurt, a style that's often called *Greek yogurt*. Draining longer makes thicker and nearly spreadable labneh.

The yogurt whey is limpidly clear and pleasantly tangy. I use it where I might otherwise use vinegar, lemon juice, or wine, when I don't want to be distracted by citrusy, pungent, or fruity flavors.

I recommend using a tangier yogurt for this recipe; very mild yogurts make for underwhelming whey. The recipe can easily be doubled, if you want to make a lot.

Start with **1 pint (500 g) good-quality whole-milk plain yogurt**. Line a strainer with four layers of cheesecloth or a very clean dish towel and set it in a larger container. Pour the yogurt into the strainer.

(Alternatively, you can cut a big square of four layers of cheesecloth, drape it in the container, and pour the yogurt in it. Gently gather the four corners of the cloth, without spilling the yogurt, and carefully twist and tie them shut in the center so they make a kind of loose hobo bundle around the yogurt, then push a wooden spoon through the gaps between the edges of the cheesecloth under the knot. Use the spoon as a crossbar at the mouth of the container to hang the bundle suspended above the bottom.)

Put the yogurt-straining setup in the fridge and let drain for 8 to 12 hours, until the yogurt in the strainer is very thick and you've collected a decent amount of clear, yellowish whey in the container. Pour the whey into a container and scrape the strained yogurt into another container. They will keep for up to 1 week in the refrigerator, tightly covered.

Yield varies; about 1 to 1¼ cups strained yogurt and ¾ to 1 cup whey per pint of yogurt

Pasta with Cherry Tomato– Yogurt Whey Sauce

This tomato sauce has a fresh acidic, clean flavor. It's also delicious over rice or other cooked grains. The recipe works especially well with Sun Gold cherry tomatoes.

Melt **2 tablespoons (30 g) cultured butter** with **1 teaspoon (6 g) sea salt** in a medium saucepan or a nonstick skillet and let it lightly brown over medium heat.

Turn the heat down to low and add **1 small chopped garlic clove**. Stir and let cook for about 30 seconds; don't let the garlic brown.

Add **1 pint (10 to 12 ounces; 275 to 350 g) stemmed Sun Gold cherry tomatoes** and turn the heat back to medium. Cover and let cook for about 2 minutes, until the tomatoes start to burst and soften. Crush them lightly in the pan with a potato masher, making sure you break each tomato open.

Add ⅓ **cup (80 g) yogurt whey** and continue to cook, uncovered, at a brisk bubble for 8 to 10 more minutes. The tomatoes will break down into a chunky sauce, and the mixture will thicken a little bit. Taste and adjust the salt as necessary.

To serve over pasta, cook **10 ounces (285 g) shaped pasta** like orecchiette, conchiglie, or campanelle. Drain when it is about 2 minutes short of how you like it done. Add the drained pasta to the whey sauce, and cook them together for 1 to 2 minutes over medium-high heat, tasting for pasta doneness and removing from the heat immediately when you reach the texture you like. The sauce will finish cooking the pasta, while soaking in some of its tangy flavors. Serve immediately.

Serves 2 to 3

Sour and Pickle-y: Lacto- fermented Plants and Vegetables

Lactic acid bacteria, as a group, don't confine themselves to dairy. They're also extremely good at finding sugars in vegetables, fermenting them and giving them a delicious, softer-than-vinegar, lactic acid tang.

Lactic acid bacteria can handle a little salt in a way that most microbes can't, so we like to give fermented vegetables a dose of it to help our microbe friends out. This comes in the form of either a liquid brine, or sprinkled over the cut vegetables until they make their own brine. Most fermented vegetables are inherently sour *and* salty.

Certain vegetables take to lactic fermentation so preternaturally you can find versions of them all over the place. There's a just about continuous, thousands-of-miles-long sour cabbage belt that starts in northern China and Korea (with the products **paocai** and **kimchi**), stretches across Russia and other Slavic regions (where it often goes by kislaya kapusta and similar names), and ends in Germany and France (as **sauerkraut** and choucroute). The belt's sour cabbage products make a brisk counterpoint to rice and tart up braised and grilled meats, stews, and dumplings.

Some of these excellent-to-ferment vegetables are actually fruits, like **cucumber pickles**, lacto-fermented in brine. Perhaps surprisingly, plums are another widely fermented fruit. Salted, pressed down, fermented, and then partially dried, plumlike Japanese ume fruit become **umeboshi**, little soft nuggets of intense fruity-salty sourness. They're so strong and complex that just an umeboshi and some rice is a deeply flavored and satisfying dish on its own. Similar salt-fermented plums and plum relatives are known as *huamei* or *li hing mui* in China, *xi muoi* in Vietnam, *kiamoy* in the Philippines, and (usually with added chile) **chamoy** in Mexico.

How Much Acid? Counting and Comparing Sour

Different sour ingredients can vary dramatically in how sour they are—a grapefruit or an orange may have plenty of sour heft to them, but the sensation of eating them is a whole different experience than the sourness of straight lime juice or vinegar. There's sour, and there's *sour* sour.

Acids, like any molecule, are measurable. As chemists are wont to do, they've actually developed several different scales for doing so. As a cook, though, I think understanding the general concepts and mechanics behind a flavor and its molecules, and being able to taste for it and think about what you're tasting, have a much more immediate importance than focusing on a lot of data about acid levels. Having gone through this chapter, I think you have a good handle on what acids are, what they do, and different places to look for iterations on the patterns of sour. Now that you've built your mind palace of sour, we can go a little deeper into numbers for annotating it.

Usually when people get started talking about acids, they go right into pH. If you'll recall, when we sense sourness, the molecule we're interacting with directly is the hydrogen (H^+) ion. pH is a measurement of how concentrated those ions are.

It's a funny scale, in two ways.

Sauerkraut: pickley, funky

Umeboshi: plummy, marzipan, shiso, salty

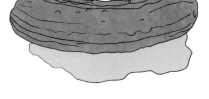
Sour Pickle: pickley, cucumbery, pleasantly musty

Chamoy: sour, fruity, spicy, salty

The brine from umeboshi is so good as a seasoning, it's decanted and packaged as umesu (literally "ume vinegar," though there's no acetic fermentation involved). It's sour, fruity, salty—delicious as a seasoning sprinkled over seaweed or vegetables.

Spicy, liquid forms of chamoy are a cultural force unto themselves, saucing corn chips and sliced fruit or jicama equally well as they do beer-based micheladas. If it's not sold near you, it's easy to acquire liter-sized bottles of it online.

Sauerkraut, kimchi, kosher dill pickles, salt-preserved lemons and limes—and nearly anything else you can find a lacto-fermented version of—mix pickled, slightly funky flavors with a rounded, mouth-filling sourness, with none of the insistent pungency of vinegar. While they're all delicious condiments and accompaniments on their own, borrowing a page from umeboshi and chamoy and using their brines as sauces and seasonings where you might otherwise use lemon juice or vinegar is a super-pro move.

Sprinkle fermented pepper brine over roasted pork or grilled chicken. Splash a bit of sauerkraut juice, kimchi juice, or the pickling liquid from any kind of fermented cabbage, turnip, or root vegetable on cooked leafy greens, fiddleheads, or green beans.

A little bit of dill pickle juice balances out a potato salad, especially if you add it to the cooked potatoes right after you drain them, giving them a chance to soak it in. Another really lovely flavor partnership is with orange vegetables. The next time you roast butternut or delicata squash (400°F and 30-ish minutes should get you in the ballpark), sprinkle a tablespoon or two of dill pickle brine (lacto-fermented, not vinegar-pickled) over them while they're still hot and serve them with a little chopped dill. (This is also delicious with cabbage kimchi juice and thinly sliced scallion, as well as with sweet potato instead of squash.)

First, it's logarithmic, kind of like decibels for judging the loudness of sound. Each step of one pH unit is ten times bigger than the last. Two things that are two pH units away from each other have a hundredfold difference in H^+ ion concentration. The pH scale is set up this way so we can talk about things with vastly different acid concentrations coherently. When we're talking about acids, we might be talking about battery acid, which is about 15 percent H^+ ions. We might also be talking about the level of acidity needed to stop mold growth in fermented foods, which is around 0.0025 percent H^+ ions. Next to 15 percent, 0.0025 percent is barely a rounding error. But it's still *very* significant in its context, so we devised a way to talk about these things being four pH units apart, rather than having a ten-thousand-fold difference.

The second quirk of pH is that it runs backward from 7, which is considered neutral. Distilled water has this pH (technically, it's 0.00001 percent, which is *actually* negligible). One pH unit away we have ten times more H^+ ions, but we count down to 6, not up to 8. A pH of 5 is ten times more acidic than a pH of 6. That vital 0.0025 percent concentration of H^+ ions to stop mold growth is a pH of 4.6. The reason we count backward is that if we're talking percentages, we're basically talking about fractions. pH 3 is greater than pH 6 in the same sense that ⅓ is greater than ⅙. pH just flips the fractions upside-down to make them easier to talk about.

It's practical to know some pH benchmarks. A pear might have a pH of 5, and 1 molecule out of every 100,000 in it is a hydrogen ion (or 0.001 percent). Lemons are closer to pH 2, or 1 in 100 (1 percent) acidic H^+ ions. Stomach acid is pH 1, or 1 in 10 (10 percent) H^+ ions. Most foods are mildly to moderately acidic, with a pH of 2 to 7.

pH is a good snapshot of how many H^+ ions are around at any specific moment in time. It's also relatively easy to measure, if you have pH paper or a pH meter. Unfortunately, it doesn't map so great onto taste, something that first became a problem in the wine industry. Sometimes, when comparing two wines, winemakers noticed that one having a higher pH than the other (so, fewer hydrogen ions and technically less acidity) would nevertheless taste sourer. Which is ass-backward if pH predicts sour taste the way we expect it to.

The trick is that only a small portion of the acid molecules in food give up their hydrogen ions at any one moment—just enough to reach a natural equilibrium point that's specific to each acid. The rest hold on to them and wait, ready to give them up if the acid balance shifts. In your mouth, this means that once you sense the H^+ ions that are unbound and free-floating, other acid molecules might be ready to release more to take their place. Which you then taste, getting a sourer experience than if the amount of H^+ ions was one-and-done, static.

The true sour-tasting acidity that a food has can be thought of as a pooled reserve of acids; the larger the pool, the more H^+ ions you taste over time, and the sourer it tastes, even if it didn't have a particularly low (and thus more acidic) pH to start. Chemists call this pooled acid reserve *titratable acidity*, because rather than measure it quickly with a pH meter, you have to carefully measure how much sodium hydroxide (also known as lye, a strong base, the chemical opposite of an acid) it takes to fully neutralize it. This process is called *titration*. As a shortcut, you'll often see *titratable acidity* written as just *acidity*, along with a percentage.

As a rule of thumb, if you see a number between 0 and 7—for instance, 3.65—that's a pH. If you see a percentage, that's titratable acidity (TA). In this chapter, I use percentages and TA.

Winemakers and juice producers measure TA, as do vinegar brewers—since for all of their ingredients, sourness is vital to get right. Wine, for which the TA determination was invented, has a TA of 4 to 8 grams per liter (0.4 to 0.8 percent),

and lemon juice—much sourer—has 50 to 80 grams per liter (5 to 8 percent). (Lemon's pH of 2 means 1 percent H^+ ions: so there are many more acid molecules in reserve than acid molecules that have given up their H^+ ions.)

Rather than take acidity levels as absolute rules, use them as a guide to help you figure out if you might need to use more or less of a new acidic ingredient where you'd usually use a different one.

Using Titratable Acidity to Mimic Sour Orange

The **sour orange** (aka the Seville, or bitter orange), a variety that combines an intense orange flavor with an acidic taste (about 3.2 percent acidity), is favored in some parts of Turkey, Iran, the Caribbean, and Mexico. Substituting a "regular" orange won't work because it's too sweet, while a lime is too sour. Combining equal volumes of sweet orange juice (1 percent acidity) and lime juice (6 percent acidity) yields a citrus juice of 3.5 percent acidity—which is close enough to sour orange's 3.2 percent for me! Use in place of sour orange juice in marinades, or to dress salads or vegetables.

Sour Orange–Marinated Pork Shoulder, Adapted to Sour Orangeless Circumstances

To make faux sour orange juice, mix ¼ **cup (60 ml) freshly squeezed orange juice** with ¼ **cup (60 ml) lime juice**.

In a food processor or with a mortar and pestle, combine **6 garlic cloves, ½ cup (120 ml) faux sour orange juice, ¼ cup (60 ml) vegetable oil, 1½ teaspoons (9 g) salt**, and **½ cup (20 g) coarsely chopped cilantro leaves**. If it's too thick, add a bit more oil—you want it to be the texture of soft mud, not entirely liquid.

Rub the paste all over a **2- to 4-pound (1 to 2 kg) boneless pork shoulder**. Marinate in the fridge overnight, then cook in a Dutch oven with a heavy, tight-fitting lid at 250°F for 4 to 6 hours, or until very tender. Serve immediately.

Serves 3 to 6

Of course, unless you're a high school chemistry teacher or a weirdo, you almost certainly don't have a titration setup ready to go in your kitchen or garage. That's okay! I've gone through the scientific literature and pulled out relevant TA concentrations.

One note, though! Most vinegar acidities are 4 to 6 percent, which seems slightly softer than lime or lemon, but because acetic acid is a really light and tiny molecule, you can pack more of them in a gram than other acid molecules (which are larger in size, but similar to one another), so each gram (or percent) of acetic acid packs a bit more punch than other acids. Figure it's about a third higher (so, 4 to 6 percent of acetic acid tastes more like 5 to 9 percent of other acids).

Average Acidities of Sour Ingredients

(As titratable acidity)

Ingredient	Titratable Acidity (TA)
Pomegranate molasses	8.00%
Lime	7.50%
Lemon	6.50%
Wine vinegar	6.00%
Sumac	5.00%
Tamarind	5.00%
Verjus	4.75%
Sour apple juice	4.50%
Rice vinegar	4.00%
Blackberry	3.75%
Black currant	2.75%
Black vinegar	2.50%
Labneh	2.35%
Sauerkraut	2.00%
Pomegranate	2.00%
Grapefruit	1.70%
Rhubarb	1.55%
Green mango	1.40%
Sour cherry	1.35%
Pickles	1.00%
Olives	1.00%
Kimchi	0.90%
Tomato	0.85%
Sweet cherry	0.75%
White wine	0.75%
Red wine	0.60%
Cherry tomato	0.40%
Mango	0.30%
Sorrel leaf	0.15%

Sweet

If salty is lifting and defining, and sour is brightening and balancing, you can think of sweet as plush and softening. It balances both sour and bitter without erasing them, making them feel richer and more luxurious. Any decent ripe tomato or peach needs lots of both sweet and sour to earn the name, and a bit of sweetness—from sugar-rich carrots or onions, sweet varieties of wine, or complex raw sugar—in a tomato sauce tastes filled out and full instead of hard and thin. The sweet-bitter

The Rules of Sweet

Sweetness is a signal for sugars, our most basic unit of biochemical energy.

Plant saps and fruits, as well as products made from them, are some of our most potent sources of sugar and sweetness.

White or cane sugar is molecules of sucrose, but other sugar types have molecules like maltose, fructose, and lactose, each with different intensities of sweetness.

As sugars are heated and processed, lots of new molecules are formed, creating sweet ingredients (like palm sugar or brown sugar) with delicious flavor complexity.

Sugars are extremely soluble in water, which can dissolve twice its own weight in sugar (for some really rich and silky syrups!).

Sweetness perceptually balances and tames sourness and bitterness.

dynamic in chocolate, coffee, and tea has helped fuel their global popularity. Sweet gets very sophisticated when it teams up with salt, adding an "oh, that's nice, what's in that?" element to chocolate chip cookies and caramel. It also happily mingles with umami, meaty, and rich dishes—barbecue, baked beans, pad thai, cheese and fruit, to name a few—making them feel richer and warmer without adding too much heft, like switching from a sweatshirt to a cashmere sweater.

Sweetness is a basic and essential sensation, signaling from our earliest time out of the womb that there is delicious, easy-to-process energy to be had. We like sweet so much we've invented an entire category of food for it—dessert. Sweetness is the flavor of celebration, luxury, and special times, and has been since the ancient world.

In the modern era, sweetness has a more complicated social history to grapple with. The sugar trade stoked early developments in global capitalism, as European colonists appropriated Indigenous land in the Americas, planted it with South Asian sugarcane, and enslaved and displaced African peoples to tend it. Sweetness and sugar have been both demonized and feared, the reward signals they set off in the brain equated to the euphoria of hard drugs and their easy calories putting them first on the list for dietary exclusion and moralizing. Biochemically, sweetness does have a special potency—the taste sensation alone is enough to trigger the release of insulin, priming the body to soak up sugar molecules before any of them actually show up in the digestive tract or bloodstream.

Why chase after sugars? Like the other tastes, sweetness links to a biological imperative (or at least, a strong suggestion). Sugars are the most straightforward biochemical fuel for lots of organisms—plants like sugarcane and maple, insects like honeybees, and omnivores like us.

Sugar and Sugar

I'm tossing around the word "sugar" like it only means one thing, but, like most interesting things, there's ambiguity there. Sugar is a sweet and crystalline substance composed of molecules like sucrose or glucose, which can be described chemically as sugars, saccharides, or carbohydrates. Sugar molecules are prodigious at linking up into larger configurations, like dangly earrings tangled up in a disorganized jewelry box. Sugars linked like this exhibit different behaviors depending on their size, so we have conventions for describing how many units any of them has. Glucose is a monosaccharide, a single sugar molecule unit. So is fructose (yeah, from the corn syrup). Sucrose, the molecule that dominates cane sugar, is a disaccharide made from a glucose linked to a fructose. There are trisaccharides, too, most of which are pretty minor culinary players. As the links get longer, we move into the realm of oligosaccharides, which, like "oligarchy," refers to not a specific number but "a few." Sugars can join up in the hundreds and thousands, making polysaccharide molecules like starches, cellulose, and glycogen, which we store as an energy reserve in our muscles and livers.

As with all our tastes, we have a dedicated receptor set up for sensing sweetness. A porelike passageway gathers the ions that make sour and salty, but our sweetness receptor is more like a catcher's mitt, specially shaped to grab on to only molecules with sugarlike structural characteristics and shapes.

Here's where sugar size becomes important! A molecule has to have the right kinds of molecular features *and* be the right size to connect with and activate the sweet receptor. Monosaccharides like glucose and fructose can do so, as can disaccharides like sucrose, maltose, and lactose. But a starch molecule, even though it's made entirely of glucoses, is way too big to fit into the receptor, so it doesn't taste sweet at all.

Sugar (Also) Loves Water

You might recall how much salt and water get along: salt is great at dissolving in water, as well as at pushing and pulling it around through osmosis. Sugar shares this affinity with water and mirrors salt's osmosis skills. Sprinkle a little sugar on cut strawberries, and the strawberry's juices

---

Impersonating Sugar

Some molecules aren't really sugars at all but, through structural quirks, can sneakily fit into the catcher's mitt of the sweet receptor and create a sweet taste sensation. For example, aspartame, saccharin, sucralose, and stevia can be thousands of times sweeter-tasting than the same amount of actual sugar, making them quite lucrative as low- or no-calorie sugar substitutes. Many of them have a unique kind of grabbiness to the sweet receptor, which doesn't actually create a categorically different type of sweetness per se but alters the "time-intensity profile" of when that sweetness peaks and levels off, which is what makes them taste definitely-sweet-but-definitely-not-sugar. Absinthe, Pernod, and ouzo have so much anise or fennel added to their distillation that there's enough of anethole (their characteristic anisey aroma molecule) to trip the sweet taste receptor. These technically unsweetened, sugar-free liquors have actual, perceptible sweetness from anethole's double-duty act.

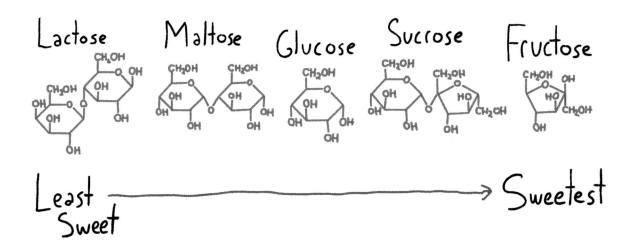

will osmotically seep out, attempting to balance the sugariness inside and outside the strawberry. We have a much higher limit on the concentration of sugar we're willing to tolerate than we do salt, so we can be quite profligate with the sugaring and still get a liquid we can use a lot of, rather than a brine we have to be careful with. Sugar-water mixtures with these can become so gravid with saccharide molecules that they take on a silky, syrupy viscosity—and the two substances play along so well that it's easy to get water to absorb and dissolve more than double, sometimes quadruple, its weight of sugar.

Sugar, like salt, also performs water-pulling osmosis on the cells of microorganisms, dehydrating them out of existence. This is what makes sugaring as viable as salting for long-term food preservation—it's how jams, jellies, and glaceed fruits last for months without spoiling.

Sweet Flavor Patterns

Sweet ingredients can bring a ton of additional flavors along for the ride. The big pattern I like to keep in mind is that those flavors tell the story of where that ingredient came from and how it was processed. Even though they all have their own character, ingredients from a similar source

(like maple and birch syrup, from tree saps) or produced in a similar way (Thai palm sugar and Latin American panela, slowly boiled down and crystallized over a fire for hours) have a kinship that makes the ocean of possibilities a little easier to navigate when choosing an ingredient for a job, or figuring out where to use a particular product.

Try This

Osmosis and Maceration

Sprinkling sugar on fruit to induce osmosis, pulling out liquids and forming a syrup, is also more practically called *maceration*. If you have an ice cream, panna cotta, cake, or anything else that could benefit from a little fruity sauciness, cut up the best fresh fruit you have on hand (strawberries into quarters or slices, cherries in half, larger stone fruit like nectarines into thin wedges) and dust the pieces with sugar, about 1 to 2 tablespoons per cup of fruit. Let it sit for half an hour (ish—1 hour won't hurt) at room temperature while the sugar and water molecules do their osmotic dance, and then pour the macerated fruit and syrup to your heart's desire.

Sweet and Sweet Alone

It's the exception that proves the rule. **White sugar** is the most boring-by-design sweet ingredient. It's the flavor of subtraction, and very good at doing the thing it is designed to do: contribute sweetness and nothing else.

White sugar is ideal for showcasing other ingredients without distraction. Think of high-flavored fruits that don't need accessorizing, just to be turned into a sorbet that transports you to the inside of the world's most neon-flavored strawberry; jam made out of special apricots or raspberries; or lemon curd that's supposed to be lemony and rich and nothing else. It's also the best option for sweet dishes that need to be very subtle or delicately milky: sweet cream ice cream, panna cotta, or a very pale and mildly flavored chiffon or sponge cake.

A bunch of free-flowing, individual crystals is the sign of a substance that's overwhelmingly only one type of molecule. (That's because a crystal is, by definition, a bunch of identical units lined up in a regimented, repeating pattern. Adding some other substance is like dumping marbles at the feet of a marching band grouped in tight formation—total chaos.)

Snowflakes are pure crystallized water, diamonds are pure crystallized carbon. Crystalline salt is almost entirely sodium chloride. And white sugar is just about entirely the disaccharide sugar molecule sucrose.

Unlike salt, there's no source to just mine crystalline sucrose out of the ground or evaporate it from seawater with minimal further processing. Instead, you've got to start with a relatively sucrose-rich source—chiefly and traditionally, the stem of the sugarcane plant, but sometimes beets—juice it, boil it down, and then use various chemical and mechanical means to separate the crystals of sucrose from everything else. Ironically, the extensive heat used to evaporate and crystallize sugarcane juice into bland-and-sweet white sugar actually creates a whole lot of new flavor molecules in the process; they just get removed. Where do they go? In a word, molasses.

Other Purified Sugars

Sucrose, in the form of white sugar, is the most common purified, refined sugar we use to cook with. Other mostly single-molecule sugars and sweeteners can be useful for specific recipes and culinary niches. Starch, the main carbohydrate in grains like corn, barley, and rice, is simply a very long chain of glucose sugars; breaking it down with enzymes or other tools releases either the individual glucoses or units of two glucoses bound together, called *maltose*. Glucose is sometimes called dextrose, or labeled as "corn sugar" or "dried corn syrup solids." It's significantly less sweet than sucrose, and is very handy in precise candy or ice cream recipes for things like controlling freezing temperature or crystallization. Maltose is the main sugar in malted barley and derivative products like dried malt extract, as well as some other sweeteners derived from grains, like brown rice syrup. The sugary component of milk is a disaccharide called *lactose*, which many adults around the world can't digest. It's also less sweet than sucrose. Since lactose is indigestible to beer yeast, too, it can survive the alcoholic fermentation process, so in pure powdered form it's used for making things like sweet and fruity beers.

Sweet and Molasses-y: Brown Sugars

The tangy, edging-on-bitter, jammy caramel intensity of molasses is all the flavors that get removed from white sugar, in concentrated form. (Quite literally: as the sugar gets more concentrated, it crystallizes in a sea of molasses, which is then removed with a centrifuge, like the spin cycle in a washing machine.) As well as being another sweet ingredient to use, molasses is responsible for the color and flavor of the many styles of **brown sugar** out there, which are generally made by adding controlled amounts of molasses back *into* white sugar, giving each crystal a little, flavorful coating.

From, roughly, lightest and mildest to darkest and most molasses-intense, brown sugars go: **turbinado** (with large, crunchy crystals), **demerara** (similar crunchy crystals as turbinado, but darker and more toffeelike flavor), **light brown sugar**, **dark brown sugar** (both of these have tiny crystals and a moist texture), and **muscovado sugar** (moist texture, even darker than dark brown sugar, and more complex).

The flavor of brown sugar is an essential part of the classic American chocolate chip cookie, and demerara syrup (2 parts demerara sugar gently heated and dissolved into 1 part water, by volume) adds delicious sweet complexity to rum drinks and just generally works well with other brown liquors, like whiskey.

In most places you're cooking with white sugar, you can usually use brown sugar for more flavor: toffees, caramel, syrups, drinks. Substituting can become a little tricky in the more precise realm of baking and pastry, because the same reactions that create brown sugar's color and flavor also make it slightly acidic, which can mess with the optimized action of baking soda and baking powder for somewhat unpredictable results. I tend to just forge ahead and try things out, maybe substituting half the white sugar instead of all of it, and prepare myself for possible sacrifices in texture in the name of flavor and, perhaps, a couple of tests to get it right.

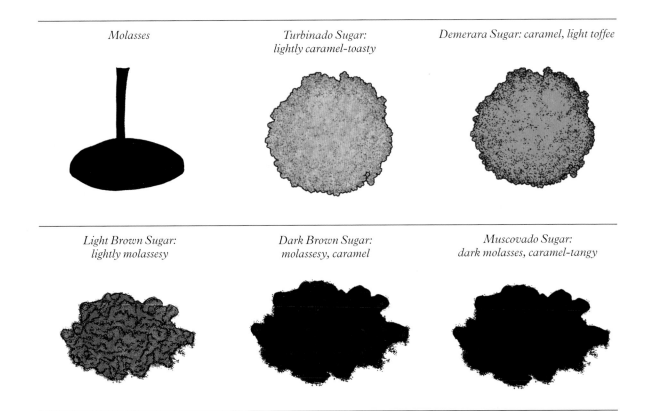

Molasses

Turbinado Sugar: lightly caramel-toasty

Demerara Sugar: caramel, light toffee

Light Brown Sugar: lightly molassesy

Dark Brown Sugar: molassesy, caramel

Muscovado Sugar: dark molasses, caramel-tangy

- Use ordinary brown sugar in dishes where you would otherwise use white sugar: add to your iced coffee, your old-fashioned, your butterscotch sauce, or use just a bit in a pan reduction or stew.
- In baking recipes that already call for brown sugar, try substituting more flavorful, unrefined sugars like panela, jaggery, or kuro sato. You'll have to grate them, but the flavor is worth it.
- Brown and unrefined sugars are more acidic than white sugar, which can affect the precision of baked recipes. To ease into experimenting, substitute brown and unrefined sugars, or syrups like honey or maple, for a quarter of the white sugar in sweet recipes—this should give you added flavor with a little less unpredictability.
- Thin honey or agave nectar up to 1:1 with boiling water to make a flavorful syrup to sweeten cocktails, iced tea, or lemonade.

Sweet, Mellow Complexity: Unrefined Sugars

If brown sugar is delicious, and we get it by taking the molasses out of sugar then putting some back in, why not just never take it out in the first place? You're right on the money. Let's enter the world of unrefined, uncentrifuged sugars. One way to think about these sugars is that they're the sugar of the people (as opposed to a refined export commodity) in the places that grow sugarcane. There are hundreds, probably thousands, of different local variations, but they're generally made by heating and evaporating cane juice until it browns and thickens, but heating it less than you would need to do to create molasses and individual crystals. From the slurrylike or pastelike state in which they exit the kettle, they're cooled and solidified into solid masses with fine-grained crystals throughout them. Rather than free-flowing granules, they come as blocks, chunks, or slabs. They're caramel-molassesy, and have a lot of individual flavor layers that tend to get cooked out of molasses: mellow earthiness, toffee, maple, smoky, buttery, or funky-floral qualities. Seriously, they're addictively complex and show the idea that sweetness is simple or boring is totally laughable.

Latin American **panela**—also known as piloncillo (Mexico), rapadura (Brazil), chancana (Peru), and papelón (Venezuela)—is cooked and formed into small loaves or cones, and often comes in the choice of either a light golden brown or darker brown level of caramelization. **Piloncillo** brings a rich, complex sweetness to nut-and-dried-chile-rich mole sauces and coffee prepared as café de olla. The drinks aguapanela and papelón con limon would be relatively boring lemonade or limeade without their namesake sugars, which give them an exciting, caramelized depth of flavor.

Panela and other unrefined sugars usually come as chunks, cakes, blocks, or as flattish, tapered cylinders. You need to grate or crush them into smaller pieces to use them, and they're much easier to incorporate by heating and dissolving into any liquids in the recipe. Sometimes you will be able to find bags of crushed jaggery, panela, or coconut sugar, with a sandy and somewhat irregular texture (since there's lots of extra stuff besides sucrose in there, they don't form big, orderly crystals).

Panela–Coconut Milk Syrup

Caramelized, slightly mapley and creamy-nutty, this coconut syrup is surprisingly versatile. Use it for sweetening iced coffee (see below), drizzling over pancakes, soaking sponge cakes, or as a sub for simple syrup in a daiquiri (page 77).

You can substitute basically any unrefined sugar here, like jaggery, palm sugar, or coconut sugar. Dark brown sugar or darker-still turbinado sugar will both work, too.

Before opening, shake **one 13.5-ounce (240-ml) can coconut milk** well. Put 8 ounces (1 cup; 2 g) of the milk ino a medium saucepan.

Crush **8 ounces/225 g panela sugar,** to yield 1⅔ cups after crushing, using a heavy mortar and pestle. Alternatively, put it in a plastic zip-top bag, wrap that in a heavy towel, and crush with a mallet until you get small-medium chunks and some sandy sugar.

Gently pour the panela into the saucepan with the coconut milk. Heat over medium-low heat, stirring occasionally to dissolve all the sugar, about 20 minutes. Turn the heat up and simmer for 5 minutes more, until slightly thickened. Remove from the heat and let cool. Store tightly covered in the fridge for up to 2 weeks.

Makes about 1½ cups

Panela-Coconut Iced Coffee

Fill a cocktail shaker (or a lidded jar) with **1 ounce (2 tablespoons; 30 ml) Panela–Coconut Milk Syrup** (see above), **6 ounces (¾ cup; 180 ml) cold-brew coffee**, and some **ice cubes**. Shake well to froth and chill, then pour into a chilled glass and serve immediately.

Serves 1

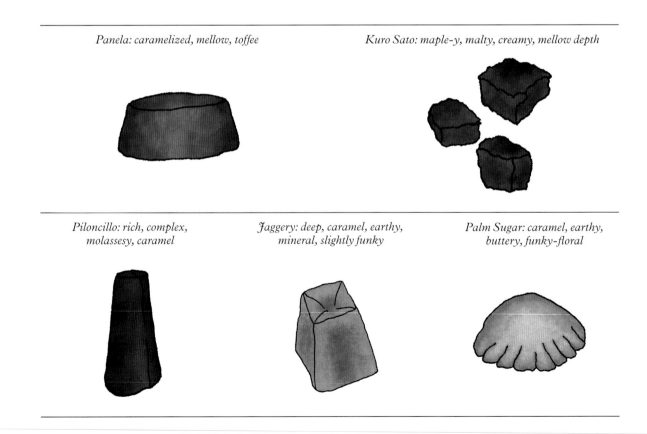

Panela: caramelized, mellow, toffee

Kuro Sato: maple-y, malty, creamy, mellow depth

Piloncillo: rich, complex, molassesy, caramel

Jaggery: deep, caramel, earthy, mineral, slightly funky

Palm Sugar: caramel, earthy, buttery, funky-floral

Panela-Coconut Daiquiri

If you want to be fancy, strain through a fine-mesh strainer, to remove the little ice shards, into a chilled glass; otherwise, just pour into the glass. A chilled coupe is traditional; I usually use a thin-walled water glass.

Fill a cocktail shaker (or a lidded jar) with **2 ounces (60 ml) Jamaican rum** (such as Appleton Estate or Smith & Cross), **1 ounce (30 ml) lime juice** (from about 1 lime), **¾ ounce (22 ml) Panela–Coconut Milk Syrup** (page 76), and **ice cubes** to fill. Shake well to chill and strain with a cocktail strainer into a glass. Serve immediately.

Serves 1

Sugarcane **jaggery**, the South and Southeast Asian cousin to panela, has a deep, caramelized, earthy taste that references molasses but is much more complex. Jaggery is closely related to gur and vellam (both from India), panutsa (from the Philippines), and namtan (from Thailand). Palm-sap-based jaggery and **gula mekala** share this earthiness, layering a buttery, almost funky-floral flavor atop it.

Complex, Unrefined Sugars from Other Plants

Sugarcane is not the only tropical plant that maintains juicy stores of sugar—many varieties of coconut, palm, and date trees do, as well. Sugar makers harvest sap directly from the living tree, often by cutting open their flowers and allowing it to drain out, then they reduce, crystallize, and dry it. Look for the name of the plant in the product name: **palm sugar**, date jaggery, and coconut sugar.

The southernmost parts of Japan are warm enough to grow sugarcane, and Japanese cuisine has its own repertoire of specialized refined and unrefined sugars. **Kokuto**, or brown sugar, is used to make malty, warmly red-brown cakes and confections in Kagoshima prefecture. The jaggery-like Okinawan rendition of kokuto has an amazingly roasty, mellow, almost salty mapley-brown-sweetness, and it's so dark it goes by the name **kuro sato** ("black sugar").

Black Sugar Pudding

Kuro sato, Okinawan black sugar, has a kind of butterscotchy quality to it, with maple-syrup-light roasted coffee notes. It's very dark (it's called black for a reason) but it's got much more complexity than molasses. I love butterscotch pudding, and this is like an even better version of that flavor.

Starch forms a gel at a higher temperature than egg does, so here you heat cornstarch and milk until it thickens, then you add eggs, which thicken with the residual heat. Brown sugars are slightly acidic, so in this recipe you minimize the amount of time you're actually heating the milk and the black sugar together without starch, to help smooth things out. Another unrefined or very dark brown sugar will also work in place of black sugar.

With both starch and eggs, this recipe technically works as a pastry cream, so you can also use it to fill cakes or donuts or whatever your heart desires.

Beat **3 large eggs** and set aside. Grate 1⅔ **ounces (50 g) kuro sato**, to yield **4 tablespoons grated**, with a grater or break up with a robust blender until chunkily powdered. Combine in a medium bowl with **3 tablespoons (23 g) cornstarch** and 7 **tablespoons (100 ml) whole milk**. Set aside.

In a medium-sized saucepan, heat ¾ **cup plus 2 tablespoons (200 ml) whole milk** with **3 tablespoons (40 g) white sugar** over medium-low heat, until steaming and just below a boil. Add the reserved milk, kuro sato, and cornstarch mixture and whisk over medium heat until it thickens (which will happen at about 170°F). When it's thick, remove from the heat and measure the temperature with an instant-read thermometer, letting cool to 170°F if it's hotter than that.

Drizzle 2 tablespoons of the hot milk-sugars mixture into the reserved eggs to gently heat them up, stirring until well combined. Whisking vigorously, pour the eggs back into the saucepan, where they will thicken the pudding from the residual heat.

Transfer to a bowl, and refrigerate to cool and set. Store tightly covered in the fridge for up to 3 or 4 days.

Serves 1 to 2

Sweet, Soft, Sticky: Saps and Syrups

All white sugar that gets birthed into existence produced syrupy molasses as a counterpart. But if you cook down sugarcane juice and stop well before crystals form, you make **cane syrup**, a delicately browned liquid form of sweetness. In the United States, it's most easy to find in sugar-producing Southern states like Louisiana, but it's also findable online. It's also a common British ingredient, under the name golden syrup.

Sugarcane is far from the only plant with a sugary sap, although other plant saps tend to have a lower concentration of sugar and a much more subtle sweetness. Boiling and reducing them like sugarcane juice similarly makes for sticky, viscous, and fluid syrup, with a lot of flavor from the original plant as well as from the cooking process. Maple trees lead the pack with **maple syrup,** full of iconic, for lack of a better word, mapley flavor and subtle acidity. Depending on when the sap was harvested, maple syrup gets darker and more intensely flavorful in tandem. While, confusingly, as of around 2014 all versions of maple syrup in the United States are called "grade A," look for "grade A dark and robust" or "grade A very dark and strong" (what used to go by grade B or C) for the most maple flavor. Birch trees can also be

tapped for sap to make into syrup. Although it's a little more difficult to find, **birch syrup** expresses really interesting and tasty fruity-berry notes when compared with maple.

With **honey**, bees do almost all of the work for us. They go forth from their hive to gorge on flower nectar. When they return, they vomit their load of nectar up, concentrate and evaporate some of its water by flapping their wings at it, then pack it into wax-encased chambers of their honeycomb for later use. There, it tends to ferment a bit with wild yeasts and bacteria, adding layers of pleasant sourness to the flavor tones that reflect the plants it came from, which can be extremely light and particularly floral in the case of clover or orange blossom honey, dark and almost savory-funky in buckwheat honey, or all kinds of colors and flavors in between.

Sweetness and Bitter, Sour, and Savory

Sweetness can dominate how we think of a food—we have "sweets" or "sweet dishes" in a way that we don't really have "salties" or "salty dishes." But as much as we sometimes choose to make sweetness the main focus, sweet is very much a team player at heart.

Sweetness is very happy to sing backup in savory dishes. Take the example of mole negro, one of the most complex of the many styles of stewed and braised moles from Oaxaca, Mexico. It has a base of toasted and burnt dried chiles long-cooked with garlic, onions, and sweet spices, and fattily enriched by lard, sesame seeds, pecans, and other nuts. And, rounding it out, ripe plantain, raisins, and drinking chocolate. It's absolutely a savory dish—very commonly cooked together with chicken and broth into a saucy stew—but it's also noticeably sweet from these sweet ingredients, a sweetness that doesn't overpower its complex

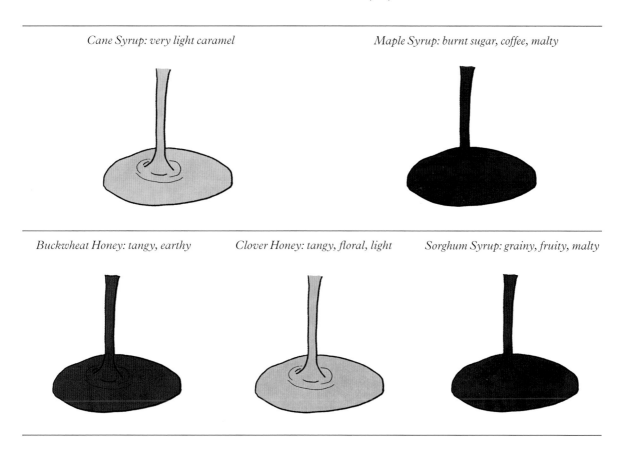

Cane Syrup: very light caramel

Maple Syrup: burnt sugar, coffee, malty

Buckwheat Honey: tangy, earthy

Clover Honey: tangy, floral, light

Sorghum Syrup: grainy, fruity, malty

spicy-and-toasted-until-burnt flavors, but cozies up to them and complements them. Sweetness chaperones and softens the edges of the spread-out, building heat, helping to moderate some of the overall richness and heft. The total flavor quality of the sweet ingredients matter, too: the sweet and fruity flavors of plantain and raisins call out some of the fruity-leathery quality in dried chiles, and the dark brown malty-sweet chocolate blends in with the depths of the long-cooked and heavily toasted flavors.

Besides mole, sweet can sneak into a dish with ingredients you'd find nowhere near a dessert: well-cooked garlic and onions give up their spicy prickle to reveal their high sugar content, as do tomatoes and carrots.

Easy Ways to Slip Sweetness into Savory Foods

- Add a tablespoon or so of brown sugar or piloncillo or 1 to 2 squares dark chocolate to chili or slow-cooked pinto beans, kidney beans, or black beans.
- Whisk a splash of dark maple syrup or an assertive honey like buckwheat honey into a vinaigrette for peppery greens.
- Dress shredded cabbage or shredded green papaya or green mango with rice vinegar with a dash of fish sauce, then grate in a sprinkling of palm sugar using a Microplane or rasp.
- Drizzle sorghum syrup or dark maple syrup over roasted carrots about 5 minutes before they're done, then top with a plop of yogurt or crème fraîche.

On a sensory level, sweetness has a lot to give besides just "more sweet." It tames and balances both bitterness and sourness, reducing their brashness and filing down their rough edges, and it's fluffed up and enhanced by umami. It tends to have a nice harmony with spiciness and fatty richness, as well as with seasoning elements like spices or fruits. A little brown sugar or a square or two of dark chocolate is often exactly what a chili or spicy bean stew needs to feel rich and in balance. A careful spoonful of white sugar added to a simple tomato sauce (white sugar, because here you're going for single-ingredient flavor clarity) can make its acidity feel beefy instead of thin and insistent.

Gastrique is essentially a cooked reduction of a sweet element and a sour element, and using it to sauce rich meats or roasted foods is a low-effort, high-payoff way to work in sweetness and braid together many of these sensory threads. It's often categorized as a primarily sour condiment, but sweet does much of the heavy lifting: keeping the sourness juicy instead of puckery, and finessing the way with any fatty, umami, meaty, or bitter notes.

Double-Apple Gastrique

A traditional French gastrique features a reduced mixture of caramelized sugar and vinegar. This version of the sauce, made with apple cider and apple vinegar, is a more solidly fruity take than the strictly textbook one. Use straight from the saucepan to finish roast chicken or pork, or drizzle it over root vegetables.

Simmer **1 cup (240 ml) apple cider** in a saucepan over medium heat until reduced to a thick, syrupy consistency (you should end up with about ¼ cup). Watch carefully so it doesn't scorch. Add ½ **cup (120 ml) apple cider vinegar**. Reduce the heat to medium-low and cook 10 to 15 minutes, until the sauce is syrupy but fluid and the vinegar flavor is tempered.

Store tightly covered in the fridge for up to 1 week.

Makes ⅓ to ½ cup

While milk naturally contains sugar in the form of lactose, it's taken to giddy heights when transformed into **sweetened condensed milk**, which is one of those nearly perfect industrial products: milk cooked to evaporate its water and concentrate its fat and proteins, mixed with its own weight in sugar, and then canned to persist in its milky, syrupy state for years and years. (Panela–Coconut Milk Syrup, page 76, is essentially a slightly lighter, plant-based version of sweetened condensed milk.) The creamy sweetness of sweetened condensed milk has made it an international hit for balancing some of humanity's favorite bitter foods, namely coffee (in Vietnamese coffee or Spanish café bombon) and tea (in Malaysia, Singapore, and Hong Kong). "A spoonful of sugar helps the medicine go down" in exactly the same manner, by suppressing how much bitterness we perceive.

Just Fruit: Highlighting and Garnishing Sweet Things

If adding sweetness to dishes can define them as sweet dishes, nowhere is sweetness more of a special highlight than in perfectly ripe fruit. Fruits' whole purpose is to accumulate enticing sugars, a reward for animals to eat them and spread around the seeds of the plant that produced the fruit. When fruit ripens, reaching its full accumulation of sugars, it also usually changes color from green to anything from light yellow to red to dark purple, and it changes the aromas it pumps out, as signals to come 'n' get it.

Really good fruit turns how we usually use sweetness on its head: instead of something we add to other things to make them sweet and harmonize with their flavors, it's the perfect center-of-the-plate food in its totally raw form, an opportunity to garnish with other flavors to harmonize with its own innate sweetness. Picture a little bit of lime zest and lime juice on sliced banana or mango; a hint of cardamom, cloves, or cinnamon sprinkled over cherries or apples; and chopped fresh herbs tossed with fruit pairings: watermelon or peach with mint, strawberries and basil, blueberries or raspberries with lemon verbena. Salt is another seasoning element that goes famously with grapefruit or melon, where it additionally helps massage away any lingering bitterness.

Plums with Shiso Salt

I really like this with a ripe but not too ripe dark-purple-skinned plum, but it is also great with red plums or greengages (with amazing green skin and flesh), if you can find them, or apricots or sour cherries (use cherry halves, rather than sliced cherries). The sweet-salt combo is even better with a little sour. If you can't source shiso, this is also tasty with thyme salt (page 42).

Cut **3 to 4 ripe plums** into 6 to 8 wedges, removing the pit. Cover with a very thin dusting of **Shiso Salt** (page 44), sprinkling sparingly across the widest part of the slices. Serve promptly on a plate.

Serves 2 to 4 as dessert

Umami

Umami is savory, brothy, meatlike richness. It fills your entire mouth, is substantial without being heavy, and comes on slow and leaves slow, too—without the attention-grabbing sharpness you can get from salty, sour, or bitter. Umami might be the most elusive taste, but nearly every cuisine enthusiastically uses it. If you like seaweed, dry-aged beef, tomatoes, miso, cheese, cured ham, mushrooms, or soy sauce, umami is what makes that favorite so delicious and satisfying.

While our other tastes have been well known for thousands of years, it took until 1909 to get a clear understanding of what umami was for, and for it to receive the name we know it by today. Kikunae Ikeda, a chemistry professor at Tokyo Imperial University, had been studying a family of kelp seaweeds called kombu. As we learned earlier, kombu is the key ingredient in dashi, the foundational stock in Japanese cuisine. And, not coincidentally, it's very savory-tasting. After steeping, boiling down, and concentrating massive quantities of kombu dashi, Ikeda was eventually able to isolate the savory-tasting molecule from everything else in the kombu: glutamate. To put a name to the flavor, he created a portmanteau from the words *umai*, meaning "delicious," and *mi*, meaning "taste."

After his discovery, Ikeda founded the company Aji-no-moto to produce and sell glutamate as a seasoning, in the form of monosodium glutamate, or MSG. Aji-no-moto is still the world's largest supplier of MSG, and it funds open-access research on umami at the Umami Information Center.

The Rules of Umami

Umami comes from glutamate (a form of glutamic acid), one of the amino acid building blocks of proteins.

Like the other tastes, glutamate is much more soluble in water than in fats.

Fermented ingredients (like soy sauce, miso, or aged cheese) are some of the best sources of umami, because the fermentation process quite literally breaks down some of their proteins into their building blocks.

Seaweeds and dried mushrooms can have tons of umami, without any fermentation needed.

You can enhance the umami in an ingredient by mixing it with another ingredient containing molecules called *ribonucleotides* (like tomatoes, fresh meats, and fish); ribonucleotides have little umami taste on their own but magnify existing umami.

Describing and Debating "Savoriness"

Prior to Dr. Ikeda's chemistry discovery, plenty of people were turned on to the idea of a "delicious," "savory" flavor, even if it was difficult to define. The Chinese term *xian wei* (which has a similar-to-umami root in "deliciousness") has been used to describe glutamate savoriness since at least the nineteenth century. If you were an intellectual-salon-going man of means in the Enlightenment, you and your gentleman-scientist buddies would be talking about "osmazome." Like "phlogiston" of the same era (a substance that supposedly saturated flammable things and was released, unchanged, when you burned them), or the "strings" in string theory physics today, osmazome was a hypothetical substance invented to explain an otherwise inexplicable natural phenomenon: What made roasted meats so damn good? We know today that this role is filled by multiple flavor sensations: umami taste, but also the browned aromas generated by the heat of roasting. (With better chemistry tools, we figured out that phlogiston was not a single substance, but actually a fast chemical reaction, between flammable molecules and gaseous oxygen. String theory remains hotly debated.)

The Science of Umami

The taste of umami comes not from a mineral (like salt) or a primary fuel source (like sweet), but from an amino acid (glutamate), one of the building blocks of proteins. Living things make and use proteins for lots of different jobs, including creating movable machinery and structure as muscles or collagen, sending or receiving signals (taste and smell receptors are this type of protein), and shepherding along all of our important biochemical reactions. Regardless of their job, proteins are all made of long, single-file strands of amino acids folded up like a necklace, with links and tangles inside the bundle holding the whole thing in place. And for stringing this necklace, there are twenty different amino acid options that make up all proteins.

The umami receptor is especially tuned to pick up on one of these: glutamic acid, specifically in the ionized form of glutamate. In addition to being a protein-builder, glutamate is also a neurotransmitter, helping carry important messages between nerve cells and shaping learning and memory. The brain can even burn it for energy, as an alternative to sugar.

Umami and MSG: Delicious and Villainized

Like saltiness in the form of pure sodium chloride, or sweetness in the form of pure sucrose, there are ingredients that are just about pure umami—namely, monosodium glutamate, or MSG. (Glutamate is technically an ion, which can float around in water by itself, but it needs a counterpart ion in solid form—thus the "monosodium," making MSG slightly salty as well as umami.)

As food molecules that attract controversy go, MSG is up there with sugars and fats, with plenty of attendant misinformation, fear, and even pseudoscience.

MSG (which can't be distinguished by the body from naturally occurring glutamate in foods) has been blamed for a rogue's gallery of ailments, including asthma, migraines, rashes, weight gain, and even neurotoxicity, starting most egregiously with something called (really) "Chinese restaurant syndrome." It was first described in a letter to the editor (note: *not* a peer-reviewed medical research report) in *The New England Journal of Medicine* in 1968 as "numbness at the back of the neck with headache, general weakness, and [heart] palpitations" after eating at Chinese restaurants. No study has actually proven that this syndrome actually exists, let alone that it's decisively caused by MSG. But a heady mix of backlash to midcentury "better living through chemistry," growing mistrust of large food and other corporations who put profits over public safety, and a not insignificant amount of anti-Chinese racism and xenophobia cemented the perception that MSG = Dangerous.

There's a whole lot of history that we don't have the time or space to delve into here, but recent research and meta-analysis has largely found that most studies that have claimed to decisively connect MSG to dangerous effects on health had methodological flaws or relied on extreme or unusual conditions that make them not very relevant for humans eating the amount of MSG you could reasonably get from foods—in other words, in normal life.

Personally, I use a lot of glutamate-rich ingredients in my cooking and also keep MSG around to sprinkle a little bit of very clean umami onto and into things.

I also think there's already too much downplaying of people's real, medical suffering in the world, and nobody should force anyone to eat something they don't want to eat. So, if your own experience leaves you convinced that you have a negative reaction to MSG, I'm not interested in proving you wrong to score facts-and-logic points. But, if you're a person who's just curious about the safety of eating glutamate, as MSG or otherwise, you should be dubious about claims that you should be universally afraid of it, because those are not backed up by any decisive science.

As an evolutionary advantage, the savory flavor of umami is a good signal for the presence of amino acids and proteins—in other words, nutritious foods that will keep our bodies and brains going. In line with this, umami increases appetite—after we taste it, we want to eat more of whatever it's in. This is a sophisticated signal, rather than a one-way switch: once we've eaten umami food in response to that appetite increase, we become satiated sooner than we would for non-umami food, and feel fuller for longer.

Umami took a while to catch on and become accepted science. A major reason is that Europeans and Americans, despite having umami elements in their food, were not well versed in recognizing its sensation. In the 1980s, sensory scientists found that while more than 50 percent of Japanese people they studied could identify umami savoriness unprompted, only around 10 percent of Americans could. This didn't mean the Americans weren't tasting it at all—in fact, after a little practice, tasting pure umami and hearing it described, pretty much everyone gets good at recognizing it (palate is about practice and paying attention, after all!). Besides this hurdle, there was some skepticism about whether we could even tell if umami was its own taste, or just an enhancer of tastes, based on chemistry and descriptions of sensation alone. What really settled it was the discovery in the year 2000 that a version of one of the brain receptors for glutamate (neurotransmitter, remember?) was also produced in the taste buds. Not only could we sense glutamate, we were sensing it as its own taste. Since then, we've figured out that we make at least three different proteins as umami taste receptors.

The primary taste receptor we use for umami is similar to the one we use for sweet: an outward-facing, catcher's-mitt-like protein with a shape that's specially tuned to fit glutamate very well, and not grab on to other things. And also like sweet, the molecules have to be both the right shape and of an overall size small enough to actually reach the "grabby" bit. Starches are full of otherwise sweet glucose, but are way too big to get near enough for the sweet receptor to grab. Proteins usually have at least 5 percent, and sometimes 20 percent, glutamate (in the form of glutamic acid) among their amino acids. These don't taste umami, though, because the bound-up glutamate can't get anywhere near the receptor, like trying to catch a beach ball with the catcher's mitt.

Our umami receptor has a feature, possibly a fluke, that creates exciting culinary opportunities. When glutamate gets near it, the catcher's mitt of our umami receptor grabs it, squeezing onto it and sending the signal for "umami!," then releases it.

The weird feature: there's a spot on the outside of the umami receptor that, once activated, makes the receptor hold on much longer when it grabs on to glutamate, like the catcher wearing the catcher's mitt got the signal not to throw the ball right away. While the receptor is holding on to glutamate, it keeps sending the umami signal—for way longer than it usually does for a single glutamate molecule.

Activating this spot on the umami receptor, hitting it with just the right molecule, makes umami things *more* umami—sometimes many times more intensely umami—without adding any actual extra glutamate. In foods, the most important molecules that activate this spot on the receptor are *ribonucleotides.*

Like amino acids, ribonucleotides are the building blocks of larger molecules. In this case,

they build ribonucleic acid or RNA, the messenger molecule that reads off the sequences in your DNA and translates them into actual proteins. Some ribonucleotides are also double-purposed as an energy storage in cells. Just like every living thing has proteins and amino acids, they also have RNA and ribonucleotides, but, like glutamate, some have way more than others.

Ribonucleotides—the ones we most care about for umami go by the acronyms IMP, GMP, and AMP—don't taste particularly umami on their own, but they beef up umami from whatever glutamate is already around. This can be a very literal beefing up: meats (from mammals, birds, or fish) are some of the richest sources of ribonucleotides—in their fresh, naked form, with no fermentation or aging necessary. It's umami synergy: the combination is greater than the sum of its parts.

In your cooking, this means you can mix a glutamate-rich ingredient with a ribonucleotide-rich ingredient and get tons of extra umami. Put another way, if you only want to use a small amount of a very strong or funky glutamate-rich ingredient in a recipe, but want more umami, you can give some oomph to the glutamate you already have by adding something with a lot of ribonucleotides, like chicken, tuna, or pork.

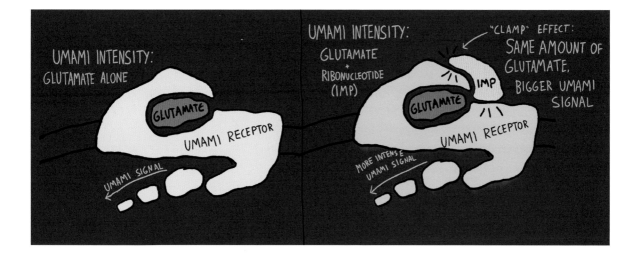

Umami Flavor Patterns

Umami makes us want to eat more of things that have it, and is also deeply satisfying. It's great at creating a feeling of richness this way—a down comforter alternative to the weighted blanket of a fattier type of richness. And since, like most taste molecules, glutamate is much happier to dissolve in water than in fat or oil, it's easy to infuse that savory-rich flavor in water-based broths, stewing, and steeping—as well as stirring it right into things, or using it as a garnish.

On your hunt for umami, the first pattern you should keep in mind: there are ingredients that naturally have a lot of free glutamate and umami taste in their raw state, and ingredients that have it because their protein has broken down into amino acids in some way.

Most naturally glutamate-rich ingredients, somewhat counterintuitively, are plants, fungi, or other vegetal and nonanimal organisms: seaweeds like kelp, shiitake mushrooms, tomatoes, green tea, even walnuts and potatoes. Meats, besides a few types of seafood, have surprisingly little free glutamate if they haven't been cured or extensively dry-aged (though they can still contribute to umami in other ingredients, with their ribonucleotides).

This brings us to the second type of glutamate-rich umami ingredients: things that start off with a lot of protein and attain umaminess when those proteins are broken down into free amino acids. These are usually aged or fermented: plant-based ingredients like miso and soy sauce, fungal ones like dried shiitakes or nutritional yeast; dairy-based ones like Parmesan, Gruyère, and blue cheeses; and animal-based ones like fish sauce, long-aged hams, and cured anchovies.

Knowing where to look for glutamate, and how it got there, gives you a preview of what other shadings of flavor you're going to get from them: deeply funky, cleanly oceanic, vegetal, and even richly fruity.

Easy Ways to Up the Umami

- Got carbohydrates (like potatoes, farro, or noodles), tomatoey stew, popcorn, wilted greens, or substantial broccoli? Finish with a shake of dry umami: finely grated Parmesan, Gruyère, or other aged hard cheeses; nutritional yeast flakes; powdered dried mushrooms (especially shiitake); flakes of nori (aonori) or kelp.
- Add a couple anchovies, a dash of fish sauce, or even a ballpark-half-teaspoon pinch of MSG to pasta sauces, braised meats, the base of a soup or stew, a pot of beans, or sautéed green vegetables when you start cooking them for umami depth.
- Make roasted vegetables rich and savory. Before roasting, toss them with one of the following mixtures (amounts given are per pound):
  - 2 tablespoons red miso mixed with 4 tablespoons melted butter (you can also apply this to chicken thighs, deliciously).
  - ½ cup finely diced dry sausage, like Italian soppressatta or Chinese lap cheong, and a drizzle of olive or other oil.
  - ¼ cup vegetable oil and ½ teaspoon powdered instant dashi.
- Make an umami-enhanced broth: a handful of ham trimmings, dried porcini or shiitake mushrooms, half a head of the sweetest green cabbage you can find, tablespoons of tomato paste, a Parmesan rind or two, several onions and/or a couple heads of garlic. Cover with water and simmer gently for as many hours as you have. Chicken carcass or beef bones optional.

Umami Super-Heavyweights

Name	What is it?	Glutamate (mg per 100 g)
Rausu kombu	Dried seaweed	3000
Marmite	Yeast extract	1960
Parmigiano-Reggiano	Aged hard cheese	1680
Kelp (general)	Seaweed	1608
Fish sauce	Fermented animal protein	1383
Nori	Seaweed	1378
Roquefort	Blue cheese	1280
Ganjang (Korean soy sauce)	Fermented soy	1264
Anchovy	Aged animal protein	1200
Douchi (fermented black soybeans)	Fermented soy	1080
Dried shiitake	Fungi	1060
Gruyère	Aged hard cheese	1050
Miso	Fermented soy	1000

Umami Heavyweights

Name	What is it?	Glutamate (mg per 100 g)
Stilton	Cheese	820
Shoyu (Japanese soy sauce)	Fermented soy and wheat	782
Walnut	Vegetable	658
Sun-dried tomato	Vegetable	650
Gouda	Processed	460
Green tea	Vegetable	450
Camembert	Processed	390
Dry-cured ham	Aged animal protein	350
Dried morel and oyster mushrooms	Fungi	310

Umami Middleweights

Name	What is it?	Glutamate (mg per 100 g)
Sake	Fermented rice	186
Aged cheddar	Cheese	180
Tomato	Vegetable	175
Bottarga	Aged salted fish roe	158
Squid	Fresh mollusc	146
Scallop	Fresh mollusc	140
Oysters	Fresh mollusc	130

Umami Welterweights

Name	What is it?	Glutamate (mg per 100 g)
Corn	Vegetable	106
Peas	Vegetable	106
Mussels	Fresh mollusc	105
Sea urchin	Fresh shellfish	103
Kimchi	Fermented vegetable	100
Potato	Vegetable	100
Garlic	Vegetable	99
Cabbage	Vegetable	94
Caviar	Salted fish roe	80
Dried porcini mushroom	Fungi	77
Soybeans	Legume	75
Crab	Crustacean	72
Duck	Fresh animal protein	69
Daikon	Vegetable	67
Onion	Vegetable	51
Spinach; carrots	Vegetable	50
Asparagus	Vegetable	49

Vegetal Umami: Plants and honorary plants

Umami is savory, meaty richness, which of course means the natural place to start is . . . vegetables? Although some of the most pronounced umami ingredients are aged, salted, and fermented foods, most vegetables have a respectable, even impressive dose of glutamate in them. (And just in case you get into an argument with a biologist, I'm using the culinary definition of "vegetable" that includes a few things that aren't, taxonomically speaking, technically plants—like mushrooms [fungi] and kelp [stamenophiles]). Instead of having heady-funky, heavy flavors, vegetable umami is light, fresh, and clean-tasting. Seaweed is an obvious example—Professor Ikeda was able to isolate umami from kelp in the first place because it has huge concentrations of glutamate. **Rausu kombu**, one of the finest forms of giant kelp grown for soup off the coast of Hokkaido, has about 3000 milligrams of glutamate per 100 grams, the highest amount in an ingredient that I can find recorded in the scientific literature. Other types of kombu kelp, **nori**, **wakame**, **sea beans,** and **red dulse** are also rich in glutamate, although in varying degrees. The umami of seaweeds is clean, oceanic, and singular. Seaweeds are also often quite salty, tasting of calcium and iodine salts along with the familiar sodium chloride.

Seaweeds' biggest umami contender is **mushrooms—shiitakes** especially, and particularly when **dried** (with about 1000 mg of glutamate per 100 g). **Oyster**, **morel**, and **porcini** (or cèpe) mushrooms also have plenty.

Lots of land plants you wouldn't necessarily clock as umami bombs have surprising amounts of glutamate in them: **green tea** and **walnuts** have an earthy-vegetal umami, and **tomatoes** have juicy flavors along with 175 mg of glutamate per 100 g, about the same amount as aged cheddar

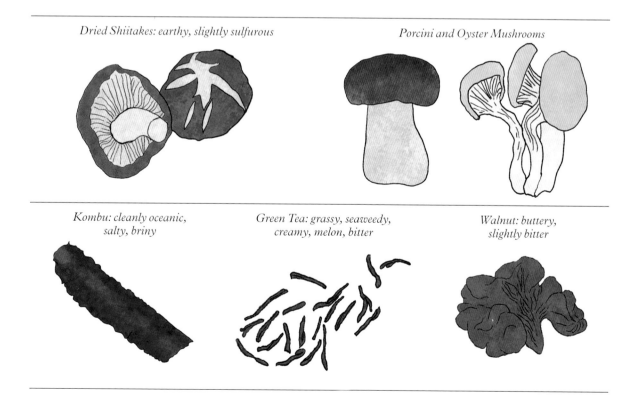

Dried Shiitakes: earthy, slightly sulfurous

Porcini and Oyster Mushrooms

Kombu: cleanly oceanic, salty, briny

Green Tea: grassy, seaweedy, creamy, melon, bitter

Walnut: buttery, slightly bitter

cheese (a savoriness that helps explain some of the fanciful variety names like beefsteak). Especially in the jelly around their seeds, tomato umami is enhanced by high levels of ribonucleotides (see "Creating Umami Synergy," page 93). Not to be overlooked for clean, earthy, and refreshing vegetal umami: **corn**, **peas**, **potato**, **garlic**, **cabbage**, **daikon**, **onion**, **asparagus**, and **spinach**.

Tomato, Pickled Shiitake, and Avocado Salad with Roasted Kelp Oil

This salad runs the spectrum of plant-based umami, with all the clean, earthy, juicy, and vegetal flavors that entails. Drying shiitakes creates even more free glutamate than fresh ones contain, and pickling them gives their earthy flavor a sharp poke from vinegar's pungency. Tomatoes are almost certainly the most umami fruit you can find, and combining the two with creamy and grassy avocado, then dressing it all in kelp oil, rounds everything out nicely.

Speaking of kelp oil, roasting the kelp and then steeping it in olive oil gives the salad a deep oceanic, nutty flavor. The credit for this technique goes to Rosio Sanchez, my former colleague at Noma, who used it to dress a hauntingly delicious dish of potatoes, roasted barley, and caviar.

Rehydrate and Pickle the Dried Shiitakes

Put **1 cup (¼ ounce; 8 g) stemmed dried shiitake mushrooms** in a medium heat-resistant bowl. Pour **3 to 4 cups boiling water** over them to cover. Steep together until the mushroom caps are soft and fully hydrated and the liquid around them is dark, at least 15 to 20 minutes and up to 1 hour. Fish out the shiitake caps and collect them in a separate bowl, saving the steeping liquid. Slice the caps lengthwise, ¼ inch thick.

Place the mushroom water in a large saucepan and bring to a boil over medium-high heat,

then turn down to a brisk simmer. Continue simmering until reduced to 1 to 1⅓ cups liquid. Add **¾ cup (180 ml) rice vinegar**, **½ teaspoon (2 ml) grated lime zest**, and the reserved sliced shiitakes to the saucepan. Bring just to a simmer, and simmer for 5 to 10 minutes, until some of the pungency of the vinegar blows off and the mushrooms have taken on some acidity. Turn the heat off, add **the juice of half a lime**, cover, and let steep and cool for 1 hour. Store the shiitake mushrooms in the fridge, in their liquid, for up to 2 weeks.

Make the Roasted Kelp Oil

Roast **3 to 4 sheets of packaged, trimmed, dried kombu** (or as much dried kombu as it takes to get **125 g, or 4.5 ounces, by weight**) on a baking sheet in a 325°F oven for 45 minutes until light golden brown, crackly, and slightly puffed. Let cool, then break up into smaller pieces with your hands. Put the kombu in a blender with **1¼ cups (300 ml) light and grassy extra-virgin olive oil** and blend until the kelp is sandy-textured and combined with the oil. Let sit for 2 hours to steep, then strain the oil from the solids through butter muslin (a tighter weave than cheesecloth, which isn't up to the task of the small kelp particles) or a very clean kitchen towel into a jar or other sealed container. (The oil can be refrigerated for 2 weeks, or store in the freezer for up to 6 months.)

Prepare the Rest of the Salad Ingredients

Slice **1 pound (450 g) very ripe tomatoes** from top to bottom, cutting them into roughly ½-inch-thick slices. Arrange the slices in a single layer on a large plate and sprinkle with **coarse sea salt,** then let sit for 15 minutes. Pour off the salted liquid they give up and reserve. Pit, peel, and cube **1 ripe avocado.**

Dress the Avocado

Combine the liquid from the tomatoes and ½ cup (120 ml) of the shiitake pickling liquid in a small bowl. Taste, and if it seems a little flat, add up to 1 tablespoon of rice vinegar. Toss the avocado cubes in this liquid, then spoon them out and reserve both the avocado and the liquid.

Assemble the Salad

On a platter or large plate, arrange the salted tomato slices, half of the pickled shiitake slices, and the drained avocado cubes in overlapping circles. Sprinkle 2 tablespoons (30 ml) of the reserved pickling liquid that you used to dress the avocado, a scant ½ cup (120 ml) of the roasted kelp oil, **a pinch of flaky sea salt**, and **freshly ground black pepper** to taste. Serve immediately.

Serves 2 to 4 as a side or starter

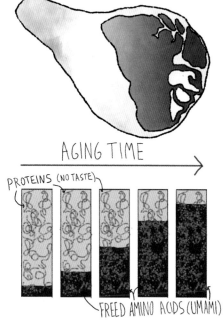

Protein x Time = Intense, Funky Umami

Umami is often easiest to spot in funky, rich, aged, and fermented foods. These are usually made from ingredients that are not particularly rich in free glutamate themselves: **cheeses** from not very umami-tasting milk; **prosciutto** from much less savory pork; **soy sauce** from comparatively bland soybeans and wheat. What do all these have in common? They start out rich in proteins, and to make them, we let them sit around for a long time, sometimes years. This combination so reliably creates umami-tasting glutamate that we can describe it as an equation: proteins x time = intense umami.

Key to this transformation are enzymes, pieces of not-quite-alive machinery that living things use to make molecular stuff. Some are for building things: taking glucoses and joining them up into starch, for example. Others are for breaking big things back down into their components, so ever thrifty cells can recycle them into new constructions, rather than starting from scratch every time. We name them in very descriptive, not very creative terms: amylases break down starch (*amylum* in Latin), and proteases, the enzymes we care about for umami, break down proteins into their individual amino acids. Which, since many proteins are made of 5 to 20 percent glutamate, translates into serious umami creation.

Some bacteria and other microorganisms go big on protease, and we've harnessed them to grease the wheels of converting protein to umami. Lactic acid bacteria produce proteases that, over time, free up impressive amounts of glutamate in aged cheeses like **Parmigiano-Reggiano**, sometimes close to 2 percent of their total weight. Culinary molds like *Aspergillus oryzae* are protease specialists, used for creating intense umami from

Soy Sauces: salty, malty, funky

Cheeses: cheesy, earthy, salty

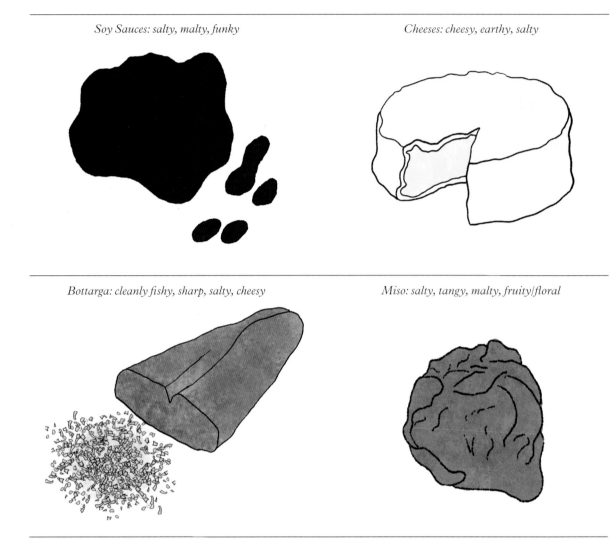

Bottarga: cleanly fishy, sharp, salty, cheesy

Miso: salty, tangy, malty, fruity/floral

Aged Hams: porky, nutty, salty

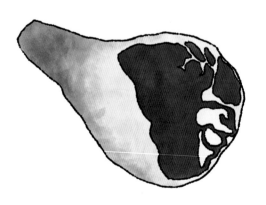

Anchovy: cleanly fishy, silky, salty

protein-rich soy in soy sauces like Japanese **shoyu** and Korean **ganjang**, and pastes and whole beans like **miso**, **doenjang**, and **douchi**. Lighter misos and soy sauces tend to have aged less, use less salt (since they don't need to be preserved as long), and often have more wheat or rice in them, all of which puts them on the lighter-umami end of the fermented soy continuum. Darker pastes and sauces are often a sign of a longer aging time, which means more time for proteases to work, and more umami. They're often also saltier, to facilitate this longer aging.

While specialized microbes are helpful, it's totally possible to harness the proteases right in the cells of once living ingredients themselves to unlock their full umami potential. When you dry-age beef or salt and air-dry a pig's leg to make ham, every single cell has proteases inside them that gradually break down the proteins around them.

And slightly grosser (but really cool!) is how we've figured out to age and ferment small fish and crustaceans to make **cured anchovies**, **fish sauces**, or **shrimp pastes**: those creatures eat even smaller creatures; things with protein in them.

So . . . they need plenty of proteases to digest them.

So . . . their digestive tracts are full of proteases.

So . . . if you salt and blend them up or even just salt and hold them, their naturally protease-rich guts will gently and gradually digest their own protein-rich bodies, creating, you guessed it, boatloads of umami-flavored glutamate.

Umami and Sour

Umami and sour may be the two most unalike tastes. Sour is bright and sharp, umami is soft and subtle. Sour enlivens and punches up whatever you pair it with; umami blankets it with lightweight richness. And rather than balancing or canceling each other out (like sweet and bitter

do), mixing them together combines all their best qualities and more, like complex harmonies generating new overtones.

It's a highly clockable flavor pattern, once you start looking for it: soy plus citrus and vinegar in ponzu sauce—umamisour. Fish sauce plus lime juice (and a little palm sugar) in a Vietnamese nước chấm—umamisour. Roman fish sauce–like garum and sour wine in oenogarum, the ancient Mediterranean's answer to ketchup—umamisour. Adobo—umamisour. Even a ripe tomato—umamisour.

Try This

Umami and sour is especially great when applied to vegetables, simultaneously enriching and enlivening them. In a ponzu mood, I combine equal parts lemon juice, grapefruit juice, and good soy sauce, and sprinkle it directly over hot broccoli, or dollop about 1 tablespoon of it into ½ cup of just-made vinaigrette for sturdy or bitter lettuces. Japanese su miso combines miso and vinegar, a fantastic idea I apply by mixing 1 part rice vinegar with 2 parts white miso (so, 2 tablespoons vinegar to ¼ cup of miso), thinning with water if necessary until creamy, then spoon over roasted carrots or grilled zucchini in a thick drizzle.

Beefing up Glutamate: Creating Umami Synergy

Looking around the list of the most glutamate-rich, umami'd-up ingredients, you might spot a conspicuous absence: Where's the beef? Surely this meatiest of meats—and meats generally—have tons of umami glutamate in them, right?

Shockingly, the answer is no: fresh meats just don't contain much glutamate. They have *some* glutamate, around 30 mg per 100 g for beef, but vegetables like garlic, cabbage, onion, even potato can have two to three times more.

This, fortunately, doesn't mean that meats, fresh or otherwise, have no role to play in umami.

Let's return for a moment to the umami receptor. It responds directly to glutamate, but the umami signal is hugely boosted when there are also ribonucleotides, like the molecules IMP and GMP, in the mix. Glutamate binds to the main part (the "active site") of the umami receptor, while ribonucleotides come in from the side and tell the receptor to hold on to glutamate and keep sending that umami signal to the brain. And meats? They're fantastic sources of ribonucleotides. Beef has a respectable 75 mg of ribonucleotides per 100 g; pork, chicken, and mackerel, around 200 mg.

Here's a second umami equation that can help put these numbers into context: 1 + 1 = 8.

Obviously, 1 plus 1 does not typically equal 8, but we're talking synergy here: the whole is greater than the sum of its parts. In the 1960s, researchers at Aji-no-moto made mixtures of glutamate and ribonucleotides at different ratios and then asked people to rate how intensely umami they were. Adding just 2 percent ribonucleotides to glutamate (so, 2 mg of IMP to 100 mg of glutamate) could double or triple the umami intensity that people tasted. This is like adding a tiny sliver of pork to an already moderately umami chunk of cheddar cheese, and making it taste as umami as a light soy sauce. **1 + 1 = 8** describes the apex of umami synergy: roughly equal quantities of glutamate and ribonucleotides make the glutamate taste about 8 times more intensely umami than the same amount of glutamate alone.

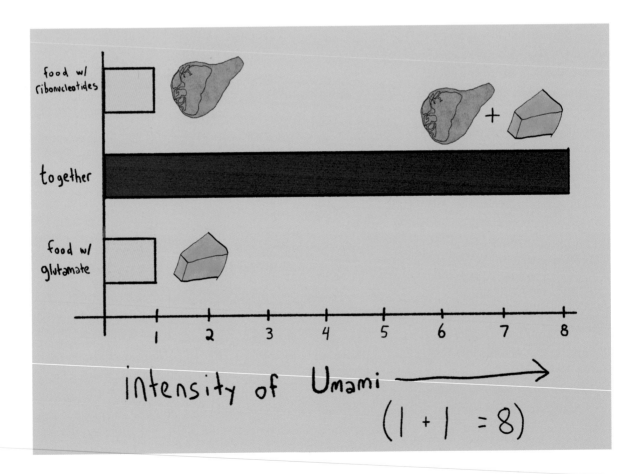

The Umami Boosters:
Ribonucleotide-Rich Ingredients

These ingredients may not have a lot of their own glutamate, but they have *tons* of ribonucleotides like IMP and GMP to boost the umami flavor of glutamate-rich ingredients.

Name	Type of Ingredient	Total glutamate (mg per 100 g)	Total ribonucleotides (mg per 100 g)
Dried anchovy	Processed animal protein	16	884
Katsuobushi	Processed animal protein	24	758
Anchovy*	Processed animal protein	1200	305
Tuna	Animal protein	16	292
Mackerel	Animal protein	36	221
Chicken	Animal protein	44	220
Pork	Animal protein	23	211
Scallop*	Animal protein	140	186
Squid*	Animal protein	146	184
Dried shiitake*	Fungi	1060	150
Prawn	Animal protein	40	87
Lobster	Animal protein	9	82
Beef	Animal protein	33	82
Crab	Animal protein	72	42
Morel (dried)*	Fungi	311	40
Sea urchin*	Animal protein	103	32
Tomato*	Vegetable	175	21

*Also a good source of glutamate.

Cooks basically figured this out a long time ago—there are many, many dishes that pair a glutamate-rich ingredient with a ribonucleotide-rich one for extra umami. For instance, shellfish and potatoes in a clambake, tomato and cheese on a pizza, kelp and katsuobushi (dried, smoked, and fermented flakes of tuna loin) in Japanese dashi broths, onions and beef broth in French onion soup. It's a finessed way of getting to your goal: rather than loading on glutamate-rich ingredients being your only option, you can pick a little of column A, a little of column B, and put a multiplier effect on flavor.

Umami synergy can help explain the savoriness of (unfermented, not highly aged, fresh) meats, which don't have particularly high levels of glutamate. If you figure 1 + 1 = 8—or, for every milligram of glutamate that you have a matching milligram of ribonucleotides for, you get 8 times the umami—then most animal proteins (beef, pork, chicken, tuna, etc.) have an umaminess somewhere between cheddar cheese and prosciutto. Each of them has tons of ribonucleotides to spare, making it most worthwhile to find glutamate-rich ingredients to join them with and further amplify.

Try This

Like kombu and katsuobushi in Japanese dashi stock, these combinations taste extremely umami because of the synergizing complementarity of glutamate-rich ingredients and ribonucleotide-rich ones: cabbage and pork; mackerel or tuna in a soy marinade; garlic and practically any meat. Others include kimchi (cabbage + garlic + dried seafood); beef broth and onion in French onion soup; and Parmesan, garlic, and anchovy on a Caesar salad.

Umami-Boosted Cacio e Pepe

Cacio e pepe, the Roman pasta dish sauced with a simple but potent mix of black pepper, cheese, and pasta water, tastes highly umami already, since it uses half a cup of intensely glutamate-rich grated Parmesan and pecorino per serving. It's one of my favorite things to eat. I like to layer in other sources of umami—with funky aged flavors (from anchovies), and toasty ones (from nutritional yeast)—to create an echoing effect, giving more complexity to the flavor, and boosting the umami tastes even further, thanks to the multiplier effect of the ribonucleotides in anchovy and nutritional yeast.

Using a Microplane or other fine grater, grate **1 cup (100 g) Parmesan or pecorino cheese** (or a fifty-fifty mix of both) and set aside.

In a small saucepan over medium-low heat, combine **4 tablespoons (60 g) butter, 1½ teaspoons (3 to 4 g) freshly ground black pepper**, and **1 minced anchovy fillet**. When the butter melts, turn off the heat, cover, and let the pepper and anchovy steep until the pasta is finished cooking.

Boil **½ pound (225 g) dried bucatini or spaghetti** in a large pot of well-salted water until fully cooked, according to package instructions. Scoop out **2 cups (450 ml) of the pasta water** and set aside, then drain the pasta and put it back in the pot. Add the anchovy-pepper mixture, **2 tablespoons (10 g) nutritional yeast flakes**, and ½ cup of the reserved pasta water to the noodles, and toss well. Gradually add the finely grated cheese to the pasta while tossing with tongs. The heat of the pasta will melt the cheese and emulsify it into the pasta water to form a sauce. Add more pasta water in tablespoon increments if it looks too thick to get a creamy texture.

Transfer to a plate and grate over a little more cheese. Serve immediately.

Serves 2

Bitter

Bitterness is the black sheep of the taste family. Like the other tastes, it senses particular molecules, but it's tuned to things that can hurt us, rather than things we want. Not everything we eat wants to be eaten, and a lot of molecules in things we use for food are toxic by design. Bitter is there to detect them. In the final, crucial moments before you swallow them down, bitter sounds the Klaxon of warning. It's the dark, shiveringly pleasant-unpleasant taste of aversion. It's the taste of latent danger.

The Rules of Bitter

Bitter is for sensing and warning about things that might be toxic or poisonous.

Since lots of things could fit this bill, bitter senses the widest range of different molecules, and we have more variations in bitter receptors than for any other taste.

Some bitter molecules or bitter foods are useful or medicinal in moderation, so we have the flexibility to learn to like them even though they still taste bitter.

This flexibility or acquired taste goes much easier when bitter is chaperoned by other flavors, both tastes and smells.

Salty, sour, and sweet are all very good at balancing, tempering, and chaperoning bitter.

Bitter foods—coffee, chocolate, endive, beer—are the most classic "acquired tastes"; literally repulsive at first contact, they're things we have to learn to like over time, if we come to like them at all. Thrill seekers that we are, we often find bitter foods (and the bitter molecules within) oddly compelling in reasonable amounts, even craveable. We deliberately ignore the danger signals to get our fix of that sick little frisson they create. These foods are not only incidentally bitter, they're enhanced by bitterness and would taste insipid or boring without it.

Used deftly, like carefully selected dissonant chords creating tension in an otherwise straightforwardly pleasant piece of music, bitter takes you straight to flavor sophistication. A grounding and deepening influence, it clarifies and awakens the palate, tempers sweetness and richness, and makes moderation satisfying. It's the single perfect ounce of strong espresso, a pile of exquisite bitter leaves, the burnt top of a crème brûlée, a strong gin and tonic or Negroni, a juicy, bracing segment of grapefruit.

The Science of Bitter

Because so many different molecules out in the world have the potential to hurt us, bitterness is the most intentionally broadly tuned of the tastes. There's no specific chemical signature, no singular mechanism in the body, that all bitter molecules share. We can't reduce them to the unifying motif of a sodium, hydrogen ion, or an amino acid—or even a general pattern, as with sugar—because toxic molecules have a radically variable range of shapes and formats.

Some molecules that create bitterness for us are formed when a chemical change happens in an ingredient, like the bitter molecules that appear in burnt, charred, or heavily browned foods. Many of them are made intentionally by plants, to protect themselves from grazing herbivores that want to eat them. In large enough doses, these bitter molecules often interfere in some ruinous way with essential processes in an animal's liver or kidneys, in its central nervous system, or in the tiny machinery inside individual cells. But even these agents of chaos don't necessarily have similar molecular shapes. Some, like caffeine, are nitrogen-based alkaloids. Others, like the bitter molecule sinigrin in mustard greens and their brassica relatives, are sulfur-based glucosinolates. Even more, like bitter limonin in citrus pith or cucurbitacin in squash and cucumbers, are triterpenes, synthesized from some of the same stuff that would otherwise go into pleasant aroma molecules.

To deal with the problem of molecules we need to sense, but that don't look like one another at all, we've rather ingeniously evolved a modular approach to bitter receptors. Their basic format is pretty similar to the sweet and umami receptors, with protein-based cubbies or pockets into which tastable molecules can nestle. But rather than one type of receptor tasked with somehow matching up to all molecules that we need to taste bitter, we make a set of about twenty-five different bitter receptor shapes, sprinkled across our bitter-sensing taste cells. Like KitchenAid mixer attachments, they're specialized for different things but hook up to the same machinery. The bitter signal sent to our brain is singularly bitter, but it can be set off by many different molecule types and shapes.

Considering all this work evolution has put into sensing and signaling toxicity and danger, are we ignoring important messages from our body when we acquire a taste for bitter foods like matcha or olives? Not exactly—it's more useful to think of bitterness as a pop-up alert than a red light, a general heads-up rather than a strict instruction. Many things we like to eat are naturally a little bitter, but we're not necessarily playing a game of toxicological chicken every time we eat them. Thinking back to what life was like for our deep ancestors, foraging and eating lots of different wild foods, a plant or another ingredient can have some bitterness in it while on balance still being a good source of energy or nutrition: think oily olives or sweet-bitter wild apples. Some bitter molecules even have innate biochemical benefits to our body, like the antioxidant effects from bitter polyphenols in tea or the antimalarial power from bitter quinine (the key bitter ingredient in tonic water, made from the bark of cinchona trees). There are plenty of times that the positives outweigh the negatives for eating something bitter. So, learning to like bitter things isn't even necessarily directed or intentional. It's part of the rather pragmatic flexibility humans have developed for bitterness: if we eat the same, unpleasantly bitter food a few times, don't die or get sick, and maybe even find out that it's a decent source of nutrition

Supertasters, Little Kids, and Bitterness

Not everyone acquires a taste for Campari or coffee. If you have a higher-than-average number of taste buds on your tongue, you also have a more acute sense of taste: everything tastes more intense to you. You're a "supertaster," and many supertasters are so sensitive to bitterness that they never warm up to it, forgoing espresso, beer, gin and tonics, dark chocolate, and even most types of nonbitter alcohol, because the bitter signal they get is so strong. Children tend to be pickier eaters for the same reason, and we all generally lose taste buds as we age.

or pleasant intoxication rather than violent toxicity, we can store memories of these experiences to guide us the next time we encounter a similar bitter food.

Bitter Flavor Patterns

Bitter can be pleasant the same way a rollercoaster can be pleasant—a brush with danger, but a controlled one. And while you could get a similar thrill from climbing into a shopping cart at the top of a hill, the greater care involved in a properly maintained rollercoaster—between the braking system, the tracks and their maintenance, lap bars and other restraints, and operator training—means the payoff, measured in consistency of experience or lack of broken collarbones, is higher, too.

Compared with the other tastes, the patterns of using bitter, the things you'll find yourself thinking about over and over, include more work in managing and balancing bitter on top of knowing where to find it and how it's shaded. Adding bitterness, but also keeping it check.

Bitter Needs a Chaperone

Just as you might look for signs of safety when lining up for a rollercoaster—functional tracks and brakes, people ahead of you in line who make it off the ride alive and unharmed—our brains pay attention to all the flavor signs that come along with bitter when eating a bitter food, in order to interpret whether to relax because it's probably a safe and fun time, or stay on guard because it's likely to be risky. These other flavors act like chaperones to bitter, smoothing its way and making polite introductions. And like the heroine of a Victorian novel, bitter simply can't be out and about without a (flavor) chaperone.

All four of our other tastes are remarkably effective at chaperoning bitter. Sweet, so much that the spoonful of sugar that helps the medicine go down has been immortalized in song. Many of the bitter drinks we drink for fun started out as medicinal tinctures, where bitter was unavoidable because the same molecules in their herbal ingredients provided beneficial bioactive effects. Any modern descendant of these drinks—say, an Italian-style amaro (Nonino, Averna, even famously extra-bitter Fernet-Branca)—has quite a bit of sugar added to it, which acts like the sugar in lemonade: it doesn't actually remove any bitter molecules, but it does reduce how strong they taste. Sour performs respectably: adding a bit of lime juice to a gin and tonic dials back the quininey bitterness without fully covering it up. And salt is a particularly effective bitter-tamer that acts directly at the source, interfering with the bitter receptor itself and dampening the signal before it even sends it to the brain.

The Italian Greyhound

This cocktail is a cousin to the salty dog, one of my favorite unpretentious drinks that improves upon the greyhound (a combination of gin and grapefruit juice) by adding a salted rim. In this recipe, the grapefruit bitterness is doubled up with bitter Campari, then brought to heel by the taming and chaperoning flavors of sour (from lime juice), salty (from that same salted rim), and sweetness (from Campari, which despite its bitterness is close to 25 percent sugar). Bartenders call this method of combining ingredients right in the glass you're going to drink it from "building" a cocktail.

Run a **cut lime** around the rim of a highball glass or tumbler. Put **¼ cup (40 g) kosher salt** in a small saucer. Invert the glass and dip the wet rim in the salt. Add **several big ice cubes** to the glass, then pour in **1 ounce (30 ml) gin**, **½ ounce (15 ml) Campari**, **3½ ounces (100 ml) grapefruit juice**, and **½ ounce (15 ml) lime juice**. Stir until combined and cold, then drink immediately.

Makes 1 drink

Besides tastes, all the other flavors in a bitter food (the ones we sense with smell) also play this chaperoning role. The blueberry or cinnamon notes in varietal coffees, the bright and juicy citrus flavors of grapefruit, the sweet-sour whisper of an 85% cacao-content dark chocolate bar—these flavors wrap themselves around bitter, buffering and tempering its intensity, softening its impact, and sending a signal that there's positive stuff other than potential toxicity happening in these foods.

Aromas are so key for this that if you don't have enough of them in a bitter food, even if you still have balancing tastes like sweet or salty, your brain responds to the lack of information by defaulting back to "bitter = dangerous." I was forcefully reminded of this principle the single time I tried to drink a Negroni, one of my favorite bitter cocktails, while suffering from a cold. A simple formula of equal parts herbal-resiny gin, fruity-tangy and sweet vermouth, and the bitter-herbaceous-sweet aperitif Campari, the Negroni's bitterness is pretty tame, just enough to be a little invigorating. But with my sinuses and nasal passages plugged up, I couldn't smell anything—so the bitter-tempering aromatic flavors (coriander and juniper from the gin, winey and caramel ones from the vermouth, citrusy-herbal ones from the Campari) were not doing their normal chaperoning jobs. Without them, the bitterness was so prominent it was practically the only thing I could taste in each shuddering sip.

Easy Ways to Chaperone Bitter

- Add salt: salted bitter rapini, obviously, but also salted chocolate, salted grapefruit, salted espresso-flavored anything. Use flaky sea salt for a textural contrast, or branch out and use an infused salt (page 41) or salty condiments like soy sauce or fish sauce.
- Add fat: olive oil, aromatic nut oils, Smoke Oil (page 210) or Roasted Kelp Oil (page 90), bacon or chicken fat, cream or crème fraîche.
- Add umami, funk, sourness, pungency with dry-cured ham or anchovies, miso, or your favorite vinegar.
- Add the aromatic suggestion of sweetness. Spices like cinnamon, cloves, cardamom, allspice, fennel; herbs like tarragon, mint, or basil all work great with bitter things that are already light, fruity, or sweet (like grapefruit or dark caramel) but also do exciting things to savory foods (like well-charred meats, broths, and bitter vegetables). Sugar doesn't even need to be there—the sweet-adjacent aromas can cover some of the chaperoning duties on their own.

Chaperoning Bitter Leaves, Roman Style

Puntarelle is like a long-and-skinny, stalky endive that's a Roman local specialty. Classically, it's cut into thin, long strips and made into an assertive salad. The dressing for puntarelle brings together almost all the bitter-tempering flavors: fatty olive oil, lots of salt, sour wine vinegar, funky-umami anchovy, and pungent garlic. If puntarelle eludes you, this dressing is a great way to dress any raw, bitter leaves, or sautéed or grilled bitter broccoli raab.

Mash together **1 small garlic clove** and **2 anchovies**, stir in **1 tablespoon (15 ml) wine vinegar**. Whisk in **3 tablespoons (45 ml) olive oil**. Add **kosher salt** and **freshly ground pepper** to taste. Store tightly covered in the fridge for up to 2 days.

Makes enough for 2 to 4 servings as a starter or side

Bitter and Vegetal

Farmers and plant breeders over the centuries have usually aimed to reduce bitterness in their crops, but there are certain delicious vegetables in which a lingering veil of bitterness is the whole point. These tend toward a pattern of distinctly and refreshingly planty flavors: leafy, earthy, grassy, bell peppery. Vegetal-bitter foods work well raw and at the beginning of the meal (like a bitter salad) to wake up the palate or pique the appetite. Raw or cooked, they're almost always tempered and chaperoned by strong flavors.

Bitter melon is a bit like a bitter, succulent zucchini with squashlike large seeds. Its primary bitter compound, momordicine, is thought to be a deterrent for egg-laying insects and it has been credited for beneficial health effects in ailments from diabetes to cancer. Since bitter melon is grown and cooked across Asia (as well as in Africa and the Caribbean), dishes that use it are like a flavor calling card for the classics of regional cuisines. It might be chaperoned by funky-umami fish sauce or fermented black beans, pungent garlic, fatty and aromatic coconut milk, or spicy chiles, all of which work wonderfully with bitter—whether you can get bitter melon where you are, or if you have other bitter vegetables to prepare.

In Europe, the chicory family produces bitter leaves in many different packages and formats: dense and compact **endive**, fluttery escaroles and **frisée**, long and crunchy **puntarelle**, and juicy **radicchio**. Accessorizing their bitterness can take a rich and fatty bent (creamy dairy on endive, bacon and its fat on frisée) as well as a more sour, vinaigrettey one. Their earthiness also plays well with contrasting sweet and fruity flavors.

Radicchio Salad with Fennel, Blood Orange, and Brown Butter

In this recipe, the dressing ingredients soften and chaperone radicchio's bitterness from several different directions. Blood orange adds sweetness, sourness, and citrusy aromas; olive oil is

fruity and fatty and further infused with anisey sweetness from the fennel seeds and toasty, malty brown butter. Rounded out with sour-pungent sherry vinegar, flaky salt, and optional resinous oregano leaves, it's all there to enhance the leaves rather than distract from them.

If you'd like a simpler preparation, you could use any really good orange in place of blood orange and segment it instead of supreming it, and combine the dressing ingredients in a mortar and pestle to crush rather than a blender—but I think it's worth it to put in the extra work if you can. I use a fast blender with a small beaker like a Magic Bullet for quick, oily flavor extractions like this. If you're worried about losing too much to the walls of your blender, you can definitely double or triple the proportions and save the extra oil in the fridge in a tight-lidded container for up to 2 weeks.

For brown butter solids, refer to A Better Brown Butter, from Heavy Cream (page 249).

Several hours before making the salad, combine **¼ cup (60 ml) good fruity olive oil**, **half of the grated zest of 1 blood orange (a scant 1 tablespoon or 5 g)**, **¼ teaspoon (1 g) fennel seeds**, and **1 tablespoon (15 g) brown butter solids** in the beaker of a powerful blender. Blend on high speed for 30 seconds or until all the fennel seeds are finely crushed. Let sit for 2 to 3 hours so the flavors infuse into the oil.

Next, peel and supreme **1 blood orange.** Hold it in your nondominant hand over a bowl to catch any juices. Carefully cut just on the inside edge of one segment. Repeat with the other side of the segment. Slide out the now easy-to-eat wedge of orange flesh and drop it in the bowl. Repeat with the other segments. Set aside.

Cut the base off **1 head radicchio** and pull the leaves off the core. Tear the leaves into slightly larger than bite-sized pieces. Wash and dry the leaves well and put them in a large bowl.

Spoon and drizzle three-quarters of the infused oil over the leaves, making sure to include

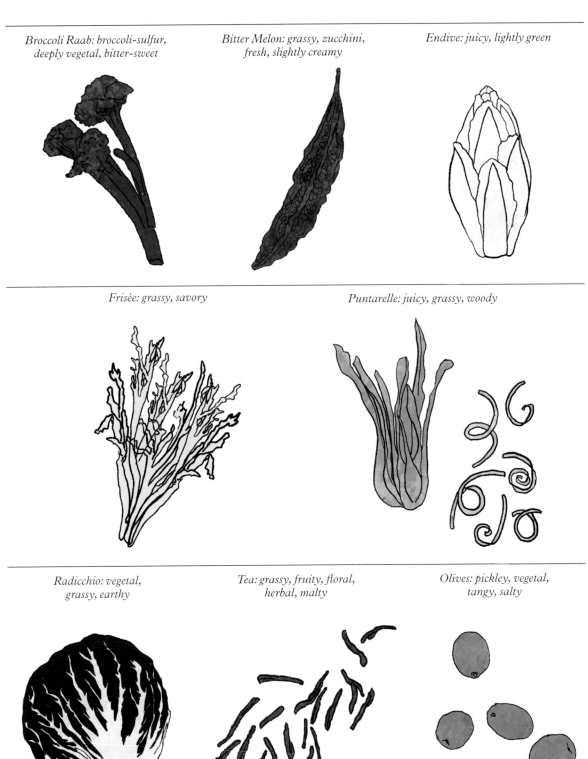

Broccoli Raab: *broccoli-sulfur, deeply vegetal, bitter-sweet*

Bitter Melon: *grassy, zucchini, fresh, slightly creamy*

Endive: *juicy, lightly green*

Frisée: *grassy, savory*

Puntarelle: *juicy, grassy, woody*

Radicchio: *vegetal, grassy, earthy*

Tea: *grassy, fruity, floral, herbal, malty*

Olives: *pickley, vegetal, tangy, salty*

any chunky bits. With your hands or tongs, gently toss the leaves. You want them all to be coated in a thin, shimmering film of oil. You can probably add the rest of the oil, but check first, so that the leaves are not swimming in it.

Sprinkle on **2 big pinches of flaky or crunchy sea salt** and toss again. There should be an even, light flurry of salt on the leaves. Eat one and see what you think—it will be bitter, but you should get distinct pops of saltiness. Next, carefully drizzle about **1 tablespoon (15 ml) sherry vinegar** (or about one-quarter as much as the oil you used). Gently toss and taste again. The vinegar should dapple the lettuce leaves.

Pour off the blood orange juice that's accumulated beneath the supremes and sprinkle on the salad. Arrange the reserved blood orange supremes on the salad, adding about **1 teaspoon (1 g) fresh oregano leaves (optional)**.

Serve immediately.

Serves 2 as a side or starter

Green tea contains natural antioxidant (and bitter-tasting) molecules called *catechins*. In black and oolong tea, which are the same plant leaves as green tea, oxidation with enzymes is encouraged by rolling and bruising, and these catechins react to form orangey-brown pigments. Green tea is heat-processed quickly to inactivate these enzymes, which preserves the green color of the leaves, the antioxidant and bitter character of the catechins, and the grassy, fresh, seaweedy flavors. These vegetal flavors come through when you steep the tea, or cook with it (in the form of powdered matcha).

Olives are technically a fruit, but a slightly bitter and unsweet one that tastes much more like a vegetable. If we're eating them whole, they're always fermented or otherwise cured. Olive fruit, leaves, and oil all have a bitterness to them that comes from a polyphenol molecule called *oleuropein*. Like many bitter molecules, it has bioactive effects: it reduces blood pressure and is a good antioxidant. Fermenting olives partially breaks down oleuropein, making it way less bitter, and the broken-down versions are still great antioxidants. Fermentation also increases their vegetal aroma and creates sour acids, which bring a bit of balance to their bitterness.

Bitter and Fruity: Citrus

Pretty much all citrus has the bitter-tasting molecule *limonin* (not to be confused with the citrusy-smelling molecule limonene!) in their pith, the white flesh between the aromatic rind and the juicy, sour-sweet flesh. While this layer is often discarded, it can also make for a very nice bitter-sour-fruity flavor experience. Orange marmalade includes strips of bitter **Seville orange** peel for flavor and its gelling pectin. The added sugar helps balance the bitterness, and the bitterness gives the whole thing sophistication. **Chinotto** can refer to a specific type of orange or the bitter soda made from it, which adds sweet-smelling cinnamon and clove for further aromatic bitter-softening. **Grapefruits** and **pomelos** have uniquely bitter flesh and juice, in addition to pith, thanks to their uniquely bitter molecule *naringin*. Salting grapefruit to tame its bitterness and make its inherent sweetness more obvious is an age-old culinary technique (one you can taste if you make The Italian Greyhound, page 99). Like chinotto, grapefruit bitterness is also well chaperoned by cinnamon and other sweet spices, a trick also used in many grapefruit-infused tiki drinks.

Naringin and limonin are both triterpenoid molecules. Citrus produces tons and tons of aromatic monoterpenoid molecules, and when three of them form like Voltron into a larger triterpenoid, they lose their aroma and, together, become bitter.

If you taste different olive oils, you'll notice that one common flavor trend in the more robust ones is "pepperiness," which is best described as a mix of spiciness and bitterness with a little something extra on top. These oils have a pleasantly throat-irritating quality most noticeable when tasted straight and on their own that some call "the cough." (Because, quite literally, it makes you cough a little.) A few years ago, researchers at the Monell Chemical Senses Center in Philadelphia (yes, it's a research institute devoted to taste and smell, and yes, I'm a huge fan) did some chemical detective work and found the molecule responsible, *oleocanthal*, is a powerful antioxidant in its own right.

Bitter and Herbal: Leaves, Flowers, and Nuts

Bitter herbs are so old-school, they're in the Old Testament (and the Torah). Where vegetables have a kind of fresh and juicy-green bitterness, bitter herbs like **wormwood**, **yarrow**, and alpine **genepi**—or more common **thyme** or **sage** in large amounts—are layered over with resiny, piney, even spiced flavors.

Bitter herbs have the biggest overlap of molecules that are both bitter-tasting and perform other biological actions in the body. So they're also one of the biggest overlaps of culinary recipes and products and medicinal ones. Traditional Chinese medicine has many of these, including **kudingcha**, a tea made from extremely bitter, saponin- and polyphenol-rich *Ilex kaushue* leaves, which are in the holly family and can reduce blood glucose and help kill cancer cells.

Herbal-bitter alcoholic infusions and liqueurs, most notably Italian **amaro**, can include dozens of bitter herbs like **wormwood**, **gentian**, and **quassia** alongside other layers of botanicals. As a *digestivo*, a kind of vestigial herbal medicine, a glass of amaro is an important finish to meals for many Italians, taking advantage of the effects many of its bitter ingredients (some triggered by the perception of bitterness, some locally by bitter molecules) have on indigestion. Postprandial needs aside, amari are also an amazing playground for different moods of herbal-bitter flavor combinations. Aroma profile can really change the vibe of the bitterness in these drinks: the more resiny and piney, the more deeply bitter and dry they taste. The more citrus, sweet herbs, and spices, the more buoyant and balanced. The unforgiving aroma profile of fernet makes it taste all the more bitter than the clove-anise flavors of (still very bitter) Underberg. Some great places to start for herbal bitterness are Campari, spicy, citrusy, colaesque Ramazzotti, and earthy Cynar, which famously includes bitter artichoke leaves.

If you're not a big amaro drinker, your reference for herbal-bitter might be **hops**. Technically flowers, these beer seasonings have resiny, floral, citrusy aromas, and when you boil them for a while, the heat flips a couple of chemical bonds

Citrus Pith: bitter, vaguely citrusy
Bitter Orange: sweetly orangey, sour
Grapefruit: piercingly fruity, fresh, herbal, sour

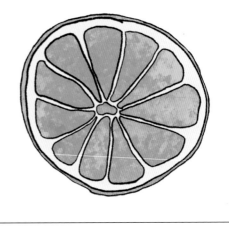

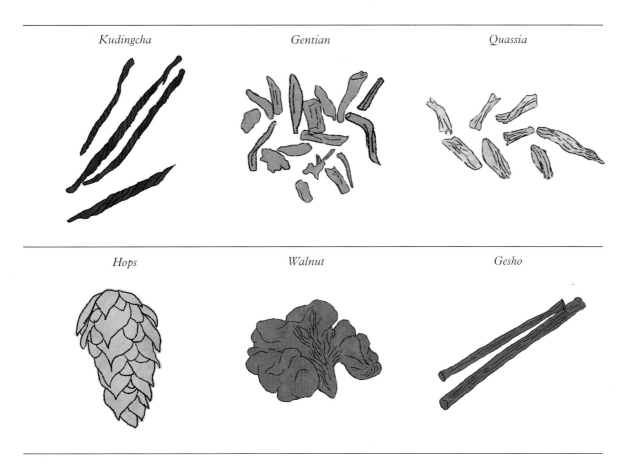

Kudingcha

Gentian

Quassia

Hops

Walnut

Gesho

in a molecule called *humulone*, creating the much more bitter *isohumulone*. The humulones protect the beer from spoiling, which is why hopped beer is so ubiquitous. Up until the fifteenth or sixteenth century, instead of hops, beers often had a mix of bitter herbs called *gruit*, which often included heather, horehound, mugwort, yarrow, bog myrtle, and juniper. Some craft brewers still use gruit in special herbal brews; many of these herbs are more commonplace now as ingredients in herbal cough drops.

Walnuts have tannic, herbal bitterness, which comes from antioxidant phenolic molecules (unlike bitter almonds, which get their bitterness from cyanide-containing amygdalin), offset by a buttery-sweet, almost mapley flavor. They're as rich-tasting as any other nut, but made into butter,

or a nut crust, or into cookies, they bring much more heft, with a more bitter edge, compared with hazelnuts, pecans, and otherwise similar nuts.

Walnut-Amaro Cake

Walnut is noticeably bitter-herbal and tannic compared with many other nuts, so this recipe leans into the bitter by adding some bitter-sweet amaro, tempered with a little extra salt. The amaro has the added benefit of essentially spicing the cake, too.

Preheat the oven to 350°F and butter a 9-inch cake pan. In a food processor, pulse **3⅓ cups (400 g) walnuts** and **3 tablespoons (30 g) cornstarch** until they're the texture of bread crumbs but not pasty. Set aside.

In a large bowl with a hand mixer or with a stand mixer, beat **5 large eggs, 1 packed cup (200 g) light brown sugar, 3 tablespoons (45 ml) Ramazotti** (or another sweet and brown amaro, like Averna or Nonino), **1½ teaspoons (8 g) baking powder**, and **½ teaspoon (3 g) salt** until the eggs are very pale yellow and rather fluffy, 6 to 8 minutes.

In a second bowl, fold the ground walnuts together with one-third of the egg mixture. (It will seem like it doesn't want to combine, but be patient and gentle so as not to deflate the whipped egg.) Continue folding in the egg mixture to the lightened nut mixture one-third at a time until fully combined.

Scrape the batter into the buttered pan and bake for 45 minutes, until the cake is puffy and somewhat matte-looking on top. A toothpick inserted in the middle should come out with wet crumbs on it, but not goop. Remove from the oven and cool for 10 to 15 minutes in the pan. Gently invert onto a rack to continue cooling.

Serve as is or with whipped crème fraîche and a sprinkle of flaky salt. Stores, well wrapped, for 3 to 4 days.

Serves 4 to 8 as dessert

Creating and Seasoning with Bitter: Browning, Caramelizing, and Burning

If you've mastered carefully accessorizing and chaperoning bitter ingredients, you might be ready to move on to the graduate-level seminar on bitter: deliberately applied bitter. The best place to start? The techniques that are so easy, you can do them by accident: **charring**, **browning**, and **burning**.

High heat kicks off a cascade of chemical reactions that create new smell molecules, golden-to-black-colored molecules, and, most important, bitter molecules. Overdo it and you'll get an acrid mess, but the list of delicious, carefully bitter, browned-burnt foods is long: roasted **chocolate** and **coffee**, **toast**, **grilled bread or sticky rice balls,** the edging-on-burnt **"leopard-spotting"** on the bottom of a good Neapolitan pizza. If you're careful about keeping well-browned and burnt spots superficial, the contrast and depth their bitterness creates make it almost like a seasoning. Charring is an essential first step applied to tomatoes and chiles for many Mexican salsas, as well as to the onions and ginger that season the broth for Vietnamese pho.

Browning, caramelizing, and burning all require a bit of carbohydrates, particularly sugar, to work well. (Technically, depending on how much sugar you have and how far you take it, you might be primarily caramelizing, burning, or browning via the Maillard reaction. But for our purposes, they all create bitterness—for way more on those processes, check out page 244.) Keep a close eye on anything really sugary or sweet, which will burn (and overdo the bitterness) superfast. Anything hard and grainy or starchy, like raw rice or other grains, raw potatoes, or coffee/cacao beans, will only get harder as it gets browner and bitterer, so plan ahead to precook these (as in a parboiled potato, steamed rice, or baked bread) before toasting/roasting/burning, or to grind them finely (like roasted coffee or cacao, or toasted rice or buckwheat) before using. Use the high heat of a heavy cast-iron pan, grill, broiler, or torch to blacken the cut surface of halved onions and the exterior of leeks, peppers, or tomatoes. Finish off cooked firm vegetables like carrots or sweet potatoes the same way. Give bananas, apricots, or peaches a quick kiss of burnt and smoky bitterness.

When used in careful, sparsely sprinkled proportions, ingredients that you have roasted the hell out of can make gorgeous seasonings, especially for foods with flavors you might otherwise use to balance out something bitter: fatty, sweet, fresh and crunchy, or umami. A prime example: garlic gets extremely bitter when burned, which is usually uniformly unpleasant, but some styles of porky, superfatty tonkotsu ramen use a drizzle of mayu, an intentionally blackened garlic oil, to deliciously balance their creamy heft.

Try **instant coffee** or **deeply browned caramel**, ¼ teaspoon at a time, to add bitter depth to chili, savory and tangy sauces, and desserts. Make Burnt Scallion Butter (page 205) and spread it over good sourdough. **Deeply brown and blacken hand-torn bread crumbs,** then sprinkle them over a crunchy salad or other vegetables, or even over ice cream, that feel like they need a little gravitas, like putting on a pair of horn-rimmed glasses before going into an important meeting.

*Spicy**

Spicy is pain. Not in the sense that life is pain, but quite literally, spicy is pain. If there's an edgy, slightly twisted element to liking bitterness, we're full-on masochists when it comes to spicy. The pedantic part of me hesitates to include spicy with the five basic tastes. That's because spicy is not technically a taste at all.

Spiciness has no taste receptor. We detect it via our sense of touch. Spicy molecules trick touch receptors—specifically, pain receptors—that usually respond to burning heat into sending those signals to the brain when there's no elevated temperature to be found. Those signals get integrated, along with smell and taste, into our holistic sense of flavor.

From a plant's perspective, spicy and bitter molecules are similar: they're an irritating line of last resort against being eaten up. They send a quick, intense sensory signal: *Get this thing out of your mouth, OMG, what is this thing? Spit it out; this isn't good for you!* Unfortunately for spicy plants, though, we humans are a bunch of pain-loving masochists who get off on the rush of endorphins flooding our endocrine systems when we get this brush with edible danger, and we go to great lengths to cultivate and cook them and their brethren.

Spicy ingredients taste great, feel great, and are totally essential in some of our most beloved foods: chile-rich salsas and kimchi, sambals and vindaloo, buzzy-spicy mapo tofu, peppered and gingered stir-fries, and sinus-clearing radishes, mustards, and wasabi.

Rules of Spicy

Spicy is sensed by pain receptors, so it's technically a touch, not a taste.

The spiciest molecules, including capsaicin, come from hot chiles, and there are chemically similar, milder molecules in black pepper, ginger, Sichuan peppercorn, and other spices.

Garlic, onion, wasabi, horseradish, and mustard all make sharp, sulfury spicy molecules that are volatile and can sting your eyes, sinuses, and mouth.

Spicy molecules are much more willing to play with fats than water, so fatty or oily cooking methods are the way to pull out and deliver the most spice.

That's a Spicy Molecule

The spiciest ingredients you can get are chiles. Chiles belong to the larger category of capsicums, and while not all of them are spicy, the spicy ones vary from barely hot to so blistering they're basically chemical weapons. Whether mild or strong, their spice comes from a molecule called *capsaicin* and its siblings, the *capsaicinoids*. Gram for gram, the capsaicinoids are the most potent pain-stimulating molecules.

---

* Not technically a taste.

Rating Spiciness

When we talk relative spiciness, we're not just making hand-waving estimates. Enter the Scoville heat unit (SHU), the answer to the question "But *how much more* will this pepper make me hallucinate the universe folding in on itself than that one?"

Technically, the Scoville scale is a measurement of the number of times you need to dilute a pepper with water to make the spiciness undetectable. More spice, more dilution necessary. If a hypothetical pepper needs to be diluted until it is 1 part pepper to 100 parts water, that's a score of 100 SHU. Pure capsaicin weighs in at 16 million SHU, and the hottest peppers (the kind bred by pepper obsessives to one-up their pepper-growing colleagues, like the Carolina Reaper) can reach 2 or 3 million SHU. Stunts aside, really spicy peppers that people actually cook with, like habañeros, have around 100,000 to 300,000 SHU.

Other spicy ingredients have molecules that are structurally similar to capsaicin, though with rather less potency. Black pepper has *piperine*. Fresh ginger has two capsaicinlike molecules: *gingerol* and *shogaol*. Sichuan peppercorns, Japanese sansho or mountain pepper, and Korean chopi pepper have *hydroxy-alpha-sanshool*, which is both spicy and buzzing, a little like carbonation but more like the electric buzz from licking a 9-volt battery.

Capsaicin, piperine, and molecules of their ilk are nonvolatile, like most taste molecules. This means that chemically they stay put in dishes, they won't waft over to you in gaseous form like the smell of garlic. (However, if you blend them in a blender, you can form a floating aerosol, a cloud of tiny liquid droplets like hairspray, which can drift like a cloud and share the experience of being essentially pepper-sprayed with anyone unlucky enough to share your kitchen.) Some plants in the onion and mustard families, including garlic, wasabi, and horseradish, take a completely different approach to spicy. They produce sulfur-based molecules with names like *allicin* and *allyl isothiocyanate* that are both spicy and truly volatile, forming pungent and highly flavored gases. In this form, they can float up into your eyes and make you cry, or up from your mouth into your nasal cavity and make your sinuses burn. (When my dad was first introducing me to sushi, he instructed me to breathe out through my mouth when I ate wasabi, which helpfully redirects the airflow to spare your nose.)

Most taste molecules are good at dissolving in water, which is why we can taste them (they dissolve in our mostly water-rich ingredients and our watery saliva to get to our taste buds). Most spicy molecules, especially capsaicin and its many brethren in peppers and spices, are much more soluble in oils and fats than in water. So water, and things made from it, like chicken broth, pick up gentler spiciness from spicy things you cook them with than oily or fatty ingredients do, because the spicy molecules have a greater affinity for fat than for water. (There is way more to read about this phenomenon of extraction, and cooking techniques that use it, starting on page 201.)

Spicy Flavor Patterns

Spicy ingredients sort neatly into four quadrants: **chiles**, with their hot-as-hell capsaicins; spicy **spices**, like black pepper and ginger, with tamer chemical relatives to capsaicin and lots of aroma; Sichuan peppercorns and their **prickly-ash fruit** cousins, with tingly, buzzy-spicy hydroxy-alpha-sanshool; and members of the **allium and mustard families**, like garlic and wasabi, that make spicy, volatile sulfur molecules.

Each quadrant embodies a particular type of spicy sensation, created by a repeating molecular pattern, and each often comes packaged with similar aromas. Chemically similar spicy molecules tend to behave in similar ways, so once you

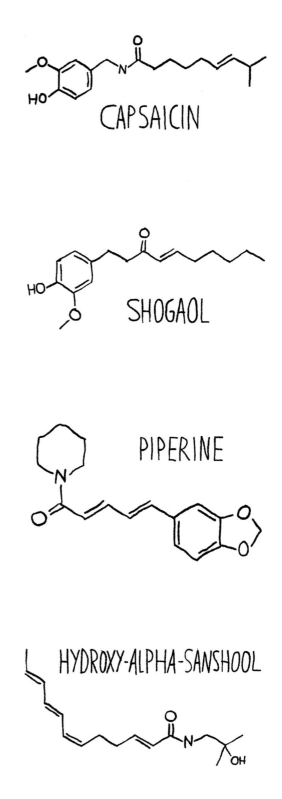

CAPSAICIN

SHOGAOL

PIPERINE

HYDROXY-ALPHA-SANSHOOL

Spicy Molecules, Ranked

Capsaicin: 16 million SHU
Capsaicinoids: 9 to 15 million SHU
Shogaol (ginger): 150,000 SHU
Piperine (black pepper): 100,000 SHU
Paradol: 100,000 SHU
Gingerol (ginger): 80,000 SHU
Alpha-sanshool (Sichuan pepper): 80,000 SHU
Zingerone (ginger): 50,000 SHU
Hydroxy-alpha-sanshool (Sichuan pepper, buzzy): 26,000 SHU

understand how spicy works for one molecule, you can bet that the others in the same quadrant behave similarly. As a root, ginger kind of resembles horseradish or onion, but if you know its spicy components act much more like those in black pepper—how they're formed, if they're volatile or nonvolatile, how stable they are, how hot they are—you can make much more confident cooking decisions.

Fruity, Smoky, Vegetal, Capsaicin-Packed Queens of Spicy: Chiles

The spiciest ingredients you can get are chiles (all species of pepper, or *Capsicum*), and they come in every level of heat from there on down to a faint whisper. They've spread around the world so well in the last several hundred years that most countries have at least one of their own traditional landrace varieties. It's thought that, since birds' pain receptors don't respond to capsaicin and they can gorge themselves on chiles with few ill effects, they were particularly invaluable in spreading chile seeds far afield during their migratory periods.

Green chiles are generally unripened forms of red or yellow chiles; they can be prickly spicier and

Molecules and Their Touch Receptors

Capsaicin feels spicy because it chemically activates a couple of touch receptors in the *transient receptor potential* (TRP) family. These touch receptors are kind of adorably specific in what they're supposed to sense: multiple types detect heat, but some only respond, very specifically, when that heat is above 109°F, while others activate only around our body temperature (98.6°F). The most important spicy touch receptor is TRPV1, which tells the brain to create a painful, burning sensation when it is triggered by heat (this is the "above 109°F one") or strong acids (like getting lemon juice in a paper cut). It's also unwittingly activated by many burning or "warm-feeling" molecules, capsaicin being one of the most potent.

Besides hot-in-here TRPV1, we also have coldness sensor TRPM8, which is also activated by not—literally—cold *menthol*, *camphor*, *eucalyptol*, and *thymol*. Minty things don't just taste like cool, they quite literally create the sensation of coldness.

In between TRPV1 and TRPM8 is TRPV3, our coziest touch sensor, the one that responds to warmth around our body temperature. *Cinnamaldehyde* and *eugenol*, the molecules responsible for the characteristic smells of cinnamon and cloves, both register their warming feeling by activating this receptor.

Other touch sensors respond to chemical irritants not as a weird side effect, but because that's their primary purpose. TRPA1, for instance, is broadly tuned to respond to *allyl isothiocyanate*, *allicin*, and other sulfur-containing spicy molecules in garlics, onions, mustards, and horseradishes; it also responds to cinnamaldehyde, THC, *curcumin*, *paradol*, and *gingerol*. *Hydroxy-alpha-sanshool* in Sichuan peppercorns and other prickly-ash fruits activates the burning-sensation receptor TRPV1, but it also turns on a totally different touch receptor that is ultrasensitive to light touches, like barely there brushes against your arm hair. By essentially turning that signal on and off in rapid succession, sanshool creates its unique buzzing feeling.

are less sweet and more vegetal than ripe chiles. Peppers have their most hot and juicy flavor when fresh, and they're delicious superthinly sliced or minced and laid on top of anything that needs that kind of spicy vibe, according to whatever your taste is as a diner. (I've seen whole fresh chiles used as a totally custom spice-level accessory: eat a bite of something, then nibble from the chile as desired.) Fermenting peppers—whether as a thicker paste like sriracha or sambal, or as a thin hot sauce—gives them delicious acidity and extra depth of flavor.

Aromatic Fermented Pepper Paste

Fermenting peppers with salt like this involves lactic fermentation, the same process that makes sauerkraut (or yogurt) sour. Wild lactic acid bacteria hang out on the skins of various plants, and a little salt and time lets them create tangy lactic acid as well as buttery, creamy, and pickley aroma notes.

This recipe works splendidly with fresh, good-quality, ripe chiles—the more flavorful the better. Depending on your tolerance for spiciness, Scotch bonnets, ripe jalapeños, cayenne peppers,

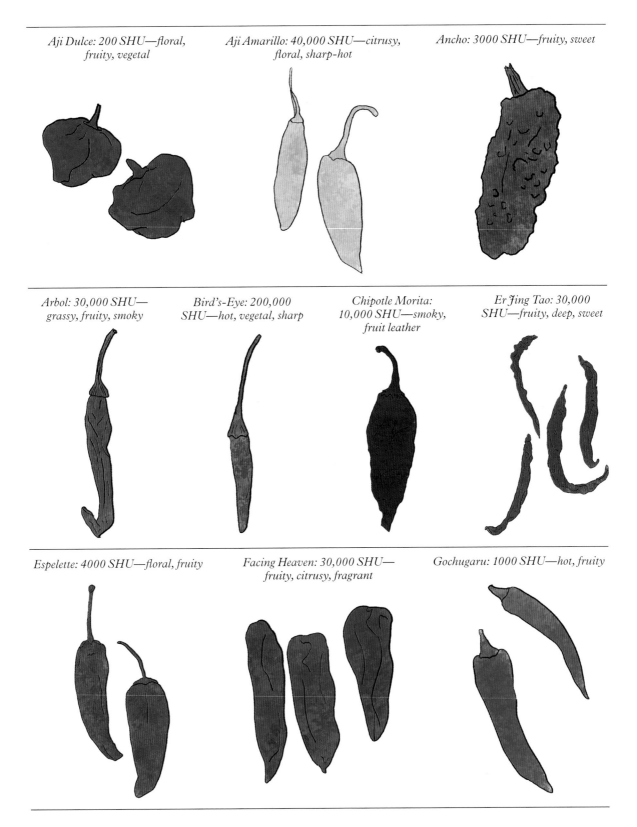

Aji Dulce: 200 SHU—floral, fruity, vegetal

Aji Amarillo: 40,000 SHU—citrusy, floral, sharp-hot

Ancho: 3000 SHU—fruity, sweet

Arbol: 30,000 SHU— grassy, fruity, smoky

Bird's-Eye: 200,000 SHU—hot, vegetal, sharp

Chipotle Morita: 10,000 SHU—smoky, fruit leather

Er Jing Tao: 30,000 SHU—fruity, deep, sweet

Espelette: 4000 SHU—floral, fruity

Facing Heaven: 30,000 SHU— fruity, citrusy, fragrant

Gochugaru: 1000 SHU—hot, fruity

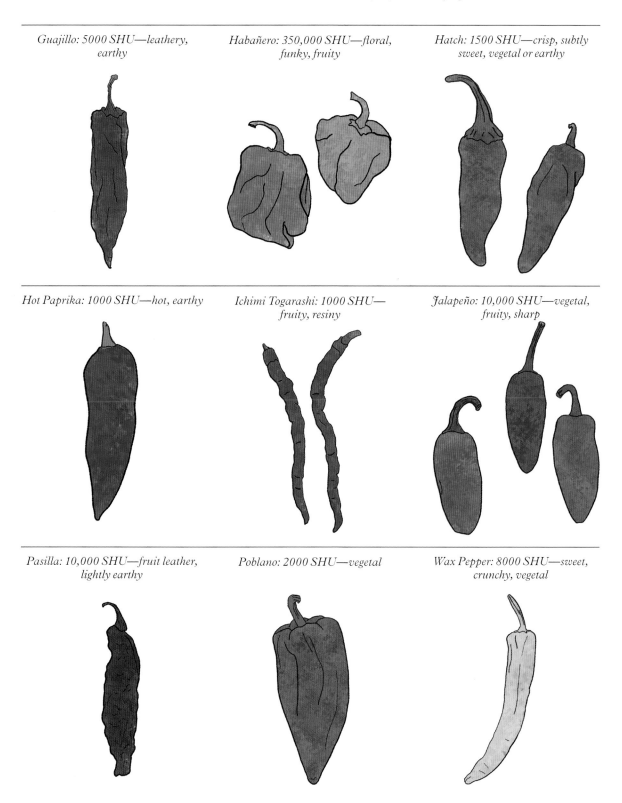

Guajillo: 5000 SHU—leathery, earthy

Habañero: 350,000 SHU—floral, funky, fruity

Hatch: 1500 SHU—crisp, subtly sweet, vegetal or earthy

Hot Paprika: 1000 SHU—hot, earthy

Ichimi Togarashi: 1000 SHU— fruity, resiny

Jalapeño: 10,000 SHU—vegetal, fruity, sharp

Pasilla: 10,000 SHU—fruit leather, lightly earthy

Poblano: 2000 SHU—vegetal

Wax Pepper: 8000 SHU—sweet, crunchy, vegetal

Easy Spicy Things

- Use a spicy condiment to balance the richness of meat: a bracing little bit of wasabi with sushi; fiery grated horseradish with roast beef; a dip in chile-infused oil for poached or fried chicken; grainy deli mustard on a hot dog; candied ginger with ham or pork belly.

- Expand your stash of chiles to cook and season with: Korean gochugaru (hot and a little fruity); hot Hungarian paprika (hot and earthy); Mexican chile-lime Tajín (hot and tangy); Turkish sun-fermented Urfa biber (dark brown, hot, and chocolatey); Calabrian flaked peperoncino (sharp-hot and bright); Basque Piment d'Espelette (hot and floral).

- For a massive upgrade on generic chile powder, pulse dried **Mexican chiles** to a rough powder in a high-powered blender (or food processor). A few classics: guajillo (medium-hot, earthy), morita (spicy and deeply smoky), ancho (fruity, sweet, and relatively mild), and pasilla (hot, fruit-leathery).

- Sprinkle a spicy powder—that chile powder you just made, a grind of black pepper, a tiny bit of Sichuan peppercorn, a mix of ground ginger and clove—on fresh fruits: mango, pineapple, peaches, nectarines, strawberries, oranges, papaya, pears, apples. . . .

- Branch out your hot sauce selection from the basics. I like bright, hot, vinegary Crystal hot sauce from Louisiana, Fly by Jing chili crisp (featuring spicy oil, flaked chiles, and crisp garlic, rather than a smooth sauce), and bottled chamoy from Mexico, a spicy-sour-fruity sauce composed of chile and fermented stone fruit.

aji amarillos, or fresh Piment d'Espelettes all work deliciously. If you're looking to experiment, unripe green chiles (like green jalapeños or poblanos) are absolutely fermentable, albeit for a much more vegetal flavor.

Shades of Chiles

Mexico is the ancestral and spiritual home of the chile, and the sheer diversity (in shape, size, flavor, and spiciness) of the chiles grown there reflects the concept that you get the most genetic variation (within a group of related organisms) the closer you are to where they originated. When approaching Mexican chiles, it's important to remember that the same variety is usually called by different names when it is fresh and when it is dried. For example, fresh poblanos go by mulato or ancho in dried form; fresh mirasols are guajillos when dried.

Stir the pepper paste into mayonnaise for sandwiches or as a dip; stir into black beans or other cooked beans; add to oil and vinegar dressing and marinate vegetables in it, especially substantial vegetables like cauliflower or green beans; dilute with olive oil or melted butter and use it to coat brussels sprouts, carrots, or sweet potatoes before roasting.

I *highly* recommend working by weight when fermenting, specifically in metric—it is very easy to keep track of ratios that way, much harder using pounds and ounces, and essentially impossible using volumes.

You may see a white film form on the top of your fermenting paste; this is harmless yeast and can be scraped off.

Fuzzy-looking mold is a sign to discard your fermentation, as are any decidedly unpleasant rotten smells. It's always better to be safe than sorry.

The most essential way to avoid spoilage is to avoid contact with air—which covering the surface with plastic wrap accomplishes pretty handily.

Wash a lidded, glass or food-safe plastic container that can hold at least 24 fluid ounces in hot and soapy water.

Wearing disposable gloves, remove the stems and most of the seeds from **1 pound (450 g) fresh chile peppers**. For greater accuracy, weigh the stemmed, seeded peppers and record their weight in grams. Calculate 3 percent of this weight (if they weigh 400 g, 3 percent is 12 g) and combine the peppers with **3 percent uniodized salt (15 g, or 1½ tablespoons kosher salt**, per pound) in a blender or food processor. Pulse until reduced to a chunky paste.

Using a very clean spatula, pack the pepper-salt paste into the container, pushing down any air pockets. Press a piece of plastic wrap loosely onto the surface of the paste and cover very loosely with the lid. The fermentation will generate gas, which will need to escape through the loose lid, so don't tighten it or it might explode. If the gas pushes the plastic wrap away from the surface, simply smooth it back down.

Ferment at room temperature for 2 to 5 weeks. Use a very clean spoon to taste-check after 1 week; there should be a mellow tanginess. If you're not sure, it is almost certainly not a bad idea to let it ferment for another week and taste again—it will get sourer, less yeasty, and generally cleaner-tasting. When it's appreciably tart and sour, you can store in the fridge, with plastic wrap pressed against the surface, and use within 4 to 6 months.

Makes 1 pound (approximately 2 cups)

Variations
Grilled Fermented Pepper Paste

If you have a charcoal or wood grill, build a hot fire and briefly grill half the peppers before fermenting, just to get a little char on them but not fully cook them. Cool, then stem and seed and proceed with the recipe.

Smoky Fermented Pepper Paste

Soak **1 or 2 seeded, stemmed, smoked dried chiles, such as chipotles,** in hot water for about 15 minutes, then drain and add to the fresh chiles and proceed with the recipe.

Other Additions to Fermented Pepper Paste

If you're looking for more flavor dimension, herbs, spices, garlic, and fermented soy (like Japanese miso or Korean doenjang) all play beautifully with chiles in lots of different dishes, and will blend and meld with the pepper paste's flavor as you ferment it. Add your selection of the following to the peppers when you puree them:

1 to 2 garlic cloves

1 tablespoon (15 ml) grated lemon or orange zest

1 tablespoon (15 ml) finely grated fresh ginger

4 or 5 allspice berries, crushed

1 tablespoon (15 ml) fresh or 1 teaspoon dried (5 ml) oregano

2 teaspoons (5 g) crushed fennel or cumin seeds

2 tablespoons (30 ml) yellow or red miso

Because chiles are essentially hollow fruits, with a relatively thin wall of flesh, they dry quickly and easily. Dried chiles are an age-old way to preserve exciting flavors way beyond a harvest season, and they're so lightweight they can be shipped almost anywhere. Cooking with dried chiles follows a few different tactics: flaking or grinding them and adding them as powder; frying and macerating in oil until spicy-crisp; or soaking (sometimes toasting first) and blending them into a paste or directly into a sauce or stew. Besides the extended shelf life, drying intensifies the flavors of chiles and accentuates earthy, floral, fruity, and leathery qualities.

When you incorporate chiles (or other spicy ingredients) in a recipe can have as much impact on spicy flavor as what kind of chile you use.

Oaxacan mole negro uses dried, then toasted, soaked, pureed, and fried chiles as a starting point, adding fruits, nuts, and other ingredients and then cooking for a long time to make a paste before any meat gets added to stew and cook. Some versions can be quite spicy, but their heat is like lying in sunbaked beach sand, intense but in a slow-building, gentle sensory-overload way. Adding spicy ingredients early on in cooking creates this gradual, more blanketing kind of burn because it gives the fat-soluble spicy molecules time to spread out and permeate everything. Get there by adding your spicy ingredients early on to long-cooked braises, stews, or curries, or rub powdered chile or chile paste onto meat before cooking it. Or use a chile oil or chile butter (read on!). (Note that this advice doesn't extend to cooking onions, garlic, horseradish, and other members of the sulfur-spicy quadrant; heat will dissipate their spicy molecules by evaporating them away and breaking them down, rather than spreading them.)

If you want spiciness to feel more like a stinging slap than a cozy weighted blanket, use chile as a garnish, like a finishing salt. Fresh, uncooked chile salsas get at this sparkly hot flavor, and so does Thai nam pla prik, with fresh, minced bird's-eye chile plus umami-funky-salty fish sauce, sweet-brown palm sugar, lime juice, and cilantro soothing the initial burn. Adding a small sprinkle of chile flakes, or minced or thinly sliced fresh serrano or habañero, will create hot, sharp bursts of spiciness, like someone is jogging over your tongue in miniature stilettos.

Chile Butter

When I want to make beans, chili, meat stews, tomato sauces, or soups spicier in an enveloping way quickly, without adding contrasting heat in the form of hot sauce or red pepper flakes, this butter gets me there. Since the capsaicin and other flavor compounds are dissolved in the butterfat, their flavor diffuses throughout the dish. For more dimensions of toasty-warm flavor, gently heat the butter when you're melting it until the milk solids start browning.

Toast **2 dried guajillo chiles** and **2 dried pasilla chiles** under a low broiler until crisp and just barely brown, about 1 to 2 minutes. Remove and discard stems and shake out and discard seeds, then crumble the chiles and reserve.

Heat **2 sticks (225 g) unsalted butter** in a small saucepan over medium heat. Stir in the **reserved crumbled chiles**, 1 teaspoon (4 g) **dried urfa biber flakes**, **6 black peppercorns,** crushed, ½ **teaspoon (2 grams) ground dried ginger**, and ¼ **teaspoon (1 g) ground cloves**.

Turn off the heat, cover, and let sit on the stove for at least 1 hour, giving the butter a stir about every 15 minutes, and gently heating when it starts to solidify. Strain through a metal tea strainer into a heatproof container. When cooled to warm, transfer to a lidded container for storage. Use within 1 month.

Makes 1 cup

Warm, Aromatic, and Spicy: Pepper, Ginger, and Other Spices

Even if you don't identify as a pain-seeking chile connoisseur, if you have even a bare-bones pantry, you probably eat something spicy at nearly every meal. I'm speaking, of course, about black pepper, the spicy, piperine-rich dried fruits of the *Piper nigrum* vine, and the final seasoning (along with salt) to countless plates of food.

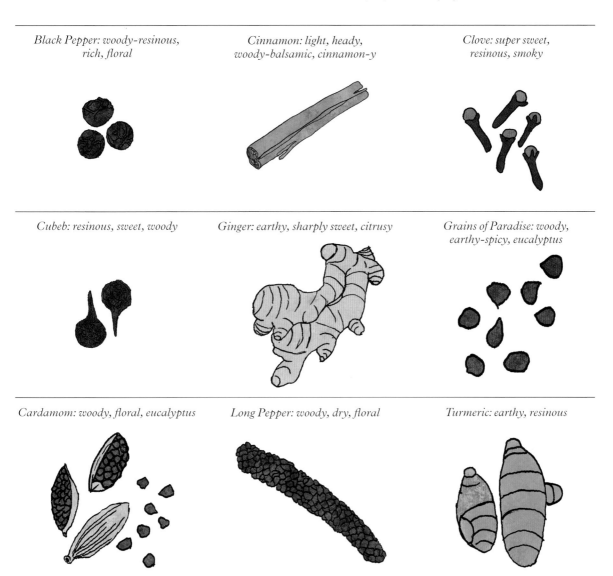

Black Pepper: woody-resinous, rich, floral

Cinnamon: light, heady, woody-balsamic, cinnamon-y

Clove: super sweet, resinous, smoky

Cubeb: resinous, sweet, woody

Ginger: earthy, sharply sweet, citrusy

Grains of Paradise: woody, earthy-spicy, eucalyptus

Cardamom: woody, floral, eucalyptus

Long Pepper: woody, dry, floral

Turmeric: earthy, resinous

Black pepper's omnipresence in European and European-influenced cooking is the hangover of a much spicier culinary past. Cooks from antiquity through the Middle Ages seasoned with copious amounts of black pepper and its other spicy relatives, especially **long pepper** and **cubeb pepper**, as well as spicy-aromatic **ginger** and less spicy, more warming and aromatic, spices like **cinnamon**, **clove**, and **cardamom**. While black pepper retained a role in savory cooking, echoes of ginger and other spices' bite mostly shows up now in sweet-spicy drinks and desserts like ginger beer and gingerbread.

Black pepper's piperine molecules are around 2000 times less spicy than pure capsaicin, but they pack into peppercorns at up to a hundred times the concentration of capsaicin in chiles—enough of a punch that chewing up a whole peppercorn can be a wheeze-inducing experience, and we usually apply them to foods as a thin, finely crushed dusting. Ginger has three different molecules that make it spicy: gingerol, shogaol, and

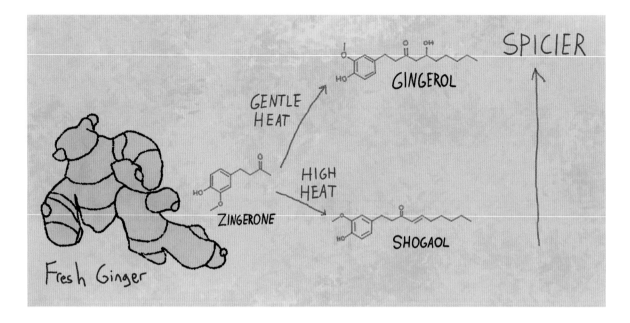

zingerone. Which of the three dominates depends on how the ginger was processed or handled.

Fresh ginger has predominantly gingerol, which is about half as spicy as the piperine in black pepper. If you apply heat to ginger, things can go one of two ways: cook it with higher heat, and the gingerol will be converted to zingerone, which is both less spicy than gingerol and has a kind of sweet-spiced aroma. If you heat ginger slowly and gently, as you would to dehydrate and grind it, you'll form shogaol, which is about three times as spicy as the original gingerol. Applying high heat to shogaol (like baking a cookie with dried ginger in a hot oven) continues transforming it, into molecules with much less spiciness.

Shifting Ginger's Spiciness

For medium-spicy ginger flavor, use it fresh and don't heat it up (or add it right at the end of cooking) and you'll taste mostly moderate gingerol. For mellower and more aromatic flavors, heat and cook the fresh ginger to convert gingerol to zingerone. For the most spicy ginger impact, use dried ginger, which is full of hot shogaol.

(If you used too much dried ginger, and it's too spicy, you can rein it back in a bit by heating it for a while.)

Ginger and pepper (black, cubeb, and long) are all spicy-hot because they have molecules, like piperine and gingerol, that act on the TRPV1 pain receptor, just like capsaicin does. So does paradol, a spicy molecule that ginger has a little of, and its cousin, melegueta pepper or grains of paradise, has a lot of. The eugenol that makes cloves taste like cloves also creates a hint of pain in this way. Many other spices have molecules that don't produce capsaicin-style burning-hotness, but can irritate another touch receptor (TRPA1) for a kind of prickly spiciness. These include cinnamon, allspice, cloves, pink pepper, and ginger relatives like turmeric, galangal, and green, black, and Ethiopian cardamom.* Even some herbs like oregano and thyme act on the TRPA1 receptor.

---

*   Just to throw us for a loop, cardamom varieties usually have a lot of resinous-spicy-smelling eucalyptol in them, which stimulates the cold-sensing TRPM8 nerve ending for an icy-hot experience.

Warm-Spicy Soda Syrup

In practice, we usually use spicy spices as much for their complex aromas as we do for spiciness. If you like spicy-aromatic ginger, why not try a heady, floral-resiny-spicy mix of cardamom, pink peppercorns, and cloves? When mixed with sparkling water, this syrup gestures in the direction of ginger beer. You can also use it to sweeten your coffee; or slice up a few firm apples or pears, add a good splash of the syrup, and simmer for 10 to 15 minutes, until softened, then pour over ice cream or crème fraîche.

In a spice grinder, grind **1 tablespoon (6 g) whole pink peppercorns**, **2 whole cloves**, and **10 cardamom pods**. Combine in a medium saucepan with **1 cup (225 ml) water** and **1 cup (200 g) sugar**. Bring to a boil, turn off the heat, cover, and let steep for 2 hours.

Once cool, strain and store in the fridge for up to 3 to 4 weeks.

For an aromatic and lightly spicy soda, mix **2 ounces (4 tablespoons; 60 ml) syrup** with **6 ounces (¾ cup; 180 ml) sparkling water**. Serve immediately over ice with **a squeeze of lemon**.

Tingly, Numbing, Buzzing Spicy: Sichuan Peppercorns and Beyond

Hot-chile spiciness is so important for Sichuan cooking, especially from cities like Chengdu and Chongqing, that it's sometimes oversimplified into a kind of synecdoche for the cuisine as a whole. The Sichuan culinary canon, which appreciates flavor patterns as much as I do, traditionally names twenty-three basic flavoring principles, many of them combinations of flavors juxtaposed together (such as fish-fragrant flavor or yu xiang wei xing, a traditional fish seasoning combining spicy fermented red chiles with salty-umami fermented beans and scallion, garlic, and ginger). If you're a fan, you've probably heard of, or at least experienced, the flavor of *málà*, or "numbing hot," an essential combination of hot chiles and buzzing, numbing **huājiāo**, or **Sichuan peppercorn**.

Sichuan peppercorn isn't actually a member of the pepper family like black pepper. It's the tiny fruit produced by prickly ash (*Zanthoxylum*) trees, which are more closely related to citruses. Along with a citrusy, floral, resinous flavor, Sichuan peppercorns are mildly spicy (in the traditional sense), and they have an astounding spicy-adjacent flavor, the *má* or "numbing" in *málà* ("numbing-hot"), which people mostly describe through metaphor: electric, like licking the terminals on a 9-volt battery. Buzzy, like a carbonated drink. Numbing, like pins and needles when your leg falls asleep.

The culprits in these numbing-buzzy-electric flavors are a family of molecules called sanshools, headed by their matriarch, hydroxy-alpha-sanshool. Like the burning of capsaicinoids, their characteristic tingly flavor comes from interacting with our sense of touch, but from mechanical, rather than temperature, sensors. Hydroxy-alpha-sanshool is thought to inhibit some of these mechanical touch receptors, explaining the numbness; at the same time, it rapidly flicks others on and off, the ones that you use to sense very light touches, like a fly landing on your skin, creating the tingle and buzz.

Many other closely related prickly ash fruits are used for spices with their sanshools and buzzy-tingly-spicy flavors, including Japanese **sansho** or **mountain pepper**, Korean **chopi**, Sumatran **andaliman**, and Nepalese **timur**. If *málà* plays up the tactile side of Sichuan peppercorn, other spice blends play more with its aromatic side, with tingly-buzzy spice as a bonus. Along with star anise and cinnamon, it's a key player in Chinese five-spice powder; with red chile, citrus peel, sesame seeds, nori, and ginger, sansho does the same in Japanese shichimi togarashi blends, which go equally well seasoning bowls of rice as of popcorn. While showing me around town, a colleague in Miyazu, Japan (north of Kyoto), brought postage-stamp-sized packets of his favorite local sansho.

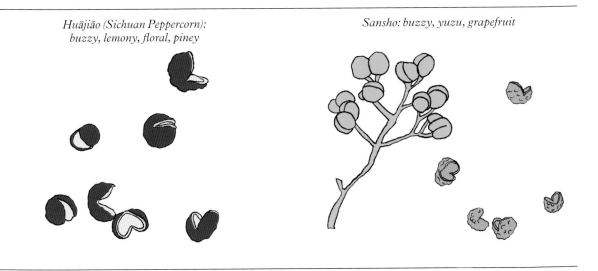

Huājiāo (Sichuan Peppercorn):
buzzy, lemony, floral, piney

Sansho: buzzy, yuzu, grapefruit

sprinkling it on grilled fish and tempura for us at dinner. It's a nice way to get into aromatic, spicy-buzzy prickly ash spices: try a light dusting of sansho, or ground Sichuan peppercorn, on fried or lightly grilled seafood or grilled vegetables, or on fresh-from-the-fryer French fries.

Fleeting, Sulfury, Sinus-Burning Spicy: Onions, Garlic, Mustard, Wasabi, Horseradish

The spiciness in **garlic, mustard seeds,** and their relatives like **horseradish**, **wasabi**, **onions,** and **radishes** has a little something extra going for it. It's certainly spicy, in a hot, fast, prickly way. If you eat enough of it at once, you'll notice the pain isn't confined to your mouth but is prickling further up in your head: your eyes, your nose, your sinuses.

The linchpin here, same as in fire and brimstone of divine punishment, is sulfur. Plants in the allium family (garlic, onions, etc.) and in the mustard family (mustard, horseradish, wasabi, radishes, etc.) create sulfur-containing molecules that are both spicy *and* volatile—in other words, gaseous, floaty, and opportunistically looking for pain receptors to menace. In onions, propanethial-S-oxide, aka "lachrymatory factor"

or-the-molecule-that-makes-you-cry; in garlic, double-sulfured allicin; in mustard-family members like radish, wasabi, and horseradish, isothiocyanates.

Like chile plants do with capsaicin, plants create these molecules to deter animals from munching on them. (Herbivores may seem quite placid and unthreatening when compared with carnivorous animals, but from the point of view of a plant, caterpillars, ruminants, and vegetable-eating humans are all apex predators.) Unlike with capsaicin, though, none of these plants (in unchewed form) actually contain any of those volatile sulfur molecules. Instead, they make them like you might prepare a two-part epoxy: right when you're going to use it, from separate reservoirs of ingredients. When a wasabi root or garlic bulb is actively being chewed, crushed, grated, or otherwise disrupted, two separate stores of molecules get mixed: big, heavy, sulfur-rich but non-aromatic molecules (glucosinolates for mustards, alliin for alliums), and enzymes that transform them into spicy tear gas (myrosinase in mustards, alliinase in alliums). Botanists call this à la minute preparation the "mustard oil bomb" (at least in mustards) and those volatile sulfur molecules are so potent, they're actually toxic to the plant as

Horseradish: earthy, woody, bitter

Wasabi: fresh, piercing, grassy

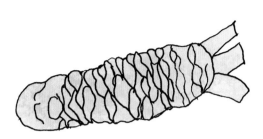

Onion: sharp, slightly sweaty

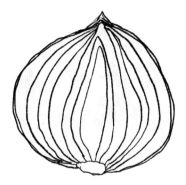

Garlic: rich, sharp, malty

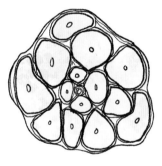

Mustard Seed: earthy, woody, tangy

Radish: sweet, dirty-earthy, crisp

well—no sense making them until they're absolutely necessary.

Since sulfury-spiciness is a factor of how much of these two starting materials come together to mix and create flavorful molecules, you can approach allium and mustard ingredients with the tactical suavity of knowing how to dial in your preferred flavor profile. For maximum spiciness, you want to crush and mix up their insides so as many enzyme molecules meet their counterparts as possible, so really grind or macerate the dickens out of them. A garlic clove, pureed, is a hell of a lot spicier than the same clove, carefully minced, for exactly this reason. And since this flavor creation uses enzymes, which are proteins and prone to thermal instability, it won't happen at all if you inactivate those enzymes. (Which will happen when you, say, heat them above 180°F or so for a little while.) Cook sulfury-spicy ingredients *before* you cut into them like this, and you'll get mostly soft sweetness and hardly any spice. Try slicing the tops off several heads of garlic, wrapping them in foil with a little olive oil, and roasting for 30 minutes at 400°F (you're welcome).

Spicy but Short-Lived

Hand-grating whole, fresh wasabi root on a sharkskin grater at a good sushi restaurant isn't just a luxurious affectation—it ensures that when you eat it, it will actually *be* spicy.

The sulfur molecules that give alliums like garlic and onions and mustard-family wasabi and horseradish their punch are as potent as they are short-lived. The innate reactivity of sulfur means that they're quick to react and rearrange to form even more new molecules, making sulfury-spicy a time-limited experience. The upside? These new molecules include things like diallyl disulfide, which make garlic aromatically garlicky instead of just spicy. The downside? That green blob you get with takeout sushi probably isn't actually wasabi, which would have lost much of its flavor in the first hour after being grated—it's more likely pasted mustard or horseradish (which stay spicy longer, though less deliciously), dyed green with perhaps a symbolic dose of wasabi at homeopathic levels.

CHAPTER 6

Smell

Sorting Out Smell and Its Flavors

A LEXICON OF WINE FLAVOR TERMS

used by Sensory Scientists + sommeliers

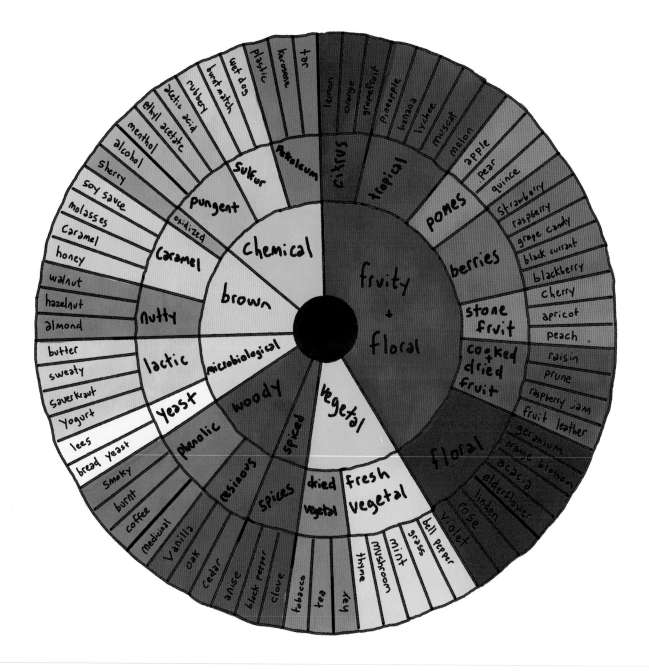

As you've read till now, tastes are pretty straight-forward sensations, and dividing taste into five categories (salty, sour, sweet, umami, bitter) to sort sensations, molecules, and ingredients is so obvious that it might be puzzling to think of how or why you'd do it some other way.

Then, turning to smell, we quickly realize that a simple system of categorization is a luxury we don't have.

Many industries where smell is important, like wine and perfume, have developed lexicons or categories for those they need to talk about the most—but there's no decisive, universal break-down like there is for taste. Smells are irreducibly multifaceted, contain many molecules mixing their signals together, and have a seemingly end-less elasticity to accommodate unique sensations. When you smell almost any ingredient you could think of—a quince, a piece of cheese—the per-ception your brain builds for you is a holistic unit of smell, overlaid with some (but not all) of the distinguishable aroma notes of its individual com-ponent molecules.

Thinking about flavor in patterns isn't just a creative philosophy, a way to apply limitations to a limitless set of possibilities to spur on ingenuity (although it is good at that). It's also how scientists approach studying flavor, particularly the parts of flavor that come from smell. They attack the delicious chaos of aroma from both ends: studying smells and flavor from the bottom up, and study-ing patterns among them from the (perceptual) top down.

If you want to know what makes a flavor com-posed of many different molecules tick, you look at it like a pile of laundry—just like you'd sort the shirts and the socks from the delicates, and the lights from the darks, you sort the smell mole-cules into their own piles, then smell the piles one by one, so you're only evaluating one kind of molecule at a time. You note down what each pile smells like, ending up with a list that looks like an annotated formula: smell molecules, amounts, and what aromas they contribute to the mix. This approach is great if you want to pinpoint where a particular note of violet or creaminess comes from in the flavor of a particular fruit—you could then go looking for that molecule in other fruits, or change how you grow or process the plant to get more of it.

The other way we study flavors—the top-down, patterns-focused approach—really em-braces the "unbridled" part of flavor: lots of molecules all doing their thing at the same time, lots of aroma qualities happening at once, odor images synergizing and condensing into wholes, ingredients with flavors that are obviously dif-ferent, in that we can tell them apart, but have more similarities with one another than they do with ingredients that are a lot more different, like oregano and rosemary vs. basil or cilantro. Rather than just throw our hands up and say, "They're all different," or compare how every single compo-nent of one flavor compares to every component of another, we invented the scientific term *multivar-iate* to encapsulate "lots of things are happening all at once, different things are changing different amounts, and we can't really separate all of it into simple parts with one unifying mechanism to ex-plain them."

So rather than try to hold every conceivable individual difference in flavors in our heads, we call upon the most chaotic and unbuttoned of the sciences to address this chaotic and unbuttoned problem: Statistics.

You may be saying to yourself—statistics, as in "There are three types of lies—lies, damned lies, and statistics." Or you may be picturing a space

A SMELL CLASSIFICATION OF PERFUMES

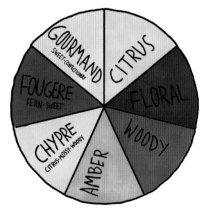

robot informing its crew that they face a precisely 90.437 percent chance of crashing into a wormhole, but there are whole branches of the field that embrace the complex, chaotic, and unbridled (or, as we call it, multivariate):

> *All the interesting worlds (physical, biological, imaginary, human) that we seek to understand* are *inevitably and happily* multivariate in nature.
> —**Edward Tufte**, *Envisioning Information*
> [emphasis added]

That's Edward Tufte, data visualization pioneer and one of my favorite statisticians.

(I realize that it may seem extremely weird to read about multiple statisticians in a book about food, but at its heart statistics is just about identifying and describing patterns in reality—so of course I love it, and it's incredibly useful for understanding flavor.)

When flavor scientists study unbridled flavors, we might start out with fifty or a hundred ways that ingredients have different flavors from one another: "This one has green flavors, but in a pronounced vegetal way; this one is also green, and a little vegetal, but more grassy; this third one is also green, maybe in between the first two in vegetalness, but the green flavors are much more minty." Or an exhaustive list of flavor molecules, with different amounts of each. It's like trying to keep a complicated picture in your head, except the picture is fifty-dimensional. We hand all that over to statistics models that compare every difference, in every molecule or flavor quality or category we track, for every ingredient we tell it to, all at once. They set off to look for patterns, and since flavor follows lots of recurring patterns, there's usually a few that are so strong, they encapsulate something like 75 percent of all the flavor differences that are going on. It's kind of like how a decent artist is able to suss out the most important parts of real-life, three-dimensional scenes and capture them in a two-dimensional drawing or painting—it's disproportionately evocative of the real thing, for something that's technically a flattened, lower-resolution interpretation. And what we get from our trusty stats model is essentially a two-dimensional map of flavor, with ingredients plotted like cities that are closer together or farther apart depending on how similar or different they are, and north, south, east, and west defined by those flavor qualities that explain the vast majority of the patterns they follow.

By thinking like a flavor scientist and focusing on these patterns and trends, you get to work with relatively tidy, understandable accounts of flavors instead of having no other option than to memorize exhaustive lists and make comparisons of comparisons of comparisons, all for the low price of setting aside the muddiest and least significant chunk of flavor differences.

From here on out, we're going to use both strategies for understanding smells and their flavors: learning some key aroma compounds that create the definitive and identifying flavor notes in ingredients, and making broad (only partially

E-2-HEXENAL

leafy, green

HEXYL ACETATE

fresh, apple, pear

HEXYL BUTYRATE

green, waxy, fruity

ETHYL-2-METHYLBUTYRATE

berry, green apple, pineapple skin

ETHYL BUTYRATE

fruity, apple, pineapple

BUTYL ALCOHOL

banana, fermented

DAMASCENONE

jammy, cooked apple, raspberry, rose

bridled) maps, clusters, and families—so we can think about ingredients in relation to one another.

Fruity: The Flavor That Says "Please Eat Me"

Fruit—packed full of sugars, balanced by acids, and soaked in aroma—is designed for you to enjoy. While other ingredients are anywhere from indifferent to actively hostile to the prospect of being eaten, fruits really, really want you to eat them. They do have an ulterior motive: the seed(s) in their centers. In exchange for their sweet and aromatic flesh, they're counting on you to leave that seed somewhere far enough away from its mother plant that it can grow and spread without competing for resources. Fruity aromas are their radio broadcast, rippling out into the ether, for you to come hither and taste.

Thinking About Fruity

Fruity flavors signal something special: an ingredient you (usually) need do nothing more to than pick and eat. You can certainly cook most of them, or assemble them uncooked in recipes like salads or drinks.

But especially at this early point in your journey to understanding flavor, I think it's useful (and fun) to approach fruity flavors like an edible catalogue of pure sensory experiences. Think of it as a low-stakes way to ease yourself into tasting and noticing, if you're not in the habit of it already, or getting very good at it, if you are. You can enjoy fruits with minimal kitchen effort, building your internal sense for flavor as you do: noticing flavor elements layering in any particular piece of fruit, noticing variations in what "fruity" can be, noticing how fruits cluster by flavor similarities within this catalogue. And all that's required to have these experiences and build this knowledge is a source of raw fruit and maybe a knife.

Fruity Experiences

Sweetness and some degree of sourness is generally a given in fruits. "Fruity" is solidly the domain of aroma. "Fruity" has lots of layers, dimensions, and variations. Any particular fruity flavor is built on a pretty generic core of fruity flavor that many fruits tend to share, with its special and unique flavor qualities layered on top.

When you taste a piece of fruit, here are some useful things to think about:

On top of general "fruitiness," sweetness, and sourness, what dimensions of fruity complexity do you notice?

Is it fruity and crisp? Fruity and rich?

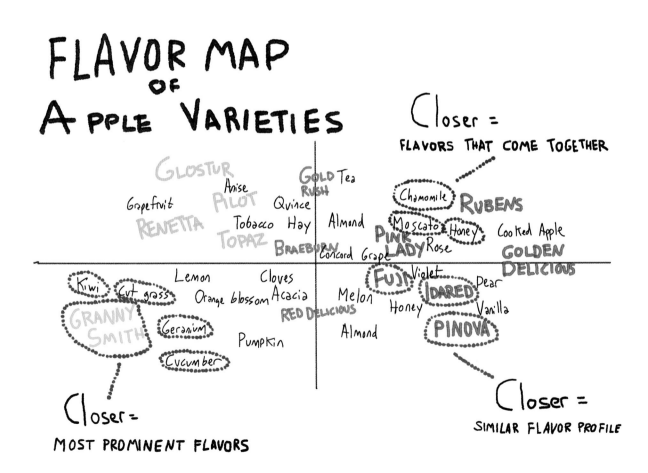

Does the fruitiness have a piercing or assertive quality to it, or is it softer and more blanketing?

What wispy or fleeting top notes can you sense? Below those, what are its deeper base notes like?

Do you get a honeyed quality or a sense of jamminess? Is it winey? Creamy? Musky? Can you notice any spiced, herbal, or floral qualities?

Elements of these combine in different ways for different fruits: green-and-creamy honeydew vs. richly juicy-creamy peaches, deep and spiced cherry vs. deep and floral raspberry, juicy and softly refreshing orange vs. juicy and piercing passion fruit.

Fruity Molecules

The aromatic base coat of "fruity" in the majority of fruits comes from molecules called *ethyl esters*, which are the plain white T-shirt of the fruity flavor world. The right blend of ethyl esters creates an "ah, yes, that's some kind of fruit" flavor; two of these are ethyl decanoate and ethyl hexanoate. Special dimensions of fruity can come from smell molecule categories like norisoprenoids (rich, honeyed, floral), more complicated esters (juicy and peary), lactones (creamy and peachy), sulfur compounds (piercing, tropical), aldehydes (orangey, melony, green), and terpenes and phenylpropenes, which have herbal, spiced, and citrusy qualities and are covered in detail in "Cumin as Chemical Weapon: Herbal and Spiced Molecules" (page 152).

All of these molecules look and behave quite differently, but we can peg them all as "fruity"— different shades or incarnations, perhaps, but obviously related. And despite studying flavor professionally for many years, I can't actually give you a good chemical reason why. Maybe we've just all learned to associate them with sweet fruit. We

Easy Ways for Fruity Fun

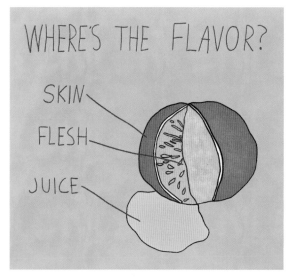

If you somehow tire of peak season fruit, simply sliced and eaten from a plate, there's several ways you can approach fruit to highlight its flavor, or to use that flavor to season something else.

- Grate the skin and use it like a spice. See "Fruity and Juicy-Fresh: Citrus" (page 134), "Apples to Oranges" (page 216), and "Herb Sauces" (page 177).
- Sprinkle the fruit flesh with sugar to macerate (page 72).
- Cook fruit flesh into a jam or marmalade. See page 226.
- Make a sorbet, like Bitter Grapefruit Sorbet (page 138).
- Make a juice, and drink or reduce into a molasses (see Fruit Molasses, page 199). Or make into a syrup or cordial (see Passion Fruit Juice Syrup, page 141) or fruit juice caramel (see Pineapple Caramel Sauce, page 141).

don't know yet what the odor images (to jog your memory: QR codes) cooked up by the olfactory bulb look like for different fruity flavors—maybe

ELEMENTS OF
FRUITY FLAVOR

BASIC FRUITY ESTERS

ETHYL BUTYRATE

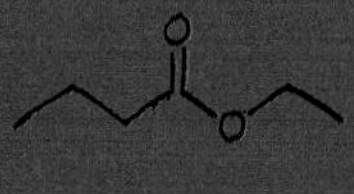

fruity, apple, pineapple

ETHYL DECANOATE

fruity, apple, grape, waxy

ETHYL-2-METHYLBUTYRATE

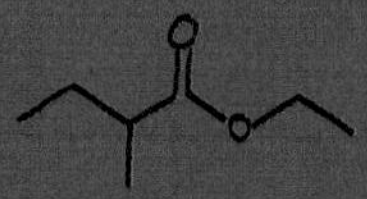

berry, green apple, pineapple skin

FANCY ESTERS

PHENETHYL ACETATE

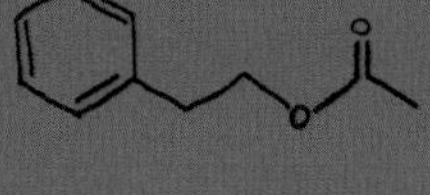

rosy-floral, honeyed

ALLYL HEXANOATE

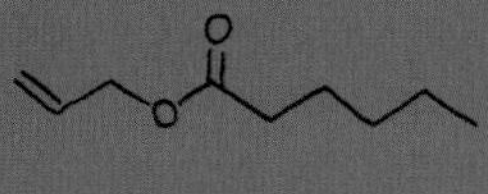

pineapple, rummy, juicy, tropical

ETHYL DECADIENOATE
"Pear Ester"

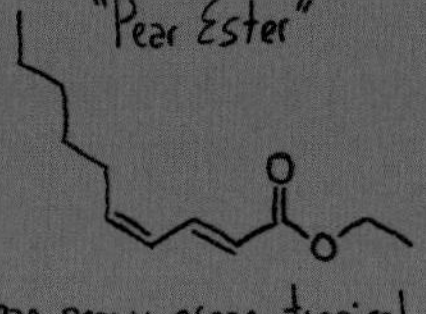

clean, pear-y, green, tropical

CREAMY, GREEN ALDEHYDES

Z-6-NONENAL

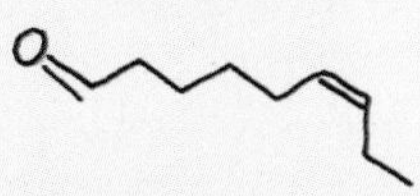

cucumber, canteloupe, vegetal

2,6-DIMETHYL-5-HEPTENAL

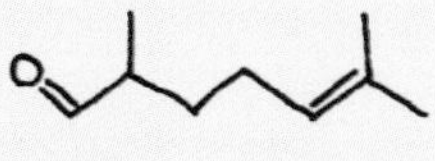

watermelon rind, floral, waxy

DECANAL

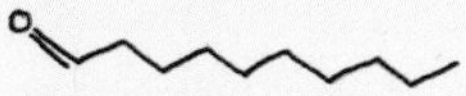

orange peel, melon, floral

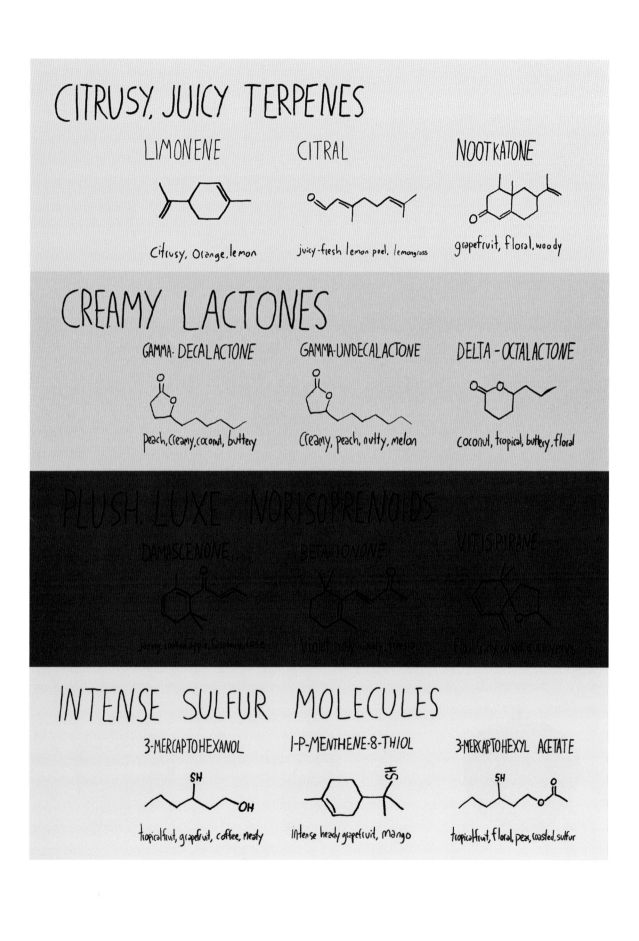

CITRUSY, JUICY TERPENES

LIMONENE
Citrusy, orange, lemon

CITRAL
juicy-fresh lemon peel, lemongrass

NOOTKATONE
grapefruit, floral, woody

CREAMY LACTONES

GAMMA-DECALACTONE
Peach, creamy, coconut, buttery

GAMMA-UNDECALACTONE
Creamy, peach, nutty, melon

DELTA-OCTALACTONE
coconut, tropical, buttery, floral

PLUSH, LUXE NORISOPRENOIDS

DAMASCENONE
jammy cooked apple, raspberry, rose

BETA-IONONE
Violet, richly woody, freesia

VITISPIRANE
floral, fruity, wood, eucalyptus

INTENSE SULFUR MOLECULES

3-MERCAPTOHEXANOL
tropical fruit, grapefruit, coffee, meaty

1-P-MENTHENE-8-THIOL
Intense heady grapefruit, mango

3-MERCAPTOHEXYL ACETATE
tropical fruit, floral, pear, roasted, sulfur

we evolved for them to be similar, even if they involve different molecules.

Fruity is a very practical endorsement for thinking in patterns; in this case, a sensory one. There's no one trait we can distill fruity down to, but we can notice how similar-but-different fruity flavors group together along with molecules like ethyl esters, norisoprenoids, and fruity terpenes.

Fruity and Crisp-Plush: The Apple Family

The regular-degular apple is the namesake (in Latin, anyway) for a group of plants within the rose family, the subtribe Malinae. Apples and the other fruits within this family are called *pomes*, and pomes tend to have a tough core with small seeds surrounded by fairly crisp and juicy flesh. **Apples'** fruity flavors start with basic fruity ethyl esters, enriched by molecules like beta-damascenone, a member of the norisoprenoid class of molecules. I think of beta-damascenone and the other norisoprenoids as the most luxurious of flavor molecules. The smell of beta-damascenone makes me picture slowly roasting the world's most deeply fruity apples and then concentrating down their golden juices with raspberries and rum over a fire of palo santo wood. Even in infinitesimally small amounts at which other molecules have no smell at all, beta-damascenone

has a deep and rich, fancy-pipe-tobacco, supernaturally fruity-floral-woody flavor, one that makes wines taste richer, fruitier, and more well balanced, gives roses seriousness and depth, and gives apples gravitas. If a generically sweet and fruity ethyl ester is the forgettable bubblegum pop of flavors, then beta-damascenone is a Mozart opera. Beta-damascenone is even more important for the flavor of cooked apples: as lighter fruity notes are boiled off, beta-damascenone is released from smell-less precursors within the raw apple by the heat.

In-the-know apple connoisseurs seek out the many heirloom, local, and specialty varieties bred over the last few hundred years that, like wine grapes, produce a staggering variety of flavor nuances. Ashmead's Kernel have fresh and pearlike flavors, Macouns are floral with sweetness balanced by a punch of sourness, Cox's Orange Pippin have notes of cherry and anise, Hidden Rose have a strawberrylike sweetness, and Orleans Reinette, citrusy nuttiness.

Hard-as-wood **quinces** are really the luxe-aroma-extreme of the pomes: layers of fruity qualities as in apples and pears, but mixed with tons of perfumed, floral-fruity, spicy complexity from norisoprenoids, caramel notes from the odd molecule it shares with caramel, and its own quincey flavor note from its own quincey

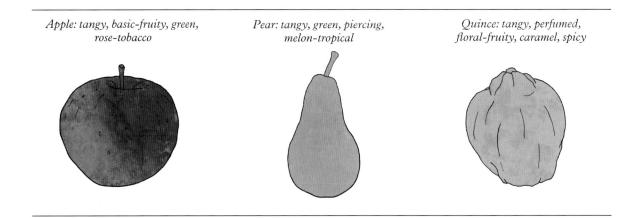

Apple: tangy, basic-fruity, green, rose-tobacco   *Pear: tangy, green, piercing, melon-tropical*   *Quince: tangy, perfumed, floral-fruity, caramel, spicy*

molecule, marmelo lactone (*marmelo* being the Japanese word for "quince"). Quinces are also prodigious producers of norisoprenoids, which give them their perfumed flavor when raw and incredible richness when cooked into compote or membrillo paste—and, it should be noted, they're inedibly tough until cooked.

Pears take the ethyl ester framework of apple flavor and multiply it with bigger, more complicated, more pear-y esters of their own. Bartlett pears (also called Williams Bon Chretien), my favorite readily available pear for eating (and also the beginnings of amazing pear brandy), have piles of long esters, especially ethyl decadienoate, which gets the nom de guerre of pear ester. It has a clean, slightly piercing, pear-y, green, slightly melony, and tropical flavor, and gives pears a lot of their characteristic peariness,

Pumping Up Tastes with Smells

Have you ever thought of a flavor, or a smell, as "sweet"? I don't mean literally sweet like sugar on your tongue—but an impression of sweetness, like caramel or vanilla, or a really ripe strawberry. This impression mostly comes down to smell, but smell isn't literally sweet, right?

Technically no, but also, kind of, yes!

You've learned to associate nonsugar flavors with sweetness through repetition: from candies, pastries, ripe fruits, etc. But, falling back into its familiar pattern of memory, contexts, and association, once you train your brain to make the connection between a smell like vanilla and a taste like sweetness, including vanilla along with sugar in something makes it taste measurably, significantly sweeter.

Scientists see this pattern over and over again: lots of fruit flavors as well as almond, caramel, vanilla, and cotton-candy ones all increase sweetness perceptually, without actually adding any more sugar.

So while these scents don't literally increase sweetness in the sense of activating more sweetness receptors on the tongue, they *do* increase a sense of sweetness in the brain, which happens when signals and memories from all the senses are pulled together to make flavor.

This special smell-taste relationship doesn't just apply to candy and desserts. Recently, we've found that cheesy aromas make foods taste significantly more salty. In theory, you could probably combine any taste and any smell, consume it regularly, and eventually train your brain so that the smell enhanced the taste and the taste enhanced the smell. These associations seem to be totally learned, so a person or a culture that doesn't usually combine, say, almonds and sweet flavors in their cuisine won't experience this taste-smell reinforcement.

I fully support experimenting on yourself, and the feedback loop of taste and smell (on top of their joining forces directly for flavor) is something that's easy to take advantage of in your cooking. If a dessert seems lacking in sweetness, you could just add more sugar—or you could add more fruity or caramelized ingredients, which would boost the sweetness without any toothachey side effects. You could get more tangy balance in a sour drink like a lemonade by incorporating a sour-emphasizing aroma like lemon zest. If your Negroni is just a little bit too bitter but you don't want to dope it up with simple syrup, adding "sweet" flavors like orange peel, cloves, or cinnamon can rein in the bitterness by increasing this sweet context. You can literally boost umami in dishes by adding fish sauce, but it will go that much further if you use a really good, aromatic-funky one. Quite the add-on to your internal Pantone of flavor.

Try This

While cooking is a must for quinces—go ahead, try biting into one and get back to me about it—it is a nice bonus choice for apples and pears. Try out any of the following for softened, fruity-caramelized pomes that are lovely with a little yogurt or cake, or even on their own. If you keep cooking until they break down all the way, no worries: that's just applesauce. Or [insert pome here] sauce. A little sprinkle of ground cinnamon, clove, allspice, and/ or cardamom with the fruit is also a nice touch.

Cut into large cubes and put in a medium pan or Dutch oven.

	Apple	Pear	Quince
Add this much sugar per whole fruit	½ to 1 tablespoon	½ tablespoon	3 tablespoons
And this much water	None	1 tablespoon	1 tablespoon
Cook over medium heat for this many minutes, or until just soft.	5 minutes	5 to 10 minutes	50 to 60 minutes

Cut in half, cut out core, and put in a medium-sized casserole or deep baking dish.

	Apple	Pear	Quince
Add this much sugar per whole fruit	To your taste; try an unrefined sugar		2 tablespoons
And this much water	None	None	½ cup, or to barely cover
Cook at 375°F for this many minutes, or until just soft	35 to 40 minutes	25 to 40 minutes	45 to 60 minutes

especially when compared to the fruity esters of apples. Pears get an extra dose of fruitiness from generally fresh-fruity and green hexyl acetate. And where apples add some depth with beta-damascenone (think Mozart opera from before), Bartlett pears make farnesene, which gives them woody, herbal, and citrusy layers in with the sweet and fruity.

Fruity and Juicy-Fresh: Citrus

Freshly squeezed lemon and lime juice are just about as bright-and-fruity-sour as you can get, but this is just one of the flavor performances the multitalented citrus family has to give. Citruses, as a group, are as hybridized and interbred as the Hapsburg dynasty, and their version of "Hapsburg jaw" is their absolute piles and piles of fresh and aromatic terpene aroma molecules (check out

THE CITRUS FAMILY TREE

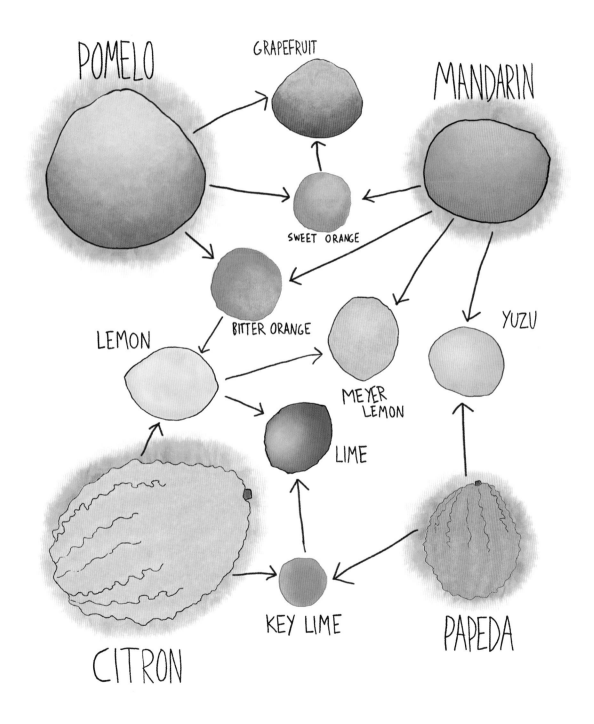

"Cumin as Chemical Weapon: Herbal and Spiced Molecules," page 153, for more on those), their incredibly aroma-rich skin, and their (usually) juicy, sour, fragrant flesh.

The selection at your local supermarket may have led you to believe that the "basic" types of citrus are the lemon, the lime, and the orange (and possibly the grapefruit)—and this may be true in one culinary sense. But all of these are actually fairly recent hybrids in a long and interconnected family tree. There are thought to be four original citrus types: the mandarin, the pomelo, the citron, and the papeda. Mandarins are usually small, sweet, and orange; pomelos are like a large, green, thick-pithed grapefruit; citrons, such as etrogs and Buddha's hands, are aromatic and relatively juiceless; papedas are usually wrinkly, aromatic, sour, or bitter fruits. Every citrus we have now—lemons, limes, oranges, grapefruits, yuzu, makrut limes, etc.—is a cross involving one or more of these ancestors.

Smell molecules usually make up such a tiny component of foods that they're invisible, but citrus skins are so packed full of them—superfresh, juicy, and even brighter than the juice—that you literally see where the flavor is by the oily droplets and films they form. Hardly anything fails to become more delicious with citrus zest: peeled and squeezed over drinks and cocktails, macerated with an equal volume of sugar to form the ultraflavorful syrup called *oleo-saccharum*, or finely grated and mixed into marinades, baked goods, and dressings. Whole, they make beautiful salt preserves, marmalades, and sugary glacés—or just great eating on their own, in the case of kumquats. Segmented—that is, cutting away the outer membranes of the segments to form delicate supremes—sour citrus are deeply assertive and sweeter ones, simply refreshing.

Nearly all citrus produce *limonene*, which despite its name does not smell specifically of lemons, but has a kind of nice, generic citrusy-orangey smell.

Along with limonene, **lemons** (the most commonly available variety is the **Lisbon lemon**) themselves have a blend of aromatic terpenes with lemony-lemon, floral, piney-resinous, and oily-woody-limey qualities. The most lemony of all, *citral*, is also the key flavor molecule in lemongrass, if you're looking for an alternate source. **Meyer lemons** get a distinct flavor, with some of the citrusy aspects of lemon, as well as spicy, herbal, thyme-leafy, slightly oxidized-citrus, woody, and cumin-like flavors.

Key limes, the original limes, are quite a bit smaller and more sour than **Persian limes**, the common supermarket variety. (Unhelpful tip: if you happen upon a nineteenth-century cocktail recipe calling for limes, Key limes are the authentic choice). Limes' distinctive flavor qualities include florals in the realm of lilac, orange flower, and lavender, as well as woody, resinous, and peppery notes. Filipino-native **calamansi** (sometimes called calamondin) has an extra-sour, extra-floral-citronella flavor, which is even more extreme in wrinkly **makrut limes** and their leaves.

Many citrus names are essentially categories, covering several varieties rather than individual fruits, and **oranges** are no exception—sweet and bright **Valencias**, softer and deeper pink-fleshed **cara caras**, stolid ordinary **navels**, berry-undertoned **blood oranges**. Oranges are one of the citrus varieties most often used specifically for their smells. *Citrus x aurantium* goes by **bitter orange**, **sour orange**, **bigarade**, **daidai**, **chinotto**, and **Seville orange**, and its sweet-and-warm-smelling peel is used as a seasoning in fresh, dried, and candied forms. English marmalade, French duck à l'orange, Belgian wheat beers, orange liqueurs like Triple Sec and curaçao, orange bitters, and spiced-bitter-orange chinotto soda from Italy all use bitter orange peel for flavor.

Grapefruits stand out from other common citrus because of their bitterness and their pretty unique aroma. That aroma is composed from

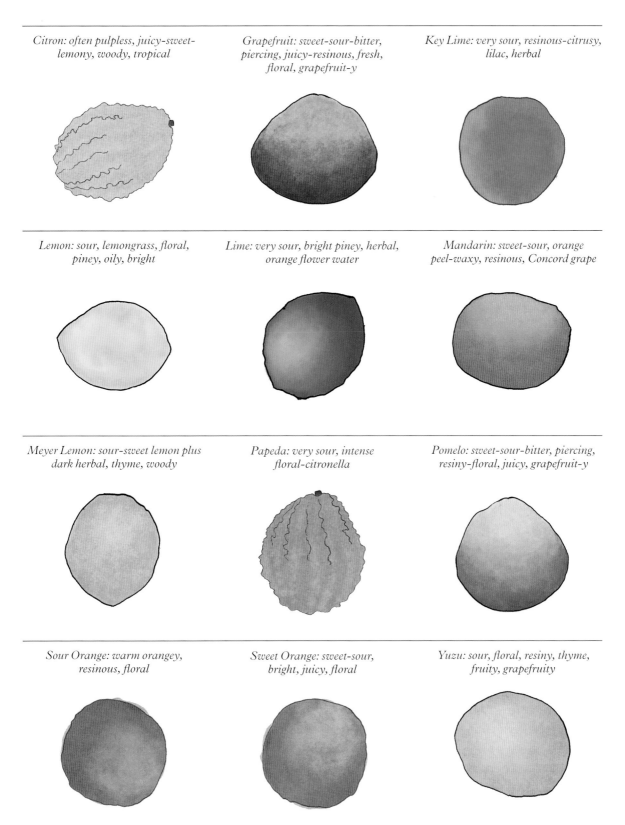

Citron: often pulpless, juicy-sweet-lemony, woody, tropical

Grapefruit: sweet-sour-bitter, piercing, juicy-resinous, fresh, floral, grapefruit-y

Key Lime: very sour, resinous-citrusy, lilac, herbal

Lemon: sour, lemongrass, floral, piney, oily, bright

Lime: very sour, bright piney, herbal, orange flower water

Mandarin: sweet-sour, orange peel-waxy, resinous, Concord grape

Meyer Lemon: sour-sweet lemon plus dark herbal, thyme, woody

Papeda: very sour, intense floral-citronella

Pomelo: sweet-sour-bitter, piercing, resiny-floral, juicy, grapefruit-y

Sour Orange: warm orangey, resinous, floral

Sweet Orange: sweet-sour, bright, juicy, floral

Yuzu: sour, floral, resiny, thyme, fruity, grapefruity

lots of familiar citrusy terpenes, plus a piercingly juicy-resinous, slightly floral, distinctly grapefruity note that comes from *nootkatone*, a special terpene, as well as a sulfur-containing molecule, *1-p-menthene-8-thiol*. We're incredibly sensitive to sulfur, and it becomes a kind of force multiplier for the flavor of whatever molecule it shows up in. Other pointedly fruity flavors from sulfur can be found in yuzu, tropical fruits, black currants, and Sauvignon blanc wines. Strong, resiny-floral grapefruity flavors are also important for grapefruit's relatives, the **oroblanco grapefruit** and the **pomelo**.

Bitter Grapefruit Sorbet

For frozen desserts to have a soft texture, they need some stabilizing action and something that keeps ice crystals very small. Ice cream gets this from proteins. In its purest form, sorbet's softness is enabled by sugar; the more sugar molecules you have dissolved in water, the harder time it has freezing, and so it stays semifrozen and soft at otherwise abnormally low temperatures. In this recipe, you add sugar to grapefruit as a syrup, but first you heat the syrup up with a little lemon juice. The citric acid from the lemon "inverts" the sucrose molecules in the syrup, breaking each of them into a glucose and a fructose, essentially doubling the antifreezing power of the sugar. Including some of the pith of the grapefruit enhances its bitterness greatly, as well as contributing pectin for extra smoothness.

This recipe works well with red grapefruits or more highly aromatic oroblanco grapefruits, which are, technically speaking, a back-crossed hybrid between a grapefruit and a pomelo rather than a true grapefruit.

Start by making syrup for the sorbet. In a large saucepan combine **2 cups (400 g) white sugar** with **200 ml (200 g) of water** and the **juice of 1 lemon**. Heat, stirring occasionally, over medium-high heat until all the sugar is dissolved, and then very gently simmer for about 20 minutes more (don't let it come to a full boil, or too much water will evaporate).

Meanwhile, wash and zest **2 small grapefruits** and then peel, reserving all. Blanch the reserved pith from a fourth of the grapefruit in 2 cups simmering water until it has absorbed some water, softens, and turns a little translucent, about 8 to 10 minutes. Remove from the saucepan, drain, lightly blot, and reserve.

Tare the weight of a blender jar on a digital scale (put the blender jar on the scale and press "tare" or "0," and the scale will now read zero, ignoring the weight of the jar and only counting what you add to it), and add the blanched grapefruit pith, **1 ounce (30 ml) lemon juice**, the 2 denuded grapefruits, and one-quarter of the reserved zest. The total weight of the mixture should be about 1 pound or 450 grams. To the blender jar, add half of this total weight in syrup (about 225 grams). Blend on high speed until everything is incorporated and broken down, 5 to 7 minutes.

Stop the motor and taste it—it should be grapefruity, sweet, tangy, and quite bitter, as well as having a slightly viscous texture. If it seems not quite sweet enough (too bitter and too sour), add syrup in tablespoon increments, tasting each time, to adjust. If it tastes not quite acidic enough (sweet enough, but too bitter), add lemon juice in tablespoon increments, tasting each time, to adjust. This will all depend on the sweetness and the acidity of your individual grapefruit.

Chill the sorbet mixture in the fridge for at least 4 hours until very cold. Spin and freeze in an ice cream maker according to the manufacturer's instructions. Store in the freezer for up to 1 month.

Makes about 1 quart (1 liter)

Fruity, Deep, and Woodsy: Berries

Strawberry is a fantastic example of a flavor that is greater than the sum of its parts. Strawberry aroma has been intensely studied for many years, driven largely by the desire to make a convincing artificial strawberry flavor. One reason that's difficult is that strawberries aren't just fruity—many fruits are not just fruity, but the seemingly innocuous strawberry combines flavor notes in a way that is weird even for fruit. Along with a backbone of standard fruity ethyl esters, strawberries combine green, pineapple, slightly cheesy, and even caramel notes.

If you smell them closely, many berries make flavor combinations in more subtle ways: for example, standard fruity plus violet (from the norisoprenoid ionone) in **raspberry**, or fruity plus clove-y, spicy, woody, and vanilla notes (from eugenol and others) in **blackberry**.

Blackberries were the first fruit I ate that I picked myself—deep in the woods of Vermont, carefully navigating thorns, zeroing in on glossy fruit that practically came apart in my fingers. Their deep, woodsy-fruity, sweet-spiced flavor lingers in my sense memories as the flavor archetype for all berries I've eaten since.

Fruity, Frisky, and Assertive: Tropical Fruits

Many tropical fruits have an especially piercing quality to their fruitiness—you sniff it and it immediately grabs your attention, kind of like onion or garlic, if that garlic quality were really delicious fruitiness. The secret ingredient, like in garlic, onion, or grapefruit, is sulfur molecules. This aroma assertiveness may act as a

Blackberry: sweet-sour, red-fruity, clove, woody, vanilla

Black Currant: sweet-sour, jammy, resiny-spiced, slightly sulfurous

Blueberry: sweet-sour, rummy-fruity, lily-lilac, spiced, green

Raspberry: sweet, red-fruity, violet, rose, jasmine, rich

Strawberry: sweet, red-fruity, pineapple, caramel, grassy, slightly cheesy

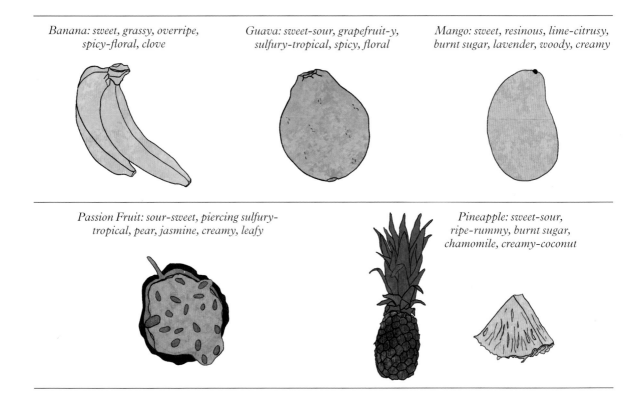

Banana: sweet, grassy, overripe, spicy-floral, clove

Guava: sweet-sour, grapefruit-y, sulfury-tropical, spicy, floral

Mango: sweet, resinous, lime-citrusy, burnt sugar, lavender, woody, creamy

Passion Fruit: sour-sweet, piercing sulfury-tropical, pear, jasmine, creamy, leafy

Pineapple: sweet-sour, ripe-rummy, burnt sugar, chamomile, creamy-coconut

kind of antifungal against the greater quantity of moldlike things that grow in higher-humidity tropical areas compared with drier or cooler regions. (A similar phenomenon happens with the color of parrots—they get brighter as you get closer to the equator, and many of those bright pigments chemically protect their feathers from mold. The more you know.) The flavors we get to enjoy because of this are some of the prime ones in **passion fruit**, **guava**, and **pineapple**.

Mango varieties have a huge range of flavors and shapes, but you may have noticed a kind of resiny note while eating any of them, especially at the transition from flesh to skin. Mangoes,

like citrus, get a lot of flavor from terpenes—especially woodsy, spiced, piney, and herbal notes, on top of their fruity and slightly juicy-caramel flavors.

Saucy Recipes for Frisky Fruits

These two recipes work smashingly with the high-acid, high-aroma flavors of tropical fruits, but they can work with any fruit you can juice: grapefruits, apples, raspberries, mango, cantaloupe, the list goes on. Even thin purees of banana, peach, or pear lend their flavors well, if you're patient about mixing and cooking them and don't mind a thicker, silkier texture. In a pinch, use high-quality bottled or frozen juice or puree.

Use these to garnish ice creams or sorbets, drizzle over black sugar pudding, mix into iced tea, especially green tea, or add to seltzer, soda, or cocktails.

Passion Fruit Juice Syrup

In a medium bowl combine the **seeds, juice, and pulp from 8 passion fruits (about 185 g)** with **¾ cup (150 g) white sugar**. Let sit for 1 hour to pull as much juice as possible from the pulp around the seeds.

Strain and keep in the fridge for up to 4 days, or freeze for longer storage.

Makes about 1 cup (250 ml)

Pineapple Caramel Sauce

Picture a caramel sauce: caramelized sugar loosened with fat and liquid. That liquid is often water and dairy. Why not use something really flavorful, like juice? Thus:

Put **½ cup (100 g) sugar** in a saucepan and add **1 tablespoon (15 ml) water.** Cook over medium heat until the syrup starts going from clear to light brown; stop when it is approximately as brown as a copper penny and wisps of smoke are just starting to blow from the top. Immediately whisk in **2 tablespoons (30 g) butter**, stirring as the mixture bubbles up and boils. Next add **2 tablespoons (30 g) heavy cream**, stir to boil off, cool, and combine. Lastly, stir in **3 tablespoons (45 ml) fresh pineapple juice**. Add **a few sprinkles salt** to taste. Eat right away.

Makes about ¾ cup caramel

Fruity and Creamy-Musky: Melons

Sad, pale, watery-flavorless chunks of dining hall honeydew and cantaloupe make it easy to forget, or never even consider, the many-layered flavor

Banana Artifacts

As a kid, it always bothered me that banana candies had a flavor that was so unlike real bananas. Like one fruity element was way overblown, and the kind of musky softness that to me screams "authentically banana-y" just wasn't there.

Later, I learned that while there are dozens of banana varieties, each with its own variations on banana flavor, the U.S. banana industry imports a monoculture of one variety, the Cavendish.

This wasn't always the case—the industry was a monoculture previously, but with a different kind of banana, the Gros Michel; the Cavendish was a variety that growers switched to in the mid-twentieth century after the Gros Michel, or Big Mike, as I like to call him, was nearly wiped out by fungal disease. "Banana flavor" has a boatload of the fruity, banana-y ester isoamyl acetate, and between the two actual banana varieties in question, guess which one makes more of that same molecule? Our old pal Mike. So now, when you taste something that's ostensibly "banana"-flavored, and you notice a mismatch to the fruit you usually eat, you're actually experiencing flavor time travel, back to when a banana was a Gros Michel and a banana candy tasted a bit more like the real thing.

delights that are melons. When they're good, you hardly need to do anything to them. If a thoughtful and sophisticated restaurant offers you a slice of melon at the end of the meal, you can look forward to a minimalist dessert experience without peer.

Cucumis melo is one species with many different-tasting varieties (kind of like wine grapes, *Vitis vinifera*, or *Brassica oleracea*, which appears in various iterations as kale, brussels sprouts, or broccoli). They're split into two categories: netted-rind, usually orange-fleshed

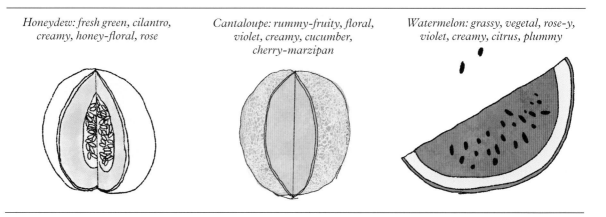

Honeydew: fresh green, cilantro, creamy, honey-floral, rose

Cantaloupe: rummy-fruity, floral, violet, creamy, cucumber, cherry-marzipan

Watermelon: grassy, vegetal, rose-y, violet, creamy, citrus, plummy

muskmelons like **cantaloupe** and **charentais**, and smooth-rind, usually greenish-fleshed melons like creamy-smelling **honeydew**. **Watermelons** are cousins (*Citrullus lanatus*), and further out in their Cucurbitaceae family, there are cucumbers (which have some green and melony flavor elements themselves), squashes, and gourds.

Melon flavor starts with lots of sweetness, low-to-medium sourness, and general ethyl ester fruitiness. Melons also have molecules called *aldehydes*, which share some shape characteristics with ethyl esters. Aldehydes create most of their melon-specific flavors—the generic "melon" quality most melons have (sort of like the generic "citrus" quality citruses get from the molecule limonene) and the super-meloniness of honeydew. Other flavors come from familiar fruity-floral-but-in-a-sophisticated-complex-way norisoprenoids and from honeyed rosy-floral phenethyl acetate, which also flavors roses and rose water.

Fruity, Creamy, and Spiced: Stone Fruits

Cherries, **plums**, **nectarines**, **peaches**, and **apricots** are united by their strongly sweet-sour tastes, and aromas, which mix fruity with notes of creamy, honeyed, almondy, floral, and spiced. They're also, as the stone fruits, members of the *Prunus* genus that are closely related enough to breed lots of delicious (and algorithmically named) hybrids: pluots, apriums, peacotums, cherrycots, nectacots, and peacharines, to name a few.

A creamy-peachy flavor that's particularly strong in peaches and nectarines comes from molecules called *lactones*—which all tend to have sweet and creamy aroma notes, though the apt name (lac-, like lactose) is sort of coincidental (early organic chemists in the mid-nineteenth century figured out how to synthesize them from lactic acid, then later found them in milk). Strawberries, coconut, oak, and flowers like osmanthus all make creamy-smelling lactones as well.

Try This

Stone fruits love to have their subtle spicy qualities played up with clove, cinnamon, or allspice. They combine surprisingly well with tiny bits of resinous herbs like thyme, savory, and rosemary, or follow Georgian or Japanese flavor combinations and play up their rich aromas with rich and refreshing fragrant herbs like tarragon and shiso. These are my favorite fruit types for baking: under or over cakes or clafoutis, crisps and crumbles, and pies—or skip the pan and wrap slices directly in pastry to make a galette.

Apricot: sweet, honey, earthy, creamy, cooked apple, floral, marzipan

Cherry: sweet-sour, berry-fruity, rosewood, clove, honey, violet, marzipan

Greengage Plum: plum plus rich, winey, resiny

Nectarine: sweet, juicy-ripe, floral, creamy, almond

Peach: sweet, ripe fruity, creamy, floral, green, marzipan

Plum: rummy-fruity, strawberry, cinnamon, buttery-creamy, floral

Almonds are also *Prunus* fruits, though their fruity part is only eaten green and immature—Persian-style, dipped in salt or stewed. The classic marzipan-almond flavor comes from a molecule called *benzaldehyde*, which both almonds and cherries have a lot of; it has a kind of maraschino cherry undertone. My favorite cherries are sour cherries. Besides their excellent acidity, they feel more deeply fruity to me than sweet cherries, and they usually have distinct clove-spiced flavors from eugenol that I find thrilling.

Besides the starring role they play in iceboxes in modernist poetry, **plums** have a smorgasbord of different culinary uses and flavors. All over Europe, plums are cooked as jams or baked into cakes and tarts, dried into prunes, and distilled into brandies like slivovitz. Georgian cuisine uses sour-fruity plum flavors in the savory-spicy sauce tkemali and, with tarragon and lamb, in the stew chakapuli. **Mirabelle**, a small, yellow, egg-shaped plum variety, has a famously coconutty flavor, from another (nonpeachy) lactone molecule. The **ume**, technically closer to an apricot but honorarily called a plum, lends its green-blush fruity-floral sourness to salt pickles like Japanese umeboshi and Vietnamese *xí muội*. I know I say this about all the fruits, but a properly ripe, nearly luminescent, ultrasweet, Rieslingy **greengage plum** is one of the finest fruit-eating experiences you can have.

Vegetal

Green, Earthy, Grounded

Vegetal flavors present something of a paradox, because while vegetables are some of my favorite foods, when it comes to unique rainbows of flavor molecules, there's not much "there" there, especially compared with fruity and herbal. Vegetables are delicious, but their vegetable-to-vegetable variations in chemistry are outweighed by those in texture.

Unlike a fruit, a vegetable doesn't particularly *want* to be eaten, and most of them require a little work on our part to make that happen: softening tough molecules like fibrous cellulose and grainy starch with heat, or accessorizing and dressing them with lively flavors to show off their best selves. The flavors you end up with in a vegetable you're going to eat give a lot of deference to texture: tender, crisp, wilted, or softly cooked.

For the same lack of desire to be eaten, vegetables haven't been pushed to develop such a varied repertoire of attractive flavor molecules as fruits have. Their broad strategy to avoid becoming dinner can be summed up in one word: green.

Shades of Vegetal

"Green" is one of the strongest through lines in vegetal flavor. In its lighter shades, it's the under-ripe edge to a green banana, the grassy aroma of a freshly cut lawn, a stalk of celery.

These vegetal flavors come from a group of aroma molecules that plant scientists call the *green leaf volatiles*, or GLVs. I think of them less as individual molecules to remember, and more like a process. GLVs don't come into existence until you slice, cut, or bite into a green plant. In response to this breach, the plant rapidly generates a bunch of molecules centered around an aldehyde called *hexenal*, which you may know as the classic aroma of freshly cut grass. It may be an invigorating and nostalgic flavor for you, but the message it's intended to send is a great disturbance in the Force—the voices of millions of bladelike leaves calling out in pain. The GLVs are like an alarm to other plants nearby inducing changes in their metabolism to get ready for trouble.

In wilted and cooked-down leaves, pea pods, or deep green bell pepper, green-vegetal takes on a darker flavor. This is the work of strong and sticky nitrogen-containing molecules called *pyrazines*.

Vegetal flavors also layer in a sense of groundedness, a stolid and comforting reminder of the ground they come from. Pyrazines can also have earthy flavors like potato skin, as well as terpenes, woody notes (like in carrots), or actual dirt flavors (the molecule geosmin, in beets). Lastly, some vegetal ingredients, especially from the mustard/radish/cabbage brassica family, get a slight spicy prickle and funky accents from sulfur molecules.

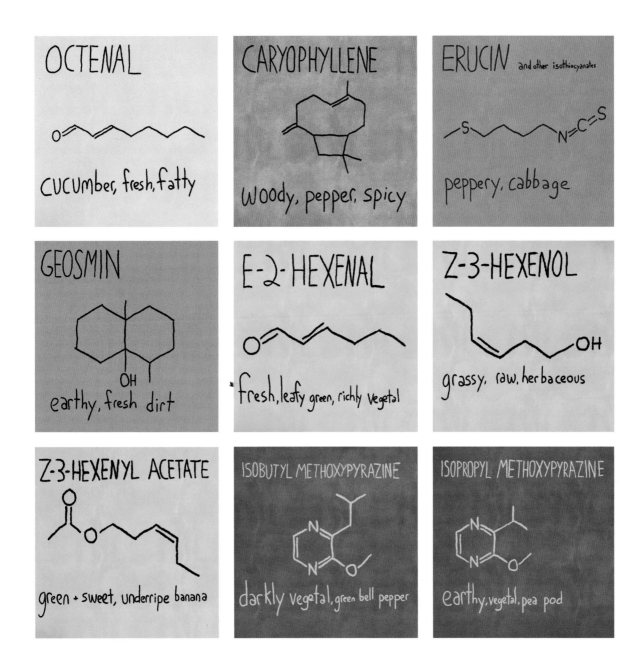

OCTENAL

cucumber, fresh, fatty

CARYOPHYLLENE

woody, pepper, spicy

ERUCIN and other isothiocyanates

peppery, cabbage

GEOSMIN

earthy, fresh dirt

E-2-HEXENAL

fresh, leafy green, richly vegetal

Z-3-HEXENOL

grassy, raw, herbaceous

Z-3-HEXENYL ACETATE

green + sweet, underripe banana

ISOBUTYL METHOXYPYRAZINE

darkly vegetal, green bell pepper

ISOPROPYL METHOXYPYRAZINE

earthy, vegetal, pea pod

Lettuce: green, vegetal, woody

Endive: light green, vegetal, bitter

Arugula: green, vegetal, peppery-spicy

Baby Greens: green, vegetal, sweet

Spinach: dark green, vegetal, earthy, slightly tangy

Watercress: green, vegetal, peppery-spicy

Purslane: green, vegetal, fresh, tangy

Radicchio: vegetal-green, woody, bitter

Cabbage: sweet, green-earthy, sulfurous

Goosefoot: spinach-y, deeply vegetal

Kale: dark green, vegetal, savory

Collards: dark green, vegetal, savory

Chard: dark green, vegetal, earthy

Mustard Greens: green, vegetal, peppery-aromatic

Amaranth: green, vegetal, earthy

Taro Leaf: green, vegetal, earthy

Dandelion Greens: green, vegetal, earthy, bitter

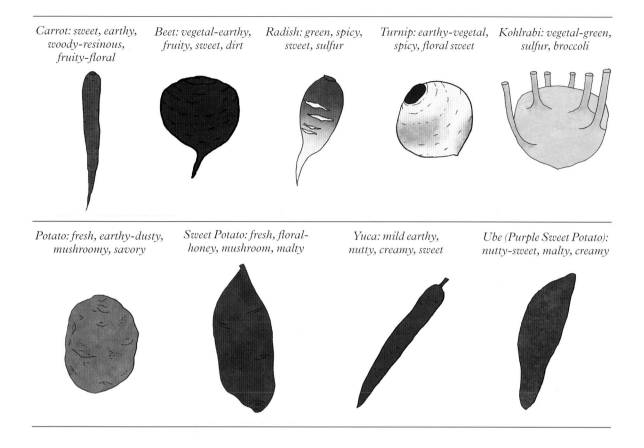

Carrot: sweet, earthy, woody-resinous, fruity-floral

Beet: vegetal-earthy, fruity, sweet, dirt

Radish: green, spicy, sweet, sulfur

Turnip: earthy-vegetal, spicy, floral sweet

Kohlrabi: vegetal-green, sulfur, broccoli

Potato: fresh, earthy-dusty, mushroomy, savory

Sweet Potato: fresh, floral-honey, mushroom, malty

Yuca: mild earthy, nutty, creamy, sweet

Ube (Purple Sweet Potato): nutty-sweet, malty, creamy

Vegetal and Earthy: Roots

When we're talking earthy, literally, it doesn't get more so than roots.

Besides the tiny, spread-out, hairlike roots that all plants use to take up water and minerals from the soil, some plants make especially supersized, swollen roots called taproots or tubers. These are underground storage organs, containing saved-up caches of water and carbohydrates to use in difficult weather.

It's the woody-with-a-hint-of-floral flavors accompanying the crunch of **carrots**, the pleasantly rootcellary earthiness of **potato**, the spot-on dirtiness of even the most well-cleaned **beet**. Rooty flavors foreground earthy-green pyrazines. Many of them, especially potatoes, hold a lot of their pleasant-earthy flavor in the skin, so naturally, keeping the skin in your recipe during serving or cooking will maintain their strength of flavor more than removing it. Carrots and beets both get distinctive qualities from a few special terpenes. Raw, these vegetal flavors project the refreshing feeling of having just pulled something out of a garden plot that you're now thoughtfully and wholesomely crunching away on. Cooked, their earthiness is more like a cozy pair of socks in front of the fire, complementing their soft, weighted-blanket texture.

In the kitchen, we can split roots into two categories based on their carbohydrate of choice, which require different cooking methods and, therefore, undergo different flavor changes. **Sugar-and-cellulose-filled roots** (like carrots, beets, and radishes) are sweet and crunchy when raw. Cooking, especially roasting, collapses their crunchy cellulose down into dense tenderness, and their sugars make them especially good at browning and caramelizing. Carrots, radishes, and daikon can show this off with extra panache when crusted with spices (see page 166).

Starchy roots (like potatoes and sweet potatoes) need to be properly cooked to turn their starch granules from gritty to fluffy or creamy. In the oven, they take on roasty-brownness well and hold on to a stodgy density. They will usually also brown well when roasted, but keep more stodgy density than crunchy-sweet roots. When boiled, starch turns creamy, and they hold on to a light veil of earthy flavor.

Vegetal and Green: Leaves

In leaves, grassy GLVs and more darkly vegetal pyrazines form the flavor backbone. Their core flavor theme is vegetal, lightly earthy green flavors, which leaves express subtle variations on, sometimes slipping in their own unique notes. Dominating all of it is a gamut of textures, from tenderest to toughest, softest to crispest. We're talking about succulent, light-green, cucumbery-almond, bitter **endives**; ultra-delicate almost creamy-green **baby greens**; silkily well-cooked and sulfurously earthy deep-green **kale** and **collards**; perky, spicy, soft, and juicy **mustard greens** and **arugula**, sweet and funky **cabbage**, earthy-fatty-mushroomy-vibing **spinach**.

Leaves, depending on age, size, and thickness, can present their best self either raw or cooked. Raw leaves will express more of the fresher green-vegetal flavors, while cooked leaves spotlight the deeper, earthier, heavier ones. There are the happy few leaves that can swing both ways: cabbage, crunchy when raw and silky when cooked, or tender specimens of mustard and spinach leaves.

Raw Salads

Raw leaves, eaten fresh and cold: a simple concept made deliciously complex by the sheer variety of raw materials available, times the many culinary philosophies of the salad. In the modern American canon, we have the iceberg wedge, the romaine-based Caesar, French-Californian-Italian-turned-bagged-mix mesclun, the arugula and olive oil, the Little Gem with Green Goddess, the various shredded marinated cabbage coleslaws, the various firm and shaved strongly flavored fennel bulbs, puntarelle, and beets.

When making salads, we have choices to make: selecting flavor first. Second: juxtaposition and layering. Are we focusing on just one kind of leaf, or building up subtle layers of flavor with several kinds? Last: chic and minimalist accessorizing—dressing a salad for the most boost and the least flavor interference. To learn how, let's dig into salad philosophies and find examples to follow.

I'd somehow assumed that the European salad tradition was a newish one, maybe summoned into existence during the Enlightenment, so I was pleasantly surprised to learn that salad eating and salad making go way back, past the Renaissance and into the medieval period and probably earlier. Retrospectively, of course that makes sense—if you're a largely agrarian society, with a cuisine

Leaf Storage: Good Handling Practices

Leaves, especially soft and delicate ones, are not long-lived ingredients. Stored with too much water, not enough water, too warm, or too cold: all of these can lead directly to wilting, which then carries on into a black, oxidized slump-mush. I take care by handling them gently, drying them well, storing with a wrapping of barely damp paper towel in a spacious zip-top bag or hard plastic container, and eating them as soon as possible.

that intimately reflects the seasons, of course you're going to go wild for any fresh green leaf that presents itself as you come out of winter. And, for as long as fairly tender greens are in season, why cook them if they taste great, together and raw?

In *The Forme of Cury*, a cookbook written by Richard II's head cook in 1390, the directions for making "salat" instruct the cook to "pick them and pluck them small with thine hand"—"them" being parsley, sage, garlic, chives, onions, leek, borage, mint, scallions, fennel, town (garden) cress, and purslane—then mix them with raw oil, then lay on vinegar and salt "and serve them forth." This, the earliest written recipe in Europe for salad, which I don't think anyone would complain about being served as a salad in 2021, just shows that lettuce (*Lactuca sativa*) isn't even a wholly necessary basis for salad—and that leafy flavor layering is baked into salad's DNA.

An Algorithm for a Minimalist but Excellent Dressing for Lettuce

Despite my appreciation for a proper vinaigrette or composed dressing, this is the way I most commonly dress a salad in real life, and it isn't even a recipe.

The idea is to get a base layer of oil all over the lettuce, to act as a little bit of a barrier from the wilting and osmotic effects of salt and vinegar, which go on after the oil primer. The flavor balance I look for is a liberal richness and aroma from the oil, then a fairly assertive level of salt, then a rounding out with sourness. I love sour foods but I think the most common failing point of a salad is a too-acidic dressing. While researching for this book, I stumbled upon essentially *The Barefoot Contessa Goes Plant Based* for early seventeenth-century Londoners, a treatise-cum-cookbook by Italian expat Giacomo Castelvetro. After throwing jingoistic shade at other

European nations for their various sins of vegetable-handling, he includes his "Sacred Law of Salads." This part, I agree entirely with: "Salt the salad quite a lot, then generously add oil to the pot, and vinegar, but just a jot." I guess not that much has changed in four hundred years.

Start with the best **olive oil** you can reasonably use, and drizzle it all over the **lettuce** (again, the best you can reasonably find) in a bowl. On the order of a thin stream, maybe 2 circular passes with the bottle. Toss and slip and move around the lettuce to get the oil layer uniform. Add more if it seems dry: you're aiming for a nice sheen over everything, and not so much that the lettuce seems sodden or weighed down, or that oil pools in the bottom. Taste a piece; there should be a pronounced oily flavor and the lettuce should feel silkenly lubricated. Next, sprinkle all over with a **fairly thin-grained finishing salt** (I am partial to flaky salts). Again, not a measurement, but take a medium-sized pinch and sprinkle half of it all over, flip the lettuce over in the bowl, then salt the rest and flip again. Do the same with maybe **five or six grinds of black pepper**. Get ready with some **vinegar**—I usually use sherry these days but a good wine vinegar, or maybe a fifty-fifty combo of lemon juice and rice vinegar, is great, too. Sprinkle it *very* sparingly over the lettuce, turning it to combine each time. You need less than you think. Taste a leaf and see if it needs more pepping up, then sprinkle in more vinegar as necessary. You might want to finish with a little more salt. Obviously, eat it immediately.

A Kaleidoscope of Cooked Greens

Cooked, leafy greens. It seems almost too obvious—"Yes, we eat greens that have been cooked. Who doesn't?"—but the technique of wilting and braising tougher leaves and stems is as much of a canvas on which cuisines play with selecting and juxtaposing leafy flavors as salads are. Flavor multiplies on top of that: flavors created by the heat and time of cooking and layered-in flavors added when you incorporate additional ingredients: the fats they're cooked in, infusions from cured meat, alliums, herbs, spices. (For my take on layering in flavors to cooked greens, see "Layered Base to Infuse Cooked Greens," page 228.)

The mix of greens known as *preboggion,* from Liguria in Italy, incorporates a wide variety of foraged and cultivated leaves. The specific content varies with the season, but can include more widely familiar plants like arugula, chard/beet greens, borage, chicory, dandelion, wild fennel, nettles, parsley, wild radicchio, and spinach; as well as local favorites like rampion, poppy leaves, dogtooth violet leaves, grattalingua (common brighteyes), leaves, pimpinella, amarago (golden fleece), cicerbita (common sowthistle), and Silene. The leaf selection is parboiled, then you have your choice of making them into a soup or stuffing them into ravioli called pansotti.

Greek cuisine calls cooked greens *horta,* or "weeds"—reflecting the use of many edible wild greens. A classic horta is Horta Vasta, or boiled dandelion greens; others might include endive, caper leaves, mustard, fennel, lamb's-quarters, wild leeks, mallow, chicory, and other leaves that seem to have once been more widely used across Europe but are found more in pockets today. *Sabzi,* the Farsi word for "herbs," covers both herbs and cooking greens in the cooked-down leafy stew ghormeh sabzi. Cooking carrot-family herbs like parsley and cilantro as if they were leafy vegetables, and adding layers of seasoning from scallions, slightly bitter fenugreek leaves, and dried black limes, ghormeh sabzi often includes rich-umami meat as well.

In South Asia, *saag* is a kind of platonic concept that accommodates many different iterations of recipes for greens cooked down with butter and seasonings into a savory, smooth braise, sometimes served and eaten on its own and sometimes as a sauce for other stewed and braised foods, like paneer or lamb. *Sarson ka saag*, an iconic Punjabi dish highlighting *sarson* (mustard greens, sometimes along with spinach and lamb's-quarters), cooked with ghee, onion, ginger, garlic, and chile, sometimes thickened with a bit of corn flour.

Callaloo is both a name for specific leafy greens (specifically amaranth or taro leaves, depending on where you are) and a few different styles of stewed dishes from the Caribbean made from leafy greens. In Jamaica, it is more typical to find callaloo, the dish, made from green amaranth leaves (callaloo, the plant) wilted, steamed and stewed down, and sometimes seasoned with tomatoes, onion, Scotch bonnet pepper, and oil. In Trinidad, callaloo (the plant) refers to taro or dasheen leaves, and the Trinidadian dish often incorporates okra, hot chiles, coconut milk, onion, garlic, and pumpkin.

Strategic Leaf Cooking

Identifying the best way to cook any given leaf can't really be found in a recipe: it depends on considering that individual leaf and its flavor and texture—your kale is not necessarily my kale, is not necessarily that guy's kale.

My mental decision tree considers how tender or tough it is, and how intense its flavor is.

Tougher leaves and more intensely deep-vegetal or bitter flavors (mature collard greens, chard, or kale) benefit from using some water in cooking, which both transfers heat extremely well, cooking and softening them more thoroughly, and dilutes the flavor a little, to stop it from getting too intense. "Some water in cooking" can look like making sure there's a deep puddle in the pan to generate a steamy environment, or cooking them directly in a pot of salted water and then straining them out.

The softer the leaf and the less earthy its vegetal flavor (for example, spinach, younger kale, mustard greens, soft beet greens), the more it can go right into a hot pan, maybe with the water from washing still clinging to it, and wilt-to-cook relatively quickly.

Try This

Like deconstructed pansotti, I like to incorporate cooked greens into pasta. For maximum efficiency, I'll heat up a pot of salted, boiling water, then simmer the greens until they're just soft and skim them out to drain. Then I boil the pasta in that same water. While that's happening, you could brown up some de-casinged pieces of sausage, dump a cup or so of crème fraîche right in to deglaze the browned parts, then add the drained greens to mingle until the pasta is finished cooking, then add that as well (and a little pasta water if it gets too thick). Finish it however your palate leads you—a few well-cracked fennel seeds or bit of chopped or pressed garlic into the creamy sauce, or a squeeze of lemon, vinegar, drained yogurt whey, or a grating of hard cheese or Dried Olive Powder (page 200) over it.

Flavors of Intensity and Defense

Like fruity flavors, the catalogue of spiced and herbal flavors—and the contributions they can make to our cooking—is one that's fun to get lost in. Resinous, green, dry and fragrant, warm, sweet, refreshing, fruity, buttery, floral, and qualities I only know how to describe by naming what they come from: clovey, black peppery, minty, thymey, cinnamony. Flavors stretching in all moods and directions as far as the eye can see.

Too intense to be the central focus of a dish, spiced and herbal flavors live to decorate, garnish, season, and enhance other ingredients. They do so boldly—culturally, they're flavor heraldry, emphatically signaling the ties between a dish and the cuisine that birthed it. A few spoonfuls of masala, proportioned to emphasize sweeter or dryly fragrant versions of spiced flavor, the exact basil variety making up a pesto, the dimension of "resinous" an herb brings to roasted pork—small amounts are all you need to tell a deep and complex flavor story in even simply cooked food.

Spiced and Herbal: The Lay of the Land

Just getting in the habit of using more herbal and spiced ingredients and flavors—seriously, any of them that strike your fancy—in your cooking will make it more delicious. If you want the easiest possible takeaway on the subject, that's it: as with fruity, the catalogue in your hands is yours to explore and find out more about what you like. With your own tasting and paying attention, you can just about choose them on the fly for a dish and not go wrong. A pasta garnished with mint, a fruit with oregano, a shortbread cookie with celery seed or turmeric, a chicken roasted with allspice, a sausage herbed with shiso—none of these fit what I think of as familiarly American cookbook cuisine vernacular, but they all have incredible potential (and, honestly, have probably been made successfully by someone, somewhere before). I'm actually having a little trouble thinking of a random combination of herbs or spicing and main ingredient that couldn't be made delicious.

I lump "spiced" and "herbal" together here not just because they're seasonings but because, if you taste closely, it's possible to find some small flavor facet of one herbal or spiced flavor somewhere in almost all of the others. And from the point of view of molecules, there's hardly any firm boundaries between the two categories. They all dip from the same pool of flavor molecules, assembling them into complex blends with lots of overlap and lots of "variations on a theme." They're all different, but some are more different than others, and their flavor borders are sort of hazy.

My own mental picture of spiced and herbal flavors is much like the maps that flavor science can pull out of this kind of messy data that I talked about at the beginning of this chapter (see page 126).

Their variations loosely group them together within the galaxy of flavor like space dust condensing into a few major nebulae.

The primary flavor regions on our star chart:

- Spiced, dry, and fragrant (black pepper and coriander)
- Spiced, earthy, and savory (cumin and saffron)
- Spiced, warm, and sweet (clove and cinnamon)
- Herbal, tender, and fragrant (basil and spearmint)
- Herbal, soft, and green (cilantro and dill)
- Herbal, resinous, and intense (oregano and sage).

I come back to these clusters or families over and over again: to improvise on a recipe I like with slightly different flavors, to figure out how to evoke a flavor feeling I have in my head, to blend them on the fly in more sophisticated ways. A simple way to think about it is that you can improvise, mix, or blend in a more subtle or familiar way by working within one cluster, or in a bolder and more complex one by choosing things that are further away, in more than one cluster: improvising in

soft and green where the original was tender and fragrant, or mixing earthy and savory with warm and sweet.

Cumin as Chemical Weapon: Herbal and Spiced Molecules

Herbal and spiced flavors make heavy use of two families of smell molecules: *terpenes* and *phenylpropenes*, which we've seen before in citrus and as minor players in a few other fruity and vegetal flavors.

To plants, terpenes and phenylpropenes have much more in common with bitter and spicy molecules than with smells like fruity: they are made for plant self-defense, as a kind of chemical weaponry.

Just about all the molecules responsible for the flavors you like in herbs and spices (flavors like clove, herbal basil, licoricey tarragon, or woody, resinous cumin) were created by evolution for similar ends: to help a plant stay alive and get an edge over other organisms. Some of them repel or intoxicate pesky, chewing insects and herbivores; some deal with environmental threats like invading molds; and some help the plant dominate and compete with other plants.

So why do we find herb and spice flavors so delicious, and so unlike chemical weapons? It's a humbling example of how the world doesn't revolve around us. A clove plant doesn't particularly give a damn if you enjoy the flavor of cloves, which is typified by the molecule *eugenol*.

Eugenol and molecules like it were already around and making their contribution to the competitive thrum of ecology and evolution, probably for hundreds of millions of years, before we ever came on the scene. We evolved in a world where plants were already making these molecules. And, like many bitter molecules, a lot of them have essentially medicinal effects (as anti-inflammatories, antibacterials, etc.) in our bodies at low doses, unlike for the insects and fungi that

(continues on page 158)

FLAVOR SUPERCLUSTERS:
HERBS AND SPICES

nutmeg

allspice

Clove

cinnamon

Fennel

Vanilla

tarragon

chervil

basil

Cilantro

dill

parsley

ginger

Shiso

Spearmint

Cardamom

peppermint

Coriander

black pepper   pink pepper

grains of
Paradise

cumin

rosemary

thyme

Sage

Marjoram

Caraway

Savory

Saffron

turmeric

Oregano

How Did Plants Figure Out How to Make Herbal and Spiced Molecules?

Herb and spice flavor molecules are "secondary metabolites" of plants: they're not strictly necessary for living, just for making life a little easier. They're great examples of nature repurposing and reusing mundane things it already makes to make new things that are complex, unique, and rather beautiful.

The terpenes, which make up half the flavor picture, are assembled like Tinkertoys from *isoprene*, one of the first secondary metabolites life ever produced (about 2.5 billion years ago). Various *isoprenoids*, molecules made from isoprene, reinforce your cell membranes, keep your skin supple, make cannabis psychoactive, make tomatoes and sunflowers colorful, and (as terpenes) make herbs, spices, flowers, citruses, and evergreen trees aromatic and flavorful.

The other flavor-molecule group of interest, the phenylpropenes, are siphoned off from a biochemical assembly line that plants invented about 500 million years ago as they left the water and colonized the land. Faced with new challenges like holding themselves up (without the buoyancy of water to support them) and protecting themselves from harsh UV sunlight (without water to filter it), they started making rearrangements and assemblages of something they made a lot of anyway because they needed it for proteins: an amino acid called *phenylalanine*. These include the *lignin* that makes tree trunks strong and woody, *polyphenol* antioxidants, and, eventually, the delicious flavors of dill, anise, cinnamon, and basil.

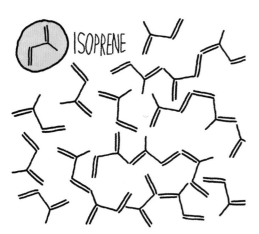

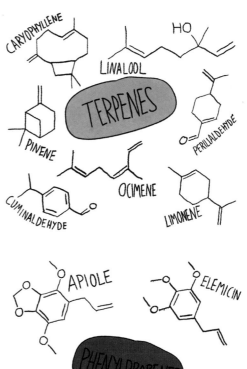

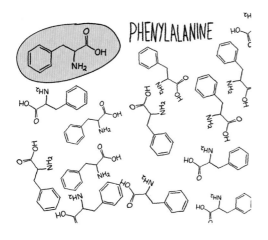

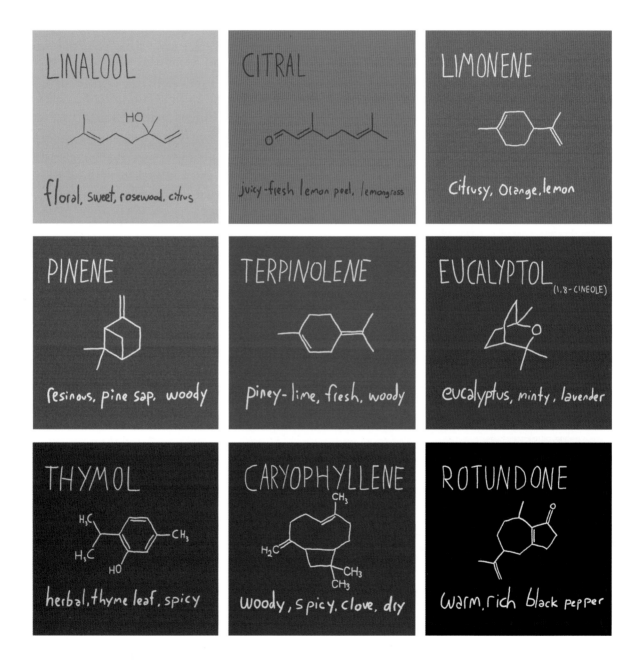

LINALOOL

floral, sweet, rosewood, citrus

CITRAL

juicy-fresh lemon peel, lemongrass

LIMONENE

citrusy, orange, lemon

PINENE

resinous, pine sap, woody

TERPINOLENE

piney-lime, fresh, woody

EUCALYPTOL (1,8-CINEOLE)

eucalyptus, minty, lavender

THYMOL

herbal, thyme leaf, spicy

CARYOPHYLLENE

woody, spicy, clove, dry

ROTUNDONE

warm, rich black pepper

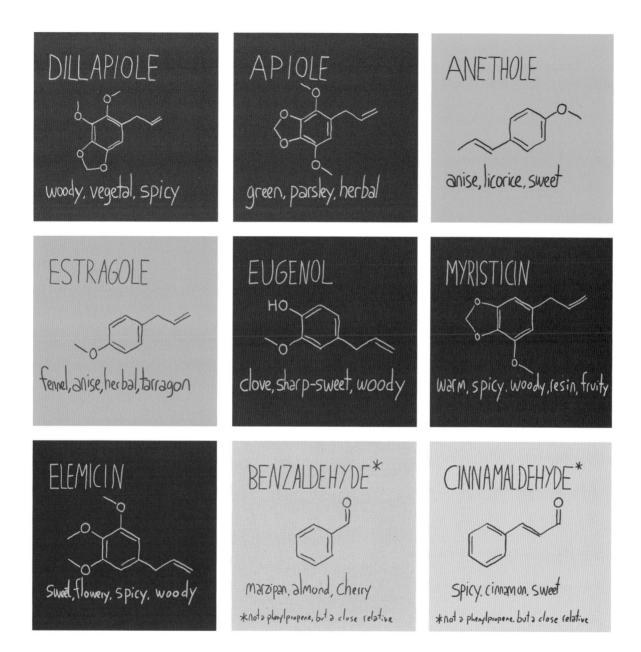

DILLAPIOLE
woody, vegetal, spicy

APIOLE
green, parsley, herbal

ANETHOLE
anise, licorice, sweet

ESTRAGOLE
fennel, anise, herbal, tarragon

EUGENOL
clove, sharp-sweet, woody

MYRISTICIN
warm, spicy, woody, resin, fruity

ELEMICIN
sweet, flowery, spicy, woody

BENZALDEHYDE*
marzipan, almond, cherry
*Not a phenylpropene, but a close relative

CINNAMALDEHYDE*
spicy, cinnamon, sweet
*Not a phenylpropene, but a close relative

they sabotage. So we adapted to be able to detect (and enjoy) them in low doses, a kind of defiant hack to get around their intended purpose.

Unlike taste molecule categories like "acids" or "sugars," there are not a lot of neat lines we can draw around terpenes and phenylpropenes: each grouping has some recognizable physical features in common, akin to curly hair, freckles, or a cleft chin—but those aren't really that useful for explaining why they have the flavors they do. So, rather than drive myself crazy with minutiae, I wrap my head around them by thinking about (you guessed it) the patterns in their flavors.

To me, terpenes are like Werner Herzog movies while phenylpropenes are like George Miller movies. Phenylpropenes, like George Miller's filmography, are respectably if not bogglingly prolific—a manageable list of a dozen or so, the most notable of which are part of a series with two wildly different vibes—*Babe*, the hero's journey of a plucky sheep-herding pig, and . . . *Mad Max*, an operatically ultraviolent dystopiad. You've got cozy, friendly, sweet-smelling phenylpropenes like warm-and-clovey eugenol and anisey estragole on the one hand, and rather intense, distinctly unsweet ones like vegetal dill apiole and woody safrole on the other.

Just like Werner Herzog's fifty-one feature films (plus fifteen short films, eighteen opera productions . . .), there are way more terpenes than you could casually keep track of. All of them are complex and multifaceted, and they are often wildly different from one another in shape or genre: feverish historical epics like *Fitzcarraldo* or *Aguirre, the Wrath of God*, horror like *Nosferatu the Vampyre*, nihilistic documentaries like *Grizzly Man*, meditative ones like *Cave of Forgotten Dreams*.

But all of them draw from a well of repeating motifs—the presence of Klaus Kinski, the folly of trying to control nature, or narration in Herzog's singular Bavaria-meets-L.A. accent—that add up to a kind of vibe that makes you think, "Yeah, that makes sense as a Herzog [slash, terpene flavor]." Terpene flavor notes range from woody to citrusy to floral to herbal to spicy to resinous to sweet to heady to deep, and each of the thousands of terpenes usually combines several of these flavors, like citrusy-floral limonene (in citrus, ginger, and dill), resinous-spicy cineole (in cardamom and basil), or woody-resinous myrcene (in black pepper and sage), for flavors that aren't necessarily just alike, but share a pattern, a vibe.

Spiced

Spices are among the most powerfully aromatic and highly flavored ingredients you can find. Usually, a few molecules out of every million is enough for a noticeable flavor—but spices don't mess around. Cloves, for instance, pack in about 20 percent of their total weight as aroma molecules. They're excessive and over the top in flavor, by a factor of about 200,000.

This aromatically flavorful excess is more or less the defining feature in calling a plant ingredient a "spice" in the first place. Whether it's a seed, flower bud, fruit, pod, root, or bark, if it's brimming with terpene and phenylpropene flavor molecules and dries well (and isn't a leaf, parts we usually categorize as herbs), we probably use it as a spice. (Though some are really smashing when fresh: ginger, galangal, green coriander seeds.)

Mapping Spiced

The landscape of spiced flavors is varied and intense, but there's a distinct strategy we can take to navigate all of it, like walking downhill until you find a stream to follow if you get lost in the woods. Carefully tasting spices, you'll notice that spiced flavors have a lot of herbal echoes in them: a strong, resinous, thyme quality in freshly ground cumin; clove notes in both allspice and basil; a shy but present black pepper flavor in rosemary. Spices tend to express sweeter or earthier extremes of flavor, with few really vegetal notes.

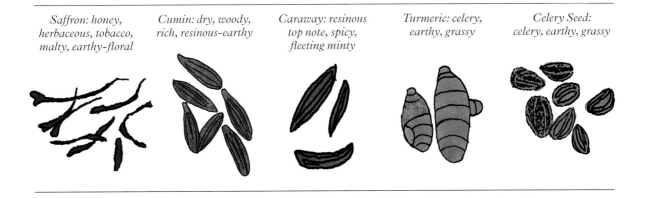

Saffron: honey, herbaceous, tobacco, malty, earthy-floral

Cumin: dry, woody, rich, resinous-earthy

Caraway: resinous top note, spicy, fleeting minty

Turmeric: celery, earthy, grassy

Celery Seed: celery, earthy, grassy

Mapping out the patterns in this flavor landscape, it's distinctly shaped by a gentle and continuous change from one direction to the other, like a mountain range on the west side of a map melting into highlands, then forest, then meadows, then beaches as we move east. The westerly extreme for spices tends to be dry, earthy, and savory, flowing through dry and fragrant on the way to warm and sweet at the eastern edge of the map. The west and central regions are more dominated by terpene flavor molecules; the east, by sweet phenylpropenes.

Individual spices cluster like points of interest on this map. The ones that are closer together, while far from identical, tend to have a more similar local flavor landscape. The further apart they are, the more they differ.

Spiced, Earthy, and Savory

Some spices are distinctly dry, like a wine without sweetness is dry, and suggest a kind of savory earthiness or woodiness. Their suite of flavor molecules covers the deepest, darkest qualities found in terpenes.

Cumin is a perfect example. When you apply it to meat or vegetables, it hits a flavor spot that is distinctly spiced, and it echoes savory or earthy flavors in those ingredients rather than contrasting them. Its dry, woody, resinous-earthy flavors set an example that other dry and savory spices

echo, and they all do amazingly crusting those meats and vegetables, going along in a braise, or adding grounded earthy flavors to grains, like a pot of rice or farro, or saucy risotto.

Cumin flavor is mostly terpenes, specifically terpenes without a lot of sweet or soft top notes. It starts with a terpene that's actually named for it, *cuminaldehyde*, with spicy, herbal, green-vegetal, and distinctly cuminy character. It gets a softening fatty-fruity spiciness from *1,3-menthadien-7-al* and *gamma-terpinene*, and rounds out with three really quintessential savory terpenes: *pinene* (resinous, woody, piney), *myrcene* (woody, celery, peppery), and *cymene* (oxidized lemon, cuminy, oregano).

Cumin is part of the carrot family, a kinship it shares with **caraway**. A pungent source of aromatic savoriness, caraway evokes dark pumpernickel Jewish rye bread, sauerkraut, and aromatic-savory alcohol like aquavit. It shares some resinous terpenes with cumin, and its key flavor component is *carvone*, a molecule that comes in two mirror-image versions. "+*carvone*," the doppelganger in caraway, creates that distinct caraway note, with spicy and subtly minty qualities. Flip to its mirror iteration, "-*carvone*," and you'll smell the characteristic flavor of spearmint. (Just like it's easier to shake someone's right hand with *your* right hand than your left hand, which are mirror images of each other, the mirror

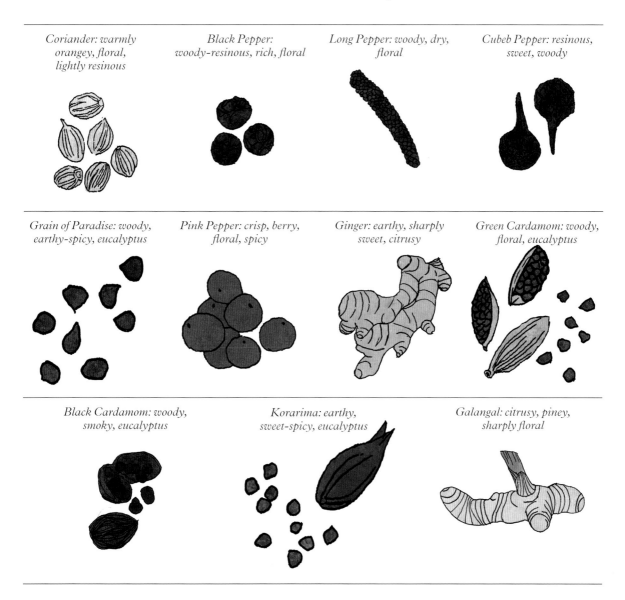

Coriander: warmly orangey, floral, lightly resinous

Black Pepper: woody-resinous, rich, floral

Long Pepper: woody, dry, floral

Cubeb Pepper: resinous, sweet, woody

Grain of Paradise: woody, earthy-spicy, eucalyptus

Pink Pepper: crisp, berry, floral, spicy

Ginger: earthy, sharply sweet, citrusy

Green Cardamom: woody, floral, eucalyptus

Black Cardamom: woody, smoky, eucalyptus

Korarima: earthy, sweet-spicy, eucalyptus

Galangal: citrusy, piney, sharply floral

versions of carvone line up to aroma receptors a little differently.)

Turmeric takes the dry-savory spice pattern in a distinctly earthy and piquant direction, as a literal from-the-ground root that has some base notes in common with its more floral-smelling cousin, ginger. **Saffron**, probably most widely used to lend rich, earthy-floral flavors to rice-based dishes from Iran and nearby countries as well as South Asia, probably gets the closest to sweetness of any of the dry and savory spices (it is the pistil of a flower, after all). It also gets

the honor of an aroma molecule named after it, *safranal*, which creates its spicy, woody, resinous-herbaceous, aromatic-tobacco flavors.

Spiced, Dry, and Fragrant

While all spices are fragrant in the technical sense, there are some for which complex, heady, sparkling fragrance *is* their main character trait. They're not usually straightforwardly sweet or straightforwardly savory, and they often combine elements of sweetness and earthiness with citrusy, woody, or resinous flavors. Like earthy and savory spices, they're full of terpenes, but they really explore the full gamut from heavy to light and fruity.

Coriander, which is the dried fruit of the cilantro plant, is a great exemplar. With just a bit of resinousness as a base (from the terpenes *pinene* and *terpinene*), it makes a hard left into sweet-floral-rosy, from *linalool* and *geraniol*, and finishes off distinctly orangey-citrusy. Like the leafy plant it comes from, it relies on *aldehydes* for some of its distinctiveness, specifically orangey-waxy *decenal* and *trans-2-dodecenal*. Coriander is great at lightening heavier spices like cumin or cinnamon in a blend, floating like a cloud of peely waxiness.

The fragrant complexity of **black pepper** is easy to forget given its sheer ubiquity in kitchens. But, if you keep a pepper grinder around to finish your food, you're coating most things you eat in a deft layer of woody-resinous base notes (*pinene*, *myrcene*, *caryophyllene*) and orange-lilac fruity-floral stuff (*limonene*, *linalool*, *alpha-terpineol*), capped off with a special, distinct "black pepper" statement from *rotundone*, a terpene that smells exactly like the core of black pepper. It's got too much light and special stuff going on to just be truly earthy and savory, but it's definitely on the driest side of fragrant. Black pepper didn't used to have such a lonely place as a basic seasoning; **long pepper**, **cubeb pepper,** and **grains of paradise** were all widely used to fill the same woody-aromatic-spicy role, with different amounts of sweeter or more resinous notes. And they're hardly unavailable today—you just have to know where to look for them.

Two cousins from the ginger family, **ginger** and **cardamom**, express fragrance, dry and woody spiciness, and floral sweetness all at once. Cardamom's primary flavor molecule, by concentration, is a terpene called *eucalyptol*. It's savory and heavy, eucalyptusy and minty, but it doesn't weigh down the flavor too much. Cardamom plays in sweet situations like spiced tea or coffee, pastries, and mulled wine like a total superstar, showing off notes of intense lemon peel, bergamot, lavender, tea, and woody roses. Ginger shares its lemon peel top note (from a terpene called *citral*), and blends in layers of earthy, sharp, sweet, and a bit resinous.

Spiced, Warm, and Sweet

Fragrant spices may have important sweet-smelling aspects, but they pale in comparison to truly sweet-spicy spices.

Warm and sweet spiced flavors are usually dominated by one, intense, distinct type of flavor molecule. Decisively clovey, slightly resinous-smoky *eugenol* is the main flavor of **cloves** and **allspice**, balsamic-woody *cinnamaldehyde* that of **cinnamon**, intensely licoricey *anethole* that of **star anise** and **fennel seed**, creamy, floral *vanillin* that of **vanilla**, sweet-peppery *myristicin* that of **nutmeg**.

Warm and sweet spices usually sport a few terpenes for flavor, but phenylpropenes do most of this heavy lifting. Toasted wood and smoke can have a sweet-spicy edge to them, too, because the woody lignin inside of it is made on the same assembly line from the same parts as phenylpropenes—just built much bigger. When heat breaks lignin down, clovey eugenol and vanillic vanillin are two of the main things it breaks down into.

The "sweet" connotation of sweet spice aromas is an interesting case. Usually sweet aromas only remind us of sweet tastes. Molecules like anethole, eugenol, and cinnamaldehyde do that, but they can also tickle at our sweet taste receptors and do, literally, taste sweet.

All these sweet-spiced flavors, especially of cloves, vanilla, cinnamon, and nutmeg, obviously work really well in distinctly sweet dishes mixed with other deliberately sweet flavors: vanilla in ice cream, fennel seeds in biscotti, allspice and cloves in pumpkin pie, cinnamon in cinnamon buns. Dessert might be the first thing you think of when you smell these spices.

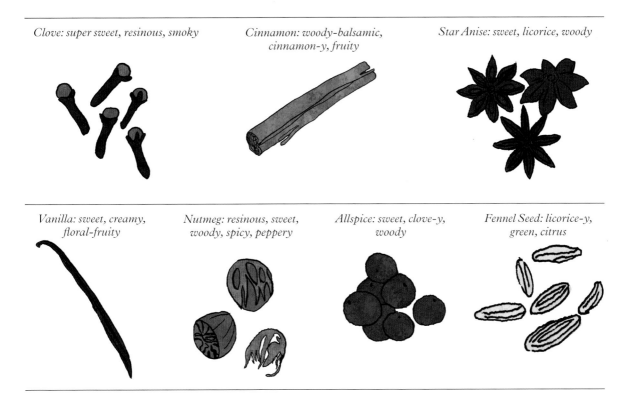

Clove: super sweet, resinous, smoky

Cinnamon: woody-balsamic, cinnamon-y, fruity

Star Anise: sweet, licorice, woody

Vanilla: sweet, creamy, floral-fruity

Nutmeg: resinous, sweet, woody, spicy, peppery

Allspice: sweet, clove-y, woody

Fennel Seed: licorice-y, green, citrus

But they also work brilliantly in savory applications, especially with meat and vegetables, and especially when they have some earthier or fragrant spices to blend and play with. Allspice in jerk chicken. Cloves in garam masala. A little cumin and a little cinnamon, together, just gently rubbed into beef or lamb before you grill or braise it become phenomenally dynamic, playing off the range of flavors in the meat and each other, and don't at all feel oversweetened or too close to dessert.

Spice Mixtures

Cumin and cinnamon are hardly the only opposite-ends-of-the-spectrum spices that work like this. Their dynamic pairing reveals an important nuance for using spices with sophistication: for as intense and complex as any spice is on its own, they get so much more interesting, and so much better, when you mix and blend several of them together.

Spice mixtures, personally, really get me going. Part of it is the idea of regarding the deeply multifaceted flavor of any spice, and then rather than treat it too preciously or austerely, kaleidoscoping it out by mixing it with other multifaceted spices, layering layer upon layer upon layer, like folding croissant or baklava dough on itself until it exponentially leaves together.

Drier spices like cumin, black pepper, or caraway get a bright lift from sweeter-inflected ones like cinnamon, cloves, or ginger, and spices that go into the "strictly for desserts" category in one place prove to be irresistible and dynamic in savory foods in another. So many countries and cuisines have been mixing spices for so long and for so many dishes according to different philosophies and ratios, that if you think you understand which spices do and don't go together from one cuisine, you'll be proven totally wrong in another. Here are a few of the most important and iconic ones, and their underlying flavor logic and uses.

Garam masala, or "hot" masala, the primary spice mixture of many north Indian cuisines, is hot not from chilies but from the warm and sweet flavors of cinnamon and cloves, and the dry and woody spicy bite of black pepper. It often also includes black or green cardamom, cumin, coriander seeds, cinnamon leaves, and sometimes rose petals, cassia, or fennel. Dry-toast, grind, and blend into dishes right before they're served.

Chai masala, or karha, spices for tea, are sometimes cardamom and ginger only; black pepper, cloves, cinnamon, star anise, fennel, coriander, rose petal, and nutmeg often make appearances. Masala chai is, as a rule, sweetened and milky; and when the milk is included in the pot that the tea and spices are heated up in, its fats and proteins help extract out more flavor from the spices. Somalian **qahwe** (coffee) is often prepared with spices, especially cardamom and cinnamon. Like the technique used for some masala chai preparations, qahwe with spices steeped all together (coffee, cardamom, cinnamon, and sugar) directly in hot milk, rather than in water alone, deepens and sweetens the overall flavor profile of the extraction. **Hawaij** is a duo of two different spice mixtures from Yemen—one savory, one sweet—that have spread geographically with the Yemeni Jewish diaspora. **Savory hawaij** is often based on turmeric (giving it an orange color), cumin, coriander, black pepper, and cardamom, with sometime additions of nigella, saffron, or caraway. Like many savory spice mixtures, it's frequently used to crust and season grilled meats and vegetables, and to season stews and soups. The sweet version of hawaij is sometimes called **hawaij for coffee** and usually has cardamom and ginger at the forefront, and cinnamon, cloves, nutmeg, fennel, and citrus peel are not unusual inclusions. Claudia Roden has collected recipes for both types of hawaij from Aden, the principal seaport in Yemen; her **savory Adeni hawaij** has coriander, cumin, cardamom, and black pepper. **Sweet Adeni hawaij** for tea (steeped without

milk) is made with cinnamon, cloves, and cardamom, with ginger added when used with coffee.

Baharat, Arabic for "spices" (with bounteous connotations from its original Persian root *bahâr*, "blossom"), is widely used as a seasoning of rice dishes, meats (as kebabs, rubbed on before grilling, or in stews), and (like hazelnut-spice **duqqah** and thyme, sumac, and sesame **za'atar**) dipped with oil as a condiment, all around the Middle East. Some normally included components are black pepper, cumin, coriander seeds, cinnamon, clove, nutmeg, and cardamom, with lots of regional and local variations, such as a very dried-mint-forward baharat from Turkey, a Gulf-style baharat with lots of dried lime and saffron, and a mostly cinnamon, rose petal, and black pepper baharat from Tunisia. Other things you might see in baharat are allspice, hot or sweet chile, and bay leaves. I've read it described as "7-spice powder," but seven is by no means obligatory.

Ras el hanout carries a vibe of prestige, meaning literally "the head of the market" or as Paula Wolfert puts it, "top of the shop"—i.e., the best spices, blended together. It's particularly associated with Moroccan and other North African cuisines, and has a lot of spices in it, sometimes marketed as ten, nineteen, twenty-six, or fifty. Intended to be particularly floral, aromatic, and sweet-leaning, ras el hanout often has rose petals, orris, coriander, and lavender as components, as well as spice mix standbys like ginger, green cardamom, nutmeg, cinnamon, clove, black pepper, turmeric, cayenne pepper, mace, and allspice. Some less commonly imported (but source-able from the right shop, or the internet) are nigella seeds, black cardamom, cubeb pepper, grains of paradise, and long pepper. Saffron, garlic, cumin, anise, mustard seed, orange peel, and fennel seeds are also not out of place in ras el hanout. It's widely used in stewed and saucy dishes, especially for high-flavored meats like lamb and game.

Traditional medieval European spice mixtures like **poudre douce, poudre fine, poudre**

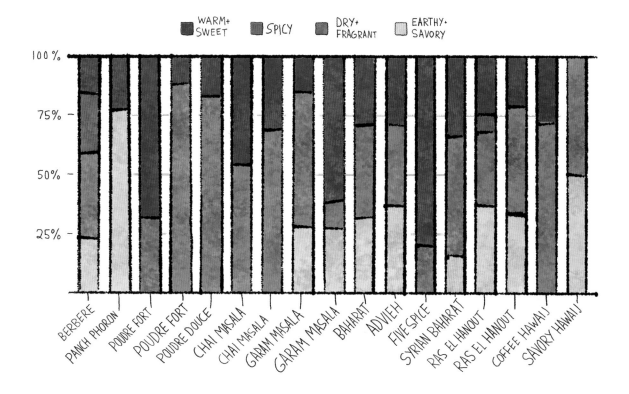

Legend: ■ WARM+ SWEET ■ SPICY ■ DRY+ FRAGRANT ■ EARTHY+ SAVORY

X-axis labels: BERBERE, PANCH PHORON, POUDRE FORT, POUDRE FORT, POUDRE DOUCE, CHAI MASALA, CHAI MASALA, GARAM MASALA, GARAM MASALA, BAHARAT, ADVIEH, FIVE SPICE, SYRIAN BAHARAT, RAS EL HANOUT, RAS EL HANOUT, COFFEE HAWAIJ, SAVORY HAWAIJ

fort, poudre Lombard (yup, European, because up until around the eighteenth century, wealthy Europeans were just as excited about spices in savory food as everyone else was, and anyone sensible still is) were often four or five spices each and often involved cloves and different types of pepper (black pepper, grains of paradise, long pepper), with different additions of ginger, cinnamon, nutmeg, and/or bay leaves. A poudre fine as described in *Le Ménagier de Paris* (a fourteenth-century "how to be a good wife" handbook that covered cooking as well as keeping hawks and getting rid of fleas) had ginger, cinnamon, cloves, grains of paradise, and sugar; another cookbook, this one from Italy (for "fine spices for all foods"), included all those as well as saffron. The same Italian book has a recipe for sweet spices or poudre douce, with ginger, cinnamon, cloves, and bay leaves.

Quatre épices ("four spices") is a French mixture of pepper, cloves, nutmeg, and cinnamon. It's possibly a holdover from similar medieval spice blends, where "sweet" spices are used with savory ingredients. Quatre épices often seasons terrines, game, or stew.

"As many versions as there are cooks" is a hard-worked war horse in food writing, but it holds true for spice mixtures. There's rarely one perfect archetype for any mixture—no "one true hawaij" or "one true baharat" (in fact, those phrases sound like a great way to start a fight), but many, many versions and variations on the *idea* or pattern of hawaij (or garam masala, baharat, ras el hanout, etc.)—variations that can still be called by the same name. Their differences could break down mostly by country, or be as granular as unique town-to-town iterations. Lots of places from North Africa through the Gulf states and into Turkey make baharat, but depending on where you are, it might go heavy on dried mint, or cinnamon, or rose petals. And, of course, many people don't blend their own spice mixtures, so the taste or aesthetic of regional manufacturers and packagers plays a big role of what any given spice mixture actually tastes like.

In the kitchen, it's a situation begging for you, an improvising cook, to go looking for patterns. For some of the spice mixtures that follow, I was thinking about combining spices from across the whole spectrum of earthy to sweet, a pattern you might see in baharat or hawaij. In Underappreciated Fragrant Spice Pumpkin Pie (page 167), I focused on the dry and fragrant middle range of the map, flavors that dominate some spice mixtures for tea or coffee, to perfume a dessert where you might more commonly taste mostly warm and sweet spices—and create an accessible way to play with some delicious and less broadly known ones like grains of paradise and korarima (or Ethiopian cardamom).

Roasted Carrots with Dry and Earthy Spices

This recipe is all about working around the earthiness and sweetness of the carrots. Serendipitously, carrots are actually closely related to three earthy spices: cumin, caraway, and celery seed. Additional relatives coriander and fennel lighten and sweeten the spice mix just a little, and our other favorite orange root, turmeric, rounds out the earthy-spicy flavors. Some of the spices will burn just a little toward the end of roasting, but the depth and bitterness this adds is actually really nice.

To make the spice mixture, mix **1 tablespoon (6 g) cumin seeds, 2 teaspoons (6 g) caraway seeds, 1 teaspoon (2 g) celery seeds, 2 teaspoons (4 g) coriander seeds**, and **1 teaspoon (2 g) fennel seeds** together in a spice grinder. Grind finely, taking care to really break down the

coriander seeds. Combine with **1 teaspoon (3 g) ground turmeric**.

Preheat the oven to 400°F. Line a large baking sheet (I use a standard restaurant-supply rimmed half-sheet pan for everything) with foil or parchment paper.

Roughly peel **2 pounds (900 g) carrots (8 to 12 medium-sized ones)** and cut into largish chunks along their length. Toss the carrots in a large bowl well coated with **3 tablespoons (42 g) melted butter, ½ teaspoon (3 g) kosher salt**, and 1½ tablespoons (12 g) of the spice mixture. If the carrots look a little patchy, add a bit more spice mix until they look evenly covered, but not encrusted.

Pour the carrots onto the baking sheet and distribute them in a single layer. Cover the whole sheet with aluminum foil and squish it down around the edges to make a relatively sealed top.

Put the sheet in the oven and cook for 15 minutes. Peel up a corner and check on the carrots—they should be steamy and moist but not totally cooked. If they still look raw, re-cover and cook for 5 to 10 minutes more.

Uncover the baking sheet and raise the heat to 450°F. Roast for 15 minutes, until the carrots are soft and have collapsed slightly, with a dry and toasty exterior. Let it go another 5 or 10 minutes if they don't look brown enough. Remove and serve.

These are delicious as is, but you could also make a quick (and carrot-family thematic) sauce by mixing **¾ cup (175 g) plain, full-fat yogurt, a small handful of dill fronds** (finely chopped), **a small handful of cilantro leaves** (finely chopped), and **½ teaspoon (3 g) sea salt**. Serve immediately.

For some real panache, plop some of the herb yogurt down on the plate like a little nest and put the carrots on top of it. Nice.

Serves 4 to 6 as a side or 2 as a main

Underappreciated Fragrant Spice Pumpkin Pie

Blended pumpkin pie spice is usually cinnamon, nutmeg, ginger, and cloves, sometimes with allspice. Cinnamon, nutmeg, cloves, and ginger were all superpopular spices in medieval and Renaissance Europe, for both sweet and savory dishes, which got me thinking about how other spices that you might have used alongside these in 1300 in Paris or Florence but aren't so widely familiar now, like grains of paradise and cubeb pepper, would be in a pumpkin pie. The answer is: really good. They're both deeply fragrant spices, with some notes in common with allspice and black pepper, but a bit fresher resinousness.

This recipe uses those two spices along with two types of cardamom: korarima (Ethiopian cardamom) and green cardamom. The result is a less sweet, very fragrant and earthy-spiced, somewhat spicy pumpkin custard, which goes well with a flaky and well-browned pie crust.

If you can't find some of the spices below, think about experimenting a little: black pepper, black cardamom, and more green cardamom are good starting points from the dry and fragrant section of the spiced spectrum; star anise, allspice, or fennel seeds are good candidates if you want to push it sweeter.

Have ready a flaky, rolled, blind-baked single 9-inch **pie crust**. Preheat the oven to 350°F.

In a spice grinder or high-powered blender, grind **2 teaspoons (4 g) cardamom seeds** (not pods), **1 teaspoon (2 g) korarima seeds**, **1 teaspoon (2 g) cubeb peppercorns**, and **½ teaspoon (1 g) grains of paradise**.

In a large saucepan, mix the ground spice mixture with **1¼ cups (300 ml) heavy cream**. Heat over medium heat until it just reaches a simmer, then remove from the heat and steep (covered) for 30 minutes.

In a large bowl, thoroughly whisk the spiced cream into **14 ounces (400 g) pumpkin puree** (either from one can of pureed pumpkin or your own roasted pumpkin). Add **3 eggs**, **1 cup (200 g) dark brown sugar**, and **1 teaspoon (4 g) salt** while still whisking.

Pour into the pie shell and bake 30 to 45 minutes, just until the center no longer sloshes when you wiggle it and the outer edges of the pie are firm. Cool to room temperature and eat. Consume within 3 days.

Serves 6 to 8 as dessert

Winter Radishes Roasted with Dynamic Spices

This dish has a similar process and flavor pattern to Roasted Carrots with Dry and Earthy Spices (page 166), but the spice mixture is much more focused on hitting a mix of savory-earthy, fragrant, and sweet notes.

Winter radishes are big (potato sized) and have the same raw wet hard crunch as their smaller, red cousins. Cook them, though, and you'll get the most silky, soft, succulent roasted vegetable you've ever had. I really like the slightly overbrowned spice crust that forms around the supersoft radishes.

To make the spice mixture, grind together **1½ teaspoons (3 g) cumin seeds**, **2 teaspoons (4 g) ground cinnamon**, **1 teaspoon (3 g) black pepper**, **1½ teaspoons (2 g) cardamom seeds** (from 2 to 3 pods), **1 teaspoon (2 g) freshly grated nutmeg**, and **1 tablespoon (6 g) pimentón** (smoked paprika). Meanwhile, preheat the oven to 425°F.

Wash and lightly peel **2 pounds (900 g) winter radishes** (or try a mix: green varieties, purple ninja, watermelon, red meat or blue meat, some smaller daikons, china rose, etc.). Cut them into rough chunks, about the size of an index and middle finger, bent all the way and scrunched together. I'll often cut the radish into thirds by cutting a wide V through the middle, so you get one middle wedge and two outer wedge shapes, and the irregularity and lack of right angles makes them easier to toss and move around.

Toss the radishes with **2 to 3 tablespoons (30 to 45 ml) halfway decent olive oil** (you can sub the oil, but go for peanut oil over a plain vegetable oil), **2 big pinches kosher salt**, and 1 tablespoon (3 g) of the spice mixture. (Add a little more oil or spices if needed to get a relatively thin, even coating. You're definitely not looking for anything pastelike or encrusted.)

Pour the radish chunks into a single layer on a rimmed baking sheet and roast for 35 to 45 minutes, rotating the baking sheet around 20 minutes in. Flip the radishes during cooking if it looks like the bottoms are cooking too fast and the tops aren't browning—this depends largely on your oven. You're looking for slightly shrunk-down, soft-middled, and golden brown, spiced outsides with some darker brown patches. Remove to a bowl or platter and serve immediately; a squeeze of lemon juice or maybe pomegranate molasses can be lightly drizzled over, at your discretion.

Serves 4 to 6 as a side or 2 as a main

Crushing and Grinding Spices

The waiter with the special task of grinding a comically gigantic pepper mill over each diner's plate of food has been a joke about pretentious restaurants for so long that seeing it coming can feel like lazy writing or an insult comparing the sophistication of your sense of humor to that of a young child—but everyone knows it for a reason. (The reason is, crushing or grinding is a really essential strategy for working with spices.)

Cut into your typical apple or grape, and you'll see that the fruit's color is concentrated in the skin, with hardly any in the flesh. In an apricot or carrot, there's a pretty even distribution of color in the whole thing. A spice's flavor molecules are more like the color of the apricot and carrot, spread pretty evenly throughout the inside and up to the surface and not concentrated in single spots.

To actually get at all the flavor a spice contains, you've got to access the flavors on the inside. Either you can use whole spices, knowing and planning to wait a very long time for these flavors to exude out of the spice and into your food, or you can break the spice open via cracking, crushing, or grinding and free them up exponentially faster.

Cracked and crushed spices work well when you're in a hurry or maybe don't need to care a lot about a very finessed texture. They're also great for including a textural element, and intense, localized pops of spice flavor: think of the crunch and briefly overwhelming flavor of the pebbly peppercorn crust on a steak au poivre, a concept (sticking semi-intact spices to foods as they're cooking) you can extend to meats in steak, roast, and whole-chicken formats, as well as to basically any roastable vegetable.

Grinding spices to a fine powder is good for quickly infusing an entire sauce or dish with spice flavor, and ensuring a smooth texture, like for spice-flavored cakes and pumpkin and other fruit pies, or stewed and braised dishes in the conceptual vicinity of curry.

The ultimate question: Do I grind myself, and when should I grind? Ground spices let their flavors diffuse quickly—but that can be into your food, or into the air. I usually (but certainly not exclusively) buy whole spices, grind and store a few tablespoons at a time, and use them up over a couple of weeks. As grinding tools, I like a bladed coffee grinder for achieving a powder texture and a heavy mortar and pestle for coarse to fine cracking. A powerful blender like a Vitamix or Nutribullet can basically do either (although, watch out, I've seen cumin and other spices pit the plastic on these). Personally, I've given up on ever successfully grinding cinnamon, turmeric, or cloves to a fine powder myself, and either use them cracked or buy them powdered.

Herbal

One of my earliest culinary memories—not just "Wow, a food I like," but "Wow, you really did something to those foods"—is from when I was around four. My chicken soup is too hot, and to cool it off, one of my parents plops in a frozen cube of cilantro pesto, courtesy of *The Silver Palate Cookbook*. It cooled down the soup, but it also permeated the whole thing with fresher than fresh, grassy-fruity, perfumed greenness. I could smell it wafting off the bowl, and it bloomed in my mouth, the small amount of herbs giving the very cooked-tasting soup a dynamism that was like running a finger up the length of a piano keyboard.

I've been pretty obsessed with herbs, their herbal flavors, and all they can do for cooking ever since. Whether they're refreshing, complex, creating a feeling of richness or cutting through stodginess—oregano in a marinade, coriander chutney, sage with pork, a sprig of mint—there's a feeling of vitality running through the whole set.

As you've probably noticed from other mentions in this book, I find herb sauces particularly compelling. Both purely for their eating pleasure—experiencing their transformative enhancements, like that soup-plus-cilantro combination—and also for the opportunities they provide to stretch my legs as a cook and experiment with patterns in flavors.

Consider pesto. Genovese-style pesto is so enrapturing that it's taken on iconic status, with its own professional consortium dedicated to it and an official recipe that calls for PDO (or Protected Designation of Origin, which is a kind of legal authentication for terroir in produce, cheese, and other agricultural products in Europe). The ingredients are Genovese basil, extra-virgin olive oil (preferably with a PDO from Liguria), PDO Parmigiano-Reggiano and pecorino, Mediterranean pine nuts, garlic, and salt.

The imperative to use Genovese basil isn't just culinary jingoism—it's to achieve a very particular flavor profile. Genovese basil is a variety with mostly herbal-resiny, green, floral, and spicy flavors, without the tarragon-licorice note most other basils have. I personally enjoy that flavor, but I also respect the confident statement that level of attention to flavor detail projects.

The original, first-documented recipe for Genovese pesto—from Giovanni Battista Ratto's 1863 book *La Cuciniera Genovese*—is focused a little more on pragmatism. It's pretty similar to the contemporary official one, with a notable difference that I've highlighted:

Take a clove of garlic, basil or *failing that marjoram and parsley*, Dutch cheese and parmesan grated and mixed together and fussy and pounded all in a mortar with a little butter until it is reduced to pulp, then untie it with fine oil in abundance.

Can pesto be Genovese pesto without that basil? I'm interested in Ratto's choice of substitutions—to me, neither parsley nor marjoram tastes *that* much like basil. But if we take a bit of a step back and look at the flavors those herbs layer together, I spot green-vegetal flavors with a little florality and spiciness in parsley and strong, herbal-resiny ones in marjoram. It's like our boy Gio took a step back from basil as a unit, thought about it as a pattern, then all he had to do was go looking for the flavors in that pattern somewhere else.

Recipes are good at what they do. They help define important cultural icons. But we don't need to only eat icons. They're not the only thing that's good. I'd rather cook something that shows off the most delicious and high-quality ingredients I can find than a technically "correct" dish with dull-tasting ingredients. To work like Giovanni, let's take a gander at what patterns we can see in the landscape of herbal flavors.

Mapping Herbal

If spiced flavor trends and varies along a spectrum, softly rolling from earthy to fragrant to sweet, we can spot one like it for herbal flavors—very green and vegetal on one end (the lush, forested coastal region to the west) and highly resinous on the other (the drier, hotter scrubland to the east). Cilantro and parsley (green, vegetal) versus thyme and oregano (resinous). The molecules that make these flavors follow a pattern, too: lots of savory phenylpropenes with the green herbs, tons and tons of intense terpenes in the resinous ones. Right in the middle, we can find herbal flavors with less intense, sweeter terpenes, as well as with some sweet and spicy phenylpropenes sometimes included for good measure: basil, tarragon, spearmint, and the like.

I'm a little dissatisfied by some of the flavor this east-west continuum doesn't explain: the

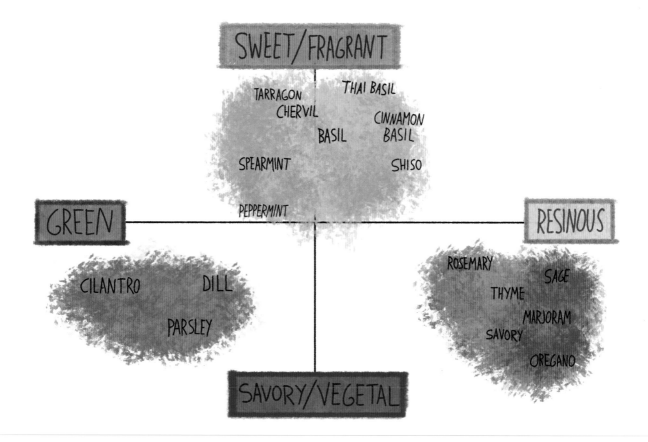

weightiness of parsley compared with cilantro, despite both being green, or the butteriness of sage compared with the uncompromising quality of oregano. So I like to pay some attention to the second, north-to-south dimension of this map. Whether green, resinous, or in between, herbal flavors also follow a pattern of range and mood from heavier and more savory to lighter, headier, and even suggestive of sweetness: savory, vegetal parsley contrasted to lighter cilantro; brawnier, more vegetal peppermint versus lighter and sweeter spearmint; intensely resinous and savory oregano versus lighter, sweeter, still-resinous rosemary.

I like to identify these groupings—rather than run through each herb one by one with less consideration of how their flavors situate them to one another—so I don't have to take on quite such a mental load keeping track of all of them.

If I were unfamiliar with shiso—but can see that it groups with other herbs as herbal, tender, and fragrant—I may not know exactly how it will taste, but its very presence in the group sets me up to anticipate powerful aromas and a mixture of sweet herbaceousness and maybe some spiciness. And shiso, does, in fact, have a powerful aroma, with citrusy-herbal qualities as well as cuminy spicy ones.

Mental maps are also useful for making quick decisions when I off-road from a recipe and start improvising. Picking out herbs near one another, from the same groups, creates more familiarity. Selecting ones further away from each other is a little bolder. And I can even mix them together, balancing or accenting the flavor effects from one group with those from another.

Herbal, Resinous, and Intense

What happens when a plant evolves in a dry, hot, and sunny environment? The ecological stakes that drove plants to develop flavorful terpene molecules in the first place—harsh ultraviolet light, desiccation, hungry grazing animals and insects—are at their highest. Accordingly, the plant absolutely packs itself full of terpenes, and develops a pattern of intense herbaceous flavors with a distinctly resinous edge, like those of oregano, thyme, and rosemary.

The strength of herbs with this flavor profile means they're best used with a light hand, as seasoning: adding them in sprinkles rather than flurries to roasted meats, roasted and sautéed vegetables, raw vegetable salads, and as small additions to sauces like chimichurri or salsa verde. These herbs are good at creating a grounded feeling of earthy-evergreen woodiness, like walking through a pine forest.

That same flavor strength also means they stand up to cooking much better than other, softer herbs, so you can confidently add them right into a pan of hot fat, or at the beginning of a long braising and stewing, without worrying that you're going to destroy all the best parts of them with heat. Most of them have a tough, waxy texture that's less prone to oxidation and wilting than softer herbs, so you can rough them up, chop them really fine, and bruise them between your fingers, to root out all their herbal flavors.

Many of these ingredients were birthed around the Mediterranean, and unsurprisingly are key flavors through lines in Greek, Provencal, and Italian cuisines, to name a few.

Oregano is the Queen of the Night of resinous and intense, with dark, camphoraceous, smoky-medicinal tones from *carvacrol*, and spiced, woody, and piney flavors from *cymene* and *terpinene*. **Thyme** lightens it up a bit with *thymol*, a sweeter, earthier version of carvacrol, plus floral linalool. **Savory** is actually two species, lighter summer savory and heavier winter savory, which have many of the same flavor molecules as thyme and oregano, and a flavor also roughly between the two. **Marjoram** is kind of a domesticated oregano, spicy-sweet with a cleaner eucalyptus-camphor flavor. **Mexican oregano** shares oregano's deep carvacrol and woody-citrusy cymene,

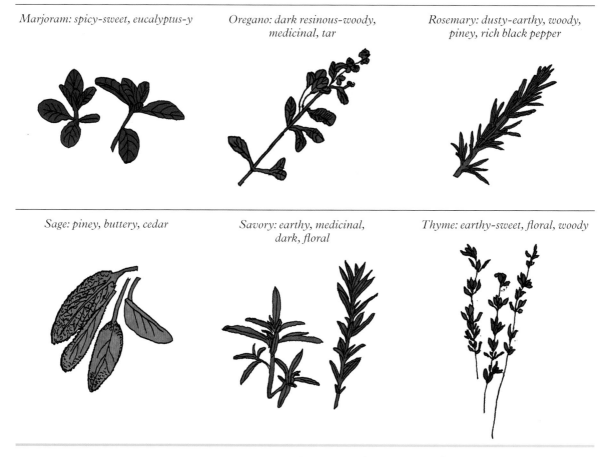

Marjoram: spicy-sweet, eucalyptus-y

Oregano: dark resinous-woody, medicinal, tar

Rosemary: dusty-earthy, woody, piney, rich black pepper

Sage: piney, buttery, cedar

Savory: earthy, medicinal, dark, floral

Thyme: earthy-sweet, floral, woody

Easy Ways to Get into Resinous Herbs

- Resinous herbs are just smashing with meaty flavors, which includes meat, but also mushrooms. Sauté some mushrooms in generous amounts of nice oil or butter, and throw in a few thyme sprigs from the beginning (maybe adding a section of rosemary branch and a smashed clove of garlic). The thyme leaves will permeate both the butter and the mushrooms, and you can just pull the little branchlets right out, as most of the leaves will pop right off during cooking. Rosemary, sage, or other larger leaves may not break down as much, but they'll infuse it wonderfully.

- Resinous herbs, if you're careful with them, are wonderful complements to fruit—in jam, a compote, or in juice, or as a superlight dusting of very finely chopped leaves right on fresh fruit. Thyme and nectarine, plum or grapefruit; rosemary and apricot or blackberry; sage with apple, pear, or pineapple; oregano and strawberry or cherry.

- These are the herbs that withstand drying with the most flavors intact—if I have more oregano, thyme, savory, or marjoram than I can use, I'll lay the sprigs out on a wire rack over a baking sheet and let them dry out for two or three days, then bag them for infinitely better herb flavor than any dried herbs you can buy in a store.

- Try thyme, summer savory, or marjoram on white bread with mayonnaise as a resinously herbed tomato sandwich.

plus fresh-resinous *eucalyptol*; it's much more closely related to verbena than oregano, as is Ethiopian **koseret**, which layers lots of floral *linalool* and lemony *geranial* over a thyme-oregano base. **Epazote** has an earthy, piney, creosote-resinous flavor that mellows out beautifully when cooked for a long time, especially with beans. **Sage** has a eucalyptusy resin note (with little of oregano or thyme's deep medicinal base layer) from *pinene*, *eucalyptol*, and *camphor*, with a sweetish, almost buttery cedar flavor from *thujone*. **Rosemary** has a pleasantly dusty-woody-earthy character from *borneol*, plus freshness from clean-piney pinene and eucalyptol, and a warm, rich, black pepper note from *rotundone*.

Herbal, Tender, and Fragrant

Herbal flavors can get intense and resinous and they can be vegetal and green, but in between, there's a distinct pattern of herbs that are simultaneously rich, nuanced, and fresh. Herbs like basil, spearmint, and shiso combine floral, resinous, woody, citrusy terpenes and often add choice softer, gentler, sweeter phenylpropenes to the mix. They lay a flavor of creamy-sweet-floral and spicy-refreshing fragrant complexity over anything you add them to. They're herbal, tender, and fragrant.

Genovese basil, the star of by-the-book pesto, gets to its iconic flavor profile (a spicy-woody core, slighty vegetal and resinous, wrapped in sweet floral and citrusy notes) with terpenes like eucalyptusy, resinous-minty *cineole*, floral-sweet *linalool*, and green-citrusy-woody *ocimene*; it also has the warm, clove-flavored phenylpropene *eugenol*. **Sweet basil** (and basils labeled simply "basil") adds a distinct licorice-anise-fennel flavor to the mix, from the phenylpropenes *estragole* and *anethole*. (Estragole is also the primary flavor in **tarragon**, as well as in lighter, more delicate **chervil**.) Other basil varieties amplify one of sweet basil's flavor dimensions: **Thai basil** goes all in on

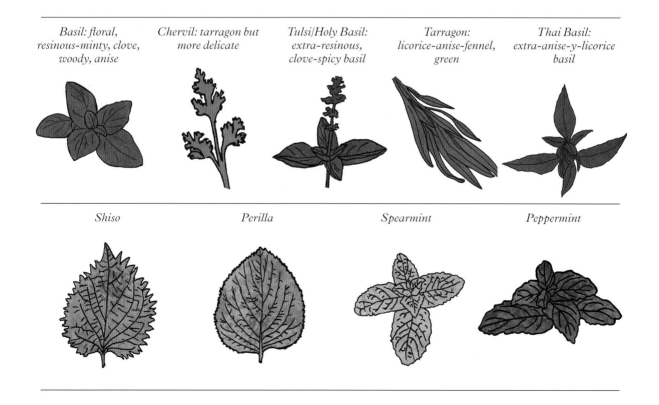

Basil: floral, resinous-minty, clove, woody, anise

Chervil: tarragon but more delicate

Tulsi/Holy Basil: extra-resinous, clove-spicy basil

Tarragon: licorice-anise-fennel, green

Thai Basil: extra-anise-y-licorice basil

Shiso

Perilla

Spearmint

Peppermint

licorice-anise flavors, **holy basil** (or tulsi) on resinous, clove, and spiced flavors, and **cinnamon basil** gets its spiced note of cinnamon from cinnamon's own key aroma molecule *cinnamaldehyde*, which is chemically close to the phenylpropenes, though not actually one itself.

Shiso is a botanical cousin to basil, and along with sister varieties of **perilla**, is like a reinterpretation of basil's sweet, spicy-grounded pattern of flavors in a different musical key. Terpenes like *perillaldehyde* give shiso cuminy spiced notes, while *caryophyllene* provides dry and woodlike ones. Aldehyde molecules like *hexenal* and *phenylacetaldehyde*, give it notable cucumber-green and floral-honeysuckle flavors, respectively.

Mints are uniquely, for lack of a better word, minty. Lighter and sweeter **spearmint** gets this from the terpene *carvone*, which comes in two mirror-image versions (like your left hand is a mirror image of your right); spearmint's is minty and fresh with faint malty caraway notes. *Limonene* gives spearmint light citrusy-orange flavors. Other types of mint are less sweet and more intense, which can feel deeply resinous and vegetal-green at the same time. **Peppermint**'s more savory mintiness comes from intense, cooling *menthol* and herbal-resinous *isopulegone*, both terpenes. Minty aroma molecules don't just suggest cold and refreshing; they interact with your cold-sensing touch receptors and create a real sensation of coolness.

Herbs that follow this pattern are pretty physically tender, with larger, floppier leaves you'll want to apply to dishes whole, roughly chopped and carpeted, or even in a thick paste, rather than the carefully doled-out flecks that befit stronger, more brooding sage or oregano.

Using them very fresh, handling them gently, and combining them with other ingredients shortly before serving show off their best qualities, especially their ability to create that fresh-and-rich cloud of flavor around a dish. It's there in a plate of hot, pesto-sauced pasta or a carefully garnished Neapolitan pizza, basil creating a flavor like a cloud of flowers dabbed with clove, sweet citrus, and just a hint of eucalyptus over it. When a shiso leaf wraps around a piece of fish sashimi or infuses fermented plum umeboshi, it imparts a clarifying fresh-bitter mintiness but also layers of rich dry cumin, soft lemon, and racy, musky, woody-spiciness. Spearmint makes a mint julep cooling and fresh, but also has a soft orangey-malty quality that lifts up the inherent vanilla-sweetness of bourbon's barrel-aged profile.

Mixed into vegetal dishes like salads, they create bursts of incandescence: you'll get mostly green-lettuce flavors as you eat, then a sudden swoon of minty, anise, lemon-cumin, or floral-spicy as you bite into an herb leaf. Besides stand-offish peppermint, they're all extremely welcome folded into hot grains and starches, like rice, earthy grains, or potatoes, or broths, soups, and stews, right before serving.

Try This

The marriage of tomato and basil is so long-standing, it would be cliché if it weren't so delicious. Tomato definitely has A Type, and I have yet to meet a fragrant herb that doesn't pair smashingly with it, taking it in wonderfully different directions. Tarragon, shiso, holy basil, a little bit of fresh spearmint . . . select whatever bunch of herbs looks best, roughly tear the leaves and mosaic them on top of an excellent sliced tomato, and you'll have a great time.

Herbal, Soft, and Green

"Green" is mostly a metaphorical way to describe flavors, but as we learned in "Vegetal" (page 144), it's incredibly evocative: cut grass, underripe banana, bell pepper. On the opposite end of the herbal flavor spectrum from resinous, the greenest, most vegetal-fresh herbal flavors are highest in members of the carrot family: dill, parsley, and cilantro. Like chartreuse, forest, and British racing green, they blend different components to arrive at their own distinct shades.

Via my second-generation-Eastern-European grandparents' house, my core flavor memories of **dill** are savory and paired with salmon, sour cream, potatoes, and cucumbers. Dill's green-plus-fragrant, savory-sweet, herb-with-a-thousand-faces routine layers vegetal phenylpropenes with terpenes, a feature of many green-herbal flavors. Dill has the aptly named *dillapiole*, green-woody and spicy, as well as warm-nutmeggy *myristicin*. More greenness and some minty qualities come in with the terpenes *alpha-phellandrene* and *carvone*, rounded out by citrusy-sweet *limonene*. **Parsley**'s dialed-down herbaceousness makes it the herb most likely to act like a leafy vegetable: combined in nearly equal volume with couscous to make refreshing tabbouleh, and sometimes as its own leafy salad, to freshen up something very heavy like roasted bone marrow. Parsley shares warm-spicy myristicin with dill, plus vegetal *apiole* (both phenylpropenes). Terpenes like *beta-phellandrene* enhance its greenness and *terpinolene* gives it a fresh, limey, and woody quality. A last layer of (fruity, waxy) green comes from the aldehyde *decenal*, also found in coriander seeds. Aldehydes are crucial for understanding **cilantro**, the greenest (in a grassy-top-note kind of way) herbal flavor. Plants make aldehydes by enzymatically snipping apart fat molecules, creating green-to-citrusy flavors that can sometimes run mushroomy, soapy, or metallic. This soapy-metallic note makes cilantro off-putting for a large minority of eaters, who tend to have a lot more copies of a smell receptor that senses aldehydes and increases sensitivity to the flavor. There's also an element of learned associations at play—soap can also have aldehydes, made from snipping fats just like plant aldehydes, and if your experience of aldehydey flavors is primarily from soap, it can unpleasantly color your experience of aldehydes in cuisine. Cilantro makes lots of (confusingly similarly named, thanks to rules chemists love and everyone else hates) aldehydes *decanal*, *decenal*, and *dodecanal*, which its fans find deliciously green, floral, orange-peely, and creamy-melony—the latter pair, of course, because orange and melons also make similar aldehydes.

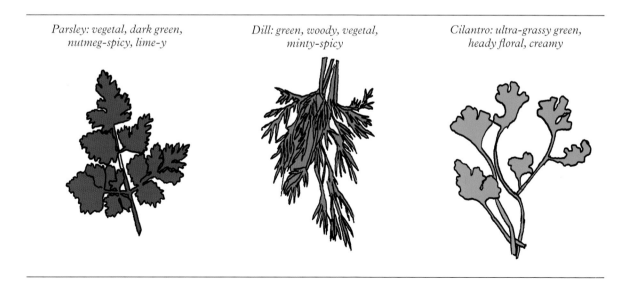

Parsley: vegetal, dark green, nutmeg-spicy, lime-y

Dill: green, woody, vegetal, minty-spicy

Cilantro: ultra-grassy green, heady floral, creamy

The Flight of Flavor Molecules

Fresh Thai basil leaves served alongside a bowl of hot, beefy pho is flavor as smell at its most exemplary. Rather than add them to the soup in the kitchen, the task is saved for you to do yourself at the last second. And once you do, you're enveloped in a cloud of sweet, spicy-anise aroma that lingers around each bite. All that flavor comes from smell molecules, flighty and light and happy to float up in gaseous form from the food they're in, and into your nasal cavity for you to smell.

Chemists call this now-I'm-chilling-like-a-solid-or-liquid-now-I'm-a-gas act *volatility*, and volatility is something you can harness if you know a little about how molecules behave.

When herbs meet soup, there's a mild repulsion between the watery soup and the herb's smellable molecules. Smell molecules are built a bit like a small, light version of an oil molecule, and they're similarly tricky to mix with water. Oil rises to float on top of water when you try to mix them; smell molecules do the same, and then continue rising in the form of a vapor.

You can send a burst of flavor flying from aromatic ingredients if you mix (or spritz) them with water, or something mostly water-based, like broth or tea. After dinner, you can do the same thing to drinks: adding a little spring water to scotch is not just an affectation—it actually loosens up the aroma molecules and makes them more smellable.

A second demonstration by basil-plus-pho: volatility also increases with temperature. You can make flavors more smellable, and more intense, by heating them up. It's a trick best used in the moments before eating, since you're effectively boiling off flavor molecules into the air. Eventually, you'll run out, and the flavor will be much more muted. This goes for some ingredients we think of as primarily taste-based: adding lemon juice to something and then continuing to heat it makes its sourness more muted, dull, and flat—but you're not losing sourness, you're losing lemony flavor from aroma molecules.

Advanced Herbal Flavors: Fragrant and Intense

Some herbs can be so intensely flavored that they push the limit of "herbal, delicate, and fragrant" right to the edge. Peppermint is aggressively fresh and minty, and Thai basil is extravagantly anise-licoricey. But they're not really resinous or piney enough to admit them into the circle of oregano and sage. Their intensity means they're usually used as a smaller garnish than other fragrant herbs—think of that handful of leaves to drop into pho, rather than thick applications of pesto. A few herbs are even more intense examples of this pattern—fragrant, highly flavored, good when used fresh and sparingly. Succulent **rice paddy herb** (or **ngo om**) is used similarly as a finishing herb, with intense lemon-dill-cumin flavors; it's delicious with fish, salads, and grains like rice. **Pápalo** is to tacos as thai basil is to pho—a final, optional garnish of a few leaves that releases lots of aroma when it contacts hot food. Its flavor is a mix of heady floral-grassy notes and heavier, woody, spicy, earthy-sweet ones. It's in the Aster family, along with tarragon and marigolds. Some younger or smaller **marigold leaves** often go by their Latin name *tagetes* when used as herbs, and show off a heady, orangey, grassy flavor even in very small amounts—a few sprigs on a beet salad, or with berries. *Tagetes minuta* or **huacatay** (sometimes called Peruvian black mint) adds flavors of, yes, mint, as well as licorice. **Lemon verbena** does the same in an intense, lemongrass-lemon-peely way.

Rice Paddy Herb (Ngo Om)

Easy Ways for Herbal, Soft, and Green Things

- Garnish with them so liberally, they're almost a leafy vegetable: a handful of chopped green herbs is never unwelcome mixed into rice or other grains like couscous or farro, approximately like tabbouleh; or vegetables like corn off the cob, boiled potatoes, avocado, squash and its melon relatives, or on top of a dressed salad.
- Add green herbs (chopped) to hot foods just before you eat them, to wilt them slightly and create a cloud of fresh aroma: hot chicken soup with dill or cilantro, or braised beef or white beans with parsley.
- Make an herb sauce (see page 180), starting with Cilantro Pesto (recipe follows).

Cilantro Pesto

This is my rendition of the cilantro pesto of my childhood. It's still just as good frozen into cubes and plopped into hot chicken or tomato soup, or sauced onto pasta or roasted vegetables.

In a blender at high speed, combine **1 bunch (about 90 g) cilantro** (washed well and dried, very thick stems removed but smaller stems included), **¾ cup (180 ml) light and grassy olive oil**, a **scant ½ teaspoon (1 g) of lemon zest**, **½ teaspoon (3 g) salt**, and a **half clove garlic**.

Push down the cilantro with a spatula, as necessary, to fully combine and break the leaves down into a smooth paste. Continue blending to fully emulsify. Add a little extra olive oil, bit by bit, if it seems more like a too-chunky paste than a thick fluid one. Turn off the blender. If storing, wait to add lemon juice to preserve the color.

Otherwise, stir in **1 tablespoon (15 ml) lemon juice** (from roughly half a lemon) to combine. Taste and add a little more salt or lemon juice, as needed.

Makes roughly 1 to 1½ cups. Use within 2 to 3 days, or freeze into cubes and store frozen in a heavy zip-top bag with all the air squeezed out, using within 3 to 4 months.

Herb Sauces

If there's one takeaway from herbal flavors, it's put more herbs on things (as in "bigger handfuls of herbs," "more kinds of herbs," and "more things that you're currently applying herbs to"). If there are two takeaways, the second one is: make herb sauces. They're an incredibly easy way to add a lot of sophisticated flavors to your food, they're very forgiving, and they're a microcosm of flavor as patterns and cooking as improvisation. Your food will be delicious, it will be hard to mess up, and you'll ease yourself into the kind of abstract thinking that's invaluable cooking and improvising intuitively.

Rolling back the clock on pesto even further than 1863, we encounter one of its deep ancestors, a Roman condiment called *moretum*, a sauce with an herb-cheese-garlic backbone and other herbal flavors layered on top. And variations on these layers abound. When he wasn't writing *Iliad* fanfiction or posthumously hanging out with Dante, the poet Virgil found some time to blog about herb sauce in verse form. His recipe involves garlic bulbs, parsley, salt, hard salty cheese, olive oil, a little vinegar (and the herb rue, which can be pretty toxic, so please don't cook with it), and a nice little flourish of coriander seeds. The Roman agronomist Columella went for something flashier in his moretum, with savory, mint, rue, coriander, parsley, chives, green onion leaves, arugula, thyme, nepeta (catmint), fresh or salty cheese, and pepper, ground together in a mortar and topped with oil, which sounds like a damn fine improvisation on the pattern of "green and resinous-herbal" that basil or parsley-plus-marjoram also carry off.

Back in the present, a census of herb sauces turns up some pretty delicious diversity. Check out the **salsa verde** of Emilia-Romagna, designed (along with mostarda) to go with the meat-heavy stew bollito misto: often parsley, garlic, capers, anchovy, olive oil, vinegar, and mustard. French **pistou** keeps Genovese pesto's basil, garlic, and olive oil, but minimalistically skips nuts or cheese, like taking off one piece of jewelry before you leave the house. Drier and more minimalist still, **gremolata** and **persillade** use finely chopped parsley and garlic, often along with citrus zest. In very German fashion, Frankfurter **Grune Sosse**, often eaten on new potatoes or fish, specifically calls for seven particular herbs: sorrel, chervil, chives, parsley, cress, salad burnet, and borage.

And instead of the cheese and oil in pesto or *moretum*, it relies on crème fraîche or another fairly loose cultured dairy, and a sieved hard-boiled egg.

Argentine/Uruguayan **chimichurri**, echoing Giovanni Battista Ratto's marjoram and parsley combo, combines parsley, oregano, garlic, olive oil, and vinegar, with added piquancy from chile. North African **chermoula** brings cumin into the mix, along with cilantro, garlic, lemon, and oil, and sometimes preserved lemons, saffron, chiles, paprika, parsley, or anchovylike salted fish. Yemeni **sahawiq** (also known as *bisbas* or *zhug)* is particularly chile-forward, and usually includes cilantro, garlic, and black cumin, sometimes with tomatoes, sardines, caraway, or cheese.

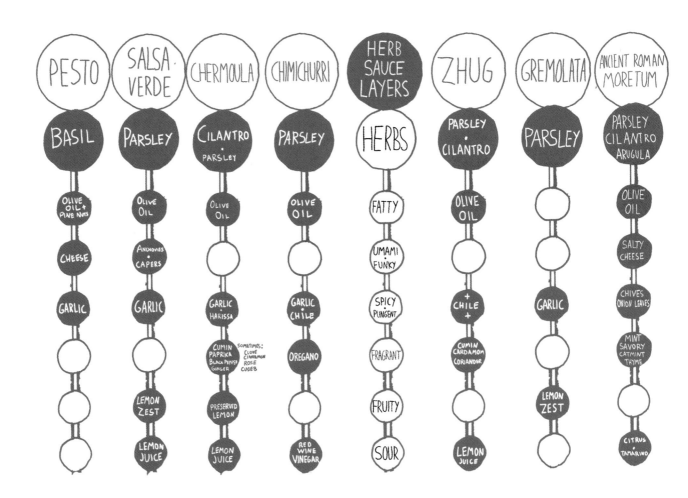

Surveying all of them, I'm noticing some patterns beyond "a sauce with herbs in it." Those herbs—green ones like parsley or cilantro, or a fragrant one like basil—form the core, along with fat in some form. Additional layers ripple out from there: something pungent and/or spicy like garlic or chile. Something with stronger aromatics, like cumin, lemon, or resinous oregano. Something a little umami-funky, like cheese or anchovies. Something sour, like vinegar or lemon juice. There's enough flexibility that some layers are absent entirely in several sauces, and still manage to taste great.

Pick out the framework (herbs-fat-pungent-umami-sour), and you have the pattern for salsa verde, which slots in parsley, oil, garlic and capers, anchovy, and vinegar. Chermoula works similarly, with cilantro and cumin, oil, garlic, and preserved lemon (and sometimes saffron, chilies, paprika, and salted fish). But you could just as happily slot in mint and tarragon alongside the parsley, lose the parsley entirely, or augment the vinegar with a kick of mustard.

Cilantro-Parsley-Tarragon Salsa Verde

I've been making this salsa verde for fifteen years, and I love how the licoriceness of the tarragon melds with the cilantro and parsley. It is really lovely served to dress a meaty, seared maitake mushroom.

In a powerful blender, combine **1 bunch (about 90 g) cilantro (**washed well and dried, large stems removed), **1 bunch (90 g) flat-leaf parsley (**washed, dried, and large and medium stems removed), **1 small bunch (10 g) tarragon (**leaves plucked from stems and stems discarded), **1 cup (240 ml) light and grassy extra-virgin olive oil, 1 tablespoon (15 g) Dijon mustard, 2 cured anchovies, 1 tablespoon (12 g) capers, 1 tablespoon (15 ml) lemon juice (**from half a lemon), and **1 peeled clove garlic**, followed by **1 teaspoon (6 g) of sea salt**, and **½ teaspoon (1 g) ground black pepper** (honestly, I usually eyeball this and adjust it at the end). Blend at medium-high speed until combined, with all herb leaves broken down and emulsified in. It should be a thick-bodied, medium-fluid sauce; add more olive oil gradually to thin out the texture if necessary. Taste and add more salt, black pepper, or lemon juice if necessary. Store for 2 to 3 days in the fridge or up to 3 to 4 months frozen.

Makes about 2 cups

Seared Maitake Mushroom with Cilantro-Parsley-Tarragon Salsa Verde

Mushrooms, though we eat them like vegetables, aren't plants at all, they're fungi. (Yes, obviously.) But what's cool about that is vegetables (and fruit) have cell walls held together by pectin, which softens and breaks down with cooking. Fungi have cell walls held together with chitin, which doesn't melt down with normal cooking. So, cooking a mushroom will collapse the many air voids inside of it, making its texture softer and denser, but the mushroom won't ever really break down to mush the way a very long-cooked vegetable will. This is why mushrooms have a sort of meatlike textural quality.

For this application, look for **2 heavy clusters of mushrooms, around ½ pound (225 g)** each. Substitute oyster or king oyster mushroom clusters if they look nicer and fresher than maitakes.

Heat two heavy-bottomed skillets or sauté pans (stainless, copper, or cast-iron will all work) over medium-high heat. When hot, add **2 tablespoons (30 ml) olive oil** to each and sprinkle the bottom of each pan with **a pinch (1 gram) salt**. Place **1 maitake mushroom cluster**, bushy top down, into the hot olive oil in each pan, and lightly press down (use a spatula, or a purpose-built weight like a light grill press or a Chef's Press, for this).

Sear for 3 to 4 minutes, until most of each mushroom's "petals" are brown and crispy. Remove any weight, flip, stem side down, and continue cooking with a little pressure on the maitake until the base is similarly seared and the mushroom is cooked and tender all the way through, 3 to 5 minutes. Remove the mushrooms from the pans and drain on paper towels. Sprinkle with **flaky salt or sea salt** and freshly ground **black pepper**.

To serve, spoon ¼ **cup cilantro-parsley-tarragon salsa verde** onto two plates, and settle the stem end of the mushroom into the sauce. Serve immediately.

Serves 2 to 4 as a starter or 2 as a
light main

Making Your Own Herb Sauce

You don't necessarily need to improvise every ingredient in a sauce; try starting conservatively, adding in or swapping one or two that strike your fancy, and see what happens. Here's some options I think have potential.

Herbs

Classics: Basil, parsley, cilantro.
Improvisations: Other members of their flavor families (green or fragrant and delicate herbs), such as spearmint, dill, tarragon, shiso.
Mining the vegetative vein a little further: Arugula, fennel, very tender carrot tops, green garlic, scallions, and lightly cooked leeks are also great.
Adding dimension with more intense herbs: Consider them as an aromatic accent to another, lighter herb, or use them for a very concentrated seasoning paste, as you might marinate porchetta in sage, rosemary, and many cloves of garlic—rosemary, thyme, oregano, sage, lemon balm, tagetes/marigold leaves, chervil, pápalo, lemon verbena.

Fat

Classic: Olive oil, pine nuts.
Improvisations: Nut oils, or double down and use both oil and whole nuts: pistachios, walnuts, hazelnuts, almonds, tahini, sunflower seeds. Animal fats: cream, yogurt, crème fraîche; schmaltz, lardo, or brown butter if you keep the sauce warm.

Sour

Classic: Wine vinegar, lemon.
Improvisations: Sour dried fruits (like sour cherries or apricots); very tangy fruits like passion fruit, underripe stone fruit, orange, grapefruit, or apple; lacto-fermented things like pickle brine, umeboshi, yogurt whey, chopped kimchi; preserved lemon.

Fragrant

Classic: A small enhancement with cumin, marjoram, or oregano.
Improvisations: Lavender, rose, citrus peel, cinnamon, allspice, fennel seeds, ginger, cloves, cardamom, pink pepper, caraway, rosemary.

Piquant

Classic: Garlic, chiles.
Improvisations: Raw or fried onions or shallots, fresh or smoked chiles, ginger, Sichuan pepper, mustard seeds.

Umami-Funky

Classics: Parmesan or other hard cheese, anchovies.
Improvisations: Aged seafood, like dried scallop or bottarga; fermented soy like miso, soy sauce, or salted black beans; aged meats

and charcuterie like prosciutto scraps, nduja, coppa.

Mining the umami-funky vein a bit further: Try raw seafood, like oysters or scallops. I leave this entirely up to your own risk tolerance and relationship with your fishmonger.

Fragrant Miso-Herb Paste

This herby paste contains many of the same flavor patterns as Genovese pesto: fragrant herbaceousness, salt, umami, piquant garlic, and nuttiness. Some of these patterns are recast: cilantro and tarragon pump up the green and anise notes; tahini throws its weight around a bit more than the usual pine nut; lemon juice adds a new zip of acidity, and miso a different dimension of fermented funk and umami than its distant cousin Parmesan.

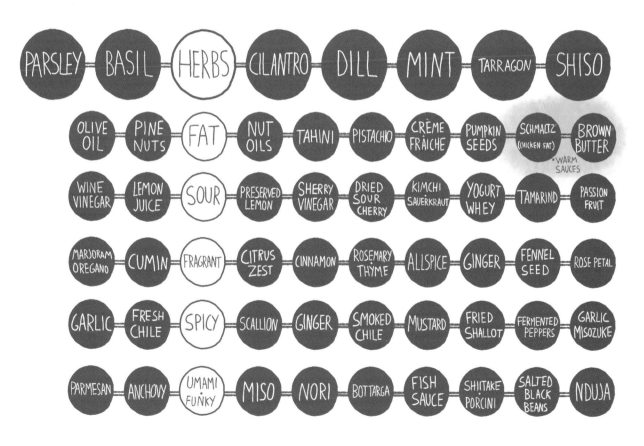

IMROVISATIONS ON HERB SAUCES LAYER BY LAYER

Toss with brown rice or roasted root vegetables (carrots, potatoes, sweet potatoes, etc.), or use as a condiment with fish or shellfish.

In a blender or mortar and pestle, combine **1 cup (12 g) lightly packed basil leaves, ¾ cup (9 g) lightly packed cilantro leaves and upper stems, ¼ cup (7 g) lightly packed tarragon leaves, 1 clove garlic, ⅓ cup (100 g) Meyer Lemon Miso** (page 276, or substitute ⅓ cup mellow, store-bought miso and the zest of 1 lemon), **2 teaspoons (10 g) tahini, ½ cup (113 g) grapeseed oil,** and **1 tablespoon (15 g) lemon juice**. Blend until smooth. Add more oil if necessary for a smooth texture. Taste to adjust salt or lemon juice. Store tightly covered in the fridge for up to 3 or 4 days.

Makes about 2 cups

Free-Associating on Herb Sauces

Shuffling through my mental Pantone, I went looking for flavors a little further afield that still fit an herb sauce pattern. I like a salsa verde with multiple herbal notes; I like the idea of funky vegetable-based additions playing a similar role to the funky cheese in pesto. Thinking about the pine nuts in pesto led to thinking about how nuts play a similar textural/flavor role in many moles, like mole negro, as well as with red peppers—hazelnuts in romesco sauce, walnuts in muhammara—and how well tangy fruits (raisins and other fruits, in many moles; tomatoes in romesco; pomegranate molasses in muhammara) work in both or all of them. Then *that* made me think of caper-raisin sauce (circling back to the funky element), which is classic on scallops or cauliflower.

So, I thought, "Let's do an herb paste where there's several herbs, there's a nut or seed component, there's a tangy component from dried fruit and a little complementary vinegar, there's garlic, and we get a funky-salty element from some kind of plant." Based on what I had around, I just felt it out and adjusted until it balanced.

Apricot, Olive, and Pumpkin Seed Herb Sauce

This is great with grilled fish, with simply cooked beans or sticky meat like short ribs or pork shoulder. You can also thin it out a bit and drizzle over charred broccoli or cauliflower.

A fruity or grassy olive oil works well here (as opposed to a more spicy or heavy one). If you can source it, consider substituting a nice pumpkin seed oil for some or all of the olive oil.

In a powerful blender, combine **a scant ½ cup (118 ml) good extra-virgin olive oil, 1¼ cups (15 g) flat-leaf parsley, 1⅔ cups (20 g) basil, 1½ teaspoons (1 g) rosemary leaves, 2 tablespoons (20 g) pumpkin seed, 3 tablespoons (30 g) dried apricots, 2 teaspoons (5 g) garlic, 1 heaping tablespoon (14 g) Castelvetrano olives** (or other high-quality, mellow olives), **½ teaspoon (2 ml) rice vinegar,** and **a small pinch (1 g) of sea salt**. Blend at high speed until all the solid ingredients are broken down and a smooth paste forms. If the paste is too thick to blend, gradually add olive oil bit by bit to loosen it up.

Taste, and adjust vinegar and salt to your liking. Store in the fridge for up to 3 or 4 days.

Makes about 1 cup

Dill, Cherry, and Pistachio Herb Sauce

Starting from the same basic pattern and tweaking components as in the previous recipe, we can get a totally different recipe that still echoes that pattern. To really hype up the flavor layers, feel free to substitute pistachio oil for some or all of the olive oil.

In a powerful blender, combine **a scant ½ cup (118 ml) good extra-virgin olive oil**, **⅓ cup (10 g) dill**, **1 teaspoon (1 g) thyme leaves**, **¾ cup (15 g) parsley**, **1 tablespoon plus 1 teaspoon (12 g) dried sour cherries**, **1½ tablespoons (14 g) pistachios**, **1½ teaspoons (4 g) garlic**, **1½ teaspoons (6 g) capers**, **½ teaspoon (2 ml) sherry vinegar**, and **a small pinch (1 g) of sea salt**. Blend at high speed until all the solid ingredients are broken down and a smooth paste forms. If the paste is too thick to blend, gradually add olive oil bit by bit to loosen it up.

Taste, and adjust vinegar and salt to your liking. Store in the fridge for up to 3 or 4 days.

Makes about ¾ cup

Meaty: Where Does the Flavor of a Steak Come From?

Most raw meat has a pretty subtle or delicate flavor, and while that can be really delicious—as beef tartare and carpaccio, lamb kibbeh nayeh, sashimi, and sushi all attest to—it's not "meaty" in the big way cooked meat is. Obviously, roasted chicken or seared beef have a lot of flavor from the browning process they go through—check out "A Whole Lot of Browning: The Maillard Reaction" (page 244) for more on that—but even meat cooked without any browning at all (poached chicken, for example) has a different, more intense flavor than raw meat. Something in raw meat, and something about the way we handle it, come together to unlock meaty flavors.

In the 1950s, meat scientists were also curious about this and decided to chase down the secret of meatiness once and for all. They combined techniques like blending, extracting with water, and filtering to separate all the kinds of molecules in a steak into categories with similar chemical properties. They ended up with a pile of washed-out muscle fibers, and what washed out from those muscle fibers was a kind of unpromising, slightly bloody-smelling clear pink liquid. Before you could say "beef fiber patty," they decided to try out just cooking these two components. Would either of them develop meaty flavor? Would both? Did they need to be cooked together?

They formed the gray, mushy pile of meat fiber proteins into a little gray meatball, and cooked it on a griddle. It gave a sad showing, not browning at all, and not developing any flavor—it just shrank and desiccated. Then they tried cooking some of the pink, water-soluble liquid they had collected. Even though there was no visible "beef" in it, just stuff *from* beef that had dissolved in water, as it heated up, boiled, and started reducing, it turned a little brownish, and conjured the smell of consommé. As they cooked it down, it smelled more and more like beef broth. When they boiled off all the water, the material that remained started smelling a bit roasted, and then uncannily like a juicy, well-seared steak.

Meats follow a pattern: most of meat, by weight, is a flavorless but texturally interesting sponge, soaked full of water (delicious, I know) with a relatively tiny amount of matter dissolved in it that actually makes meat, well, meaty.

Meat is greater than the sum of its parts—these meaty-flavored components only became obvious after combining with heat, which activates other types of less flavorful molecules dissolved in the raw meat. Meat is also way more delicious than a pile of gray and tasteless fibers and a little bit of flavored goop—splitting it apart to make it simpler helps explain it, but the full appeal of meaty doesn't really come into focus unless they're together, the fibers giving structure and texture to the flavors.

Facilitating Meaty Molecules

More than other flavors coming from smell, selecting for meaty flavor requires some inference on your part, and some stewarding to fully activate. A chemist might describe raw meat as being full of "precursors"—molecules that are special not so much for what they currently are, but for what they can be transformed into under the right conditions. You can't taste or smell the meaty flavors you might ultimately achieve in a piece of raw meat. But if you understand what they come from and how they're made, you can look for the signs—color, water content, information on breed, age, and diet—then make sure you're handling the meat in a way that maximizes them.

Besides the fibrous proteins that give meat structure, the stuff that soaks its spongelike structure, the meat juice, also has proteins in it. One of them is *myoglobin*, which, like similarly named hemoglobin, carries oxygen. Hemoglobin carries oxygen for blood, myoglobin carries it within muscles. It holds on to oxygen with a little atom of iron at its center. When they get hot enough, the iron and the oxygen from myoglobin work together to chew up *fatty acids* from the muscle cells' membranes, releasing smaller, smellable molecules. Dark meat has more myoglobin (that's why it's dark!) than white meat, which means more flavor-creating reactions during cooking.

During the same heating process, enzymes in the meat juice break down other floating proteins and amino acids into similarly smaller, smellable molecules. There are two amino acids made with sulfur—*methionine* and *cysteine*—and when that sulfur gets carried over into these smellable molecules, it does the same thing it does for flavor molecules in garlic and tropical fruits: makes them super, intensely smellable, so a really small amount makes a really meaty flavor.

Exactly how these reactions shake out depends on how a given animal made its muscles and what it filled them with. Species matters—which is why chicken, beef, and lamb have different flavors—but so does diet. Vegetation is turned into somewhat different fatty acids than grains, which iron and oxygen break down differently, which is why pastured or grass-fed meat tastes different than corn-fed. The animal's fat deposits, in and around the muscle, contribute to flavor, too—they can break down into flavors, but they also act as a reservoir for flavors and precursors that aren't so water-soluble, including (since you are what you eat) plant aroma molecules from the animal's diet.

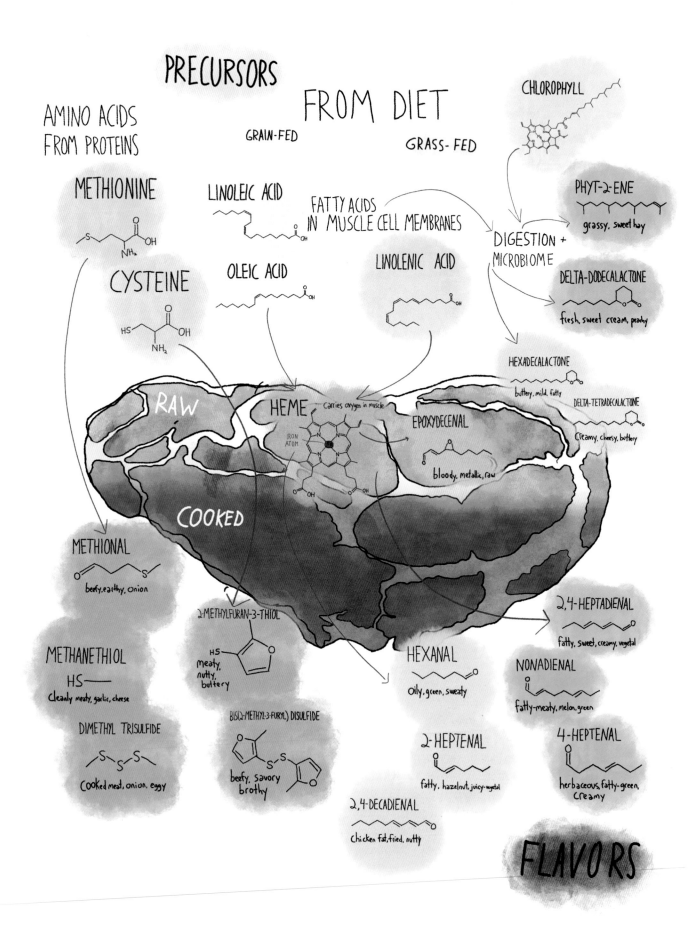

PRECURSORS

FROM DIET

AMINO ACIDS
FROM PROTEINS

GRAIN-FED

GRASS-FED

CHLOROPHYLL

METHIONINE

LINOLEIC ACID

FATTY ACIDS
IN MUSCLE CELL MEMBRANES

DIGESTION +
MICROBIOME

PHYT-2-ENE

grassy, sweet hay

CYSTEINE

OLEIC ACID

LINOLENIC ACID

DELTA-DODECALACTONE

fresh, sweet cream, peachy

HEXADECALACTONE

buttery, mild, fatty

DELTA-TETRADECALACTONE

Creamy, cheesy, buttery

RAW

HEME   Carries oxygen in muscle

IRON
ATOM

EPOXYDECENAL

bloody, metallic, raw

COOKED

METHIONAL

beefy, earthy, onion

2-METHYLFURAN-3-THIOL

HS

meaty,
nutty,
buttery

HEXANAL

Oily, green, sweaty

2,4-HEPTADIENAL

fatty, sweet, creamy, vegetal

NONADIENAL

fatty-meaty, melon, green

METHANETHIOL

HS—

Cleanly meaty, garlic, cheese

DIMETHYL TRISULFIDE

S—S—S

Cooked meat, onion, eggy

BIS(2-METHYL-3-FURYL) DISULFIDE

beefy, savory
brothy

2-HEPTENAL

fatty, hazelnut, juicy-vegetal

4-HEPTENAL

herbaceous, fatty-green,
creamy

2,4-DECADIENAL

chicken fat, fried, nutty

FLAVORS

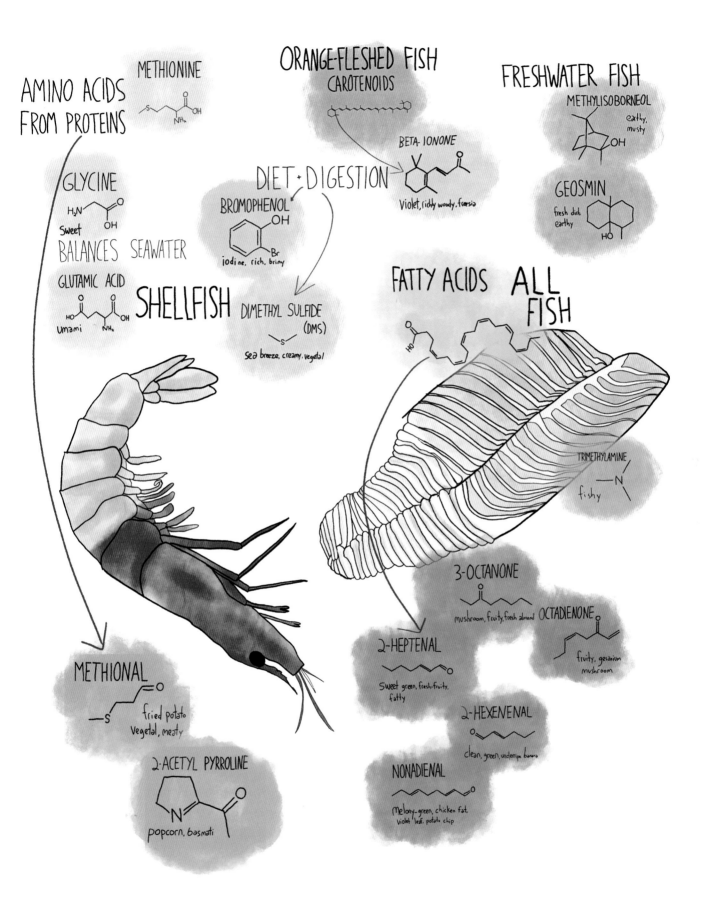

As the temperature gets higher, enzymatic reactions cease, and, at the surface of the meat where the heat is strongest, the browning Maillard reaction takes place between *free sugars* and *free amino acids*, creating characteristically browned and roasted flavors. Exactly which cooked and browned flavors will be produced depends on variations in the meat you are cooking, leading to the obvious differences in roasted chicken and roast beef flavors.

Aquatic Meaty: Fish and Shellfish

I'm using "meat" broadly in this chapter. Fish flavor is a special case, because fish muscle is a special muscle. Compared with mammals and birds, fish have odder stuff dissolved in their muscles. Saltwater fish use various nitrogen-containing molecules, including amino acids, to help their cells balance the saltiness of seawater. Shellfish use a lot of the specific amino acids *glycine* and *glutamate* to do this, which have sweet (glycine) and umami (glutamate) flavors. Saltier waters mean more amino acids are needed to balance the cells, which can make for sweeter and more umami-rich meat—like an oyster or clam from the fully salty waters of a cove versus the more brackish ones nearer to where a river empties into the sea.

Fish have more *unsaturated fat* than land animals, and it oxidizes differently than saturated fat. Fish fatty acids can chemically oxidize, like the fatty acids we met in the steak, but will also meet up with enzymes and create specific green- and sometimes metallic-smelling molecules (members of that exclusive club from cilantro, leafy vegetables, and melons: the *aldehydes*).

Redder-fleshed fish, like tuna, do a lot more intense swimming than other types and have more myoglobin (like mammalian dark meat has) to oxygenate their muscles, which creates more meaty flavors when cooked than in lighter-meat fish.

Fish have different diet-related molecules dissolved in their flesh, like oceanic-smelling *bromophenol* from algae and seaweed. Bottom-feeders like catfish have more earthy *geosmin*, and orange-colored fish get their tint from dietary *carotenoids*, which can break down into *norisoprenoids* (which you might recall in apples and quinces have really lovely luxe fruity-floral-fine tobacco flavors—see page 132).

Less deliciously, the nitrogen compounds that fish produce as waste products can hang around in the muscle. They don't have much flavor, but after slaughter, the clock is ticking: eventually, they break down into molecules that smell fishy (*trimethylamine*) and ammonia-scented—which is one of the reasons you want to shop for well-handled fish that hasn't been sitting around and has been kept very cold. If you've ever seen a sake or citrus marinade on fish, it does more than just make it taste like sake or citrus—the nitrogen volatiles are sensitive to acid, which alters them slightly and makes them much stickier to water. The volatiles don't actually go away, they just stop floating into the air where we can smell them.

Handling Meat for Full-Spectrum Meatiness

Cooking methods themselves can enhance, change, or dilute the flavors of main ingredients. Techniques like blanching, simmering, or boiling involve the addition of water. The ever present tendency toward balance means that some of the ingredients' flavors will diffuse out into the water, and some of the water will diffuse into the ingredients, inherently diluting their flavor a little. This can be an essential part of the dish, as for the poached whole chicken in Hainanese chicken rice or khao man gai—and the chicken broth that results from the poaching.

On the other end of the spectrum, the intense dry heat of searing or roasting will evaporate water away, intensifying flavors but also transforming them through the Maillard reaction and pyrolysis (see "Pyrolysis: Charring and Burning," page 239). Think a rib roast, standing in the intense heat of the oven, or carrots sizzling and browning on a naked baking sheet.

I sometimes find myself reflexively going to the formula Dry heat + Browning = Flavor, which is true but, like all simple doctrines, is not the only truth. Between poaching and searing, you can cook ingredients entirely in their own, self-contained juices, neither adding nor removing significant amounts of water. You don't lose any flavor molecules, but you're also handling the meat so all of its flavor precursors convert and contribute to flavor, not just the ultrabrowned ones.

Sealing the cooking vessel, or approximating doing so, helps achieve this equilibrium. Think: the humid heat of a piece of lamb cooked slowly using only its own moisture, or carrots quasi-steamed in their own water content. Generally, the cooking vessels used for these tasks have hot spots, which is actually a bonus: a little bit of browning adds complexity to gentle texture and purity of flavor, which are both delicious and comforting.

Techniques like cooking en papillote, in clay or dough crusts, wrapping in leaves or seaweed, in bladders, or underground vessels or pits (like Mexico's lamb barbacoa, cabrito, or cochinita pibil) all follow this approach. In a kitchen without these capabilities, you can achieve the same kind of moisture retention in a cast-iron Dutch oven with a heavy, well-fitting lid, or even a deep casserole or baking dish tightly covered in foil.

The Meatiest Slow-Cooked Meat

Cooking in its own moisture, water-equilibrium cooking: it's applicable for leafy greens, root vegetables, brussels sprouts, and all manner of tougher, slowly cooked meats. Here, I focus on beef and lamb, but it works equally well with tough cuts of pork, and probably other species of mammal, though I haven't tested those extensively with it.

Select a tougher, connective tissue-rich cut, preferably in one or two large pieces: brisket, short ribs, chuck, oxtail, shank, shin, bottom round, neck, whatever cut "stew meat" is. If you can't find a big piece of chuck or neck, this will also work with small pieces of meat, or a smaller total quantity, such as 1 to 1½ pounds.

Preheat the oven to 275°F, and prepare **2 to 3 pounds (1 to 1½ kg) of beef or lamb (see headnote)** by lightly sprinkling it with **kosher salt** and **freshly ground black pepper**. Resist the urge to sear the meat, or to add a lot of aromatics or liquid to the pot; that's a different recipe.

Put the meat in a heavy, cast-iron Dutch oven with a tight-fitting lid, and cover, making sure it's seated correctly and not off-kilter, which would allow moisture to escape and dry out the meat.

Cook, checking after 90 minutes and adding a splash of water if there are no juices visible, until the meat appears fully cooked, not very browned, is starting to separate at the fibers slightly, and is very easy to cut into, about 2 to 3 hours. If desired at this point, uncover the pot and put back in the oven to lightly brown for 10 to 15 minutes, turning the heat up to 350°F.

Remove the meat and slice into chunks. I recommend eating it with a little Cilantro-Parsley-Tarragon Salsa Verde (page 179) or green herb tahini (page 181), or topped with very lightly cooked Cherry Tomato–Yogurt Whey Sauce (page 64), as a relish.

Serves 2 to 4

Variation
The Meatiest Slow-Cooked Meat with Butter, Onion, and Spices

Prepare **Meatiest Slow-Cooked Meat**, as above, but add **half a yellow or white onion** (thinly sliced), **1 tablespoon (8 g) of your favorite ground spice mixture**, and **3 tablespoons (42 g) butter** when you put the meat in the pot.

The Meatiest Slow-Cooked Meat Crusted With Herb Paste

Prepare **Meatiest Slow-Cooked Meat**, as above. When fully cooked, remove from the pot (separate from any liquid fat and juices in the pot) and cool at room temperature for 1 hour before cooling fully in the refrigerator.

Take the cooled, slow-cooked meat and place it in a large bowl. Pour over it **1 to 1½ cups** Cilantro-Parsley-Tarragon Salsa Verde (page 179), **Fragrant Miso-Herb Paste** (page 181), or **Cilantro Pesto** (page 177). Turn the meat to coat in whatever herb paste you choose.

Heat a heavy, cast-iron or stainless-steel skillet on the stove over high heat and turn on your extractor hood or other ventilation. When the pan is very hot, transfer the meat from the bowl (allow the excess herb paste to drip off back into the bowl) to the pan. Sear the meat for 4 to 5 minutes on each side, browning it and forming an herby crust. You want the meat inside to be warmed through, but not so heated that it overcooks. (Alternatively, sear the meat under a very hot broiler, or on a very hot grill, 3 to 5 minutes per side.)

Slice or shred the meat while it is hot and serve immediately.

PART 3

The Third Law of Flavor

Flavor Can Be Concentrated, Extracted, and Infused

What do pomegranate molasses, almond milk, and chili oil all have in common?

All three are products of transformation—quite literally, flavors changing their physical forms. Pomegranate molasses starts as a whole pomegranate, with all its flavor intensified and concentrated into a syrup by cracking it open, juicing it, and boiling it. Almond milk starts as almonds blended intensely with water, extracting out fat, protein, and almondy flavors in the process, and leaving behind relatively flavorless solids. Chili oil pulls and spreads out the tightly packed spicy and aromatic molecules inside dried chilis into a bath of warm oil, like unspooling and draping a bolt of tightly wrapped fabric.

They're all examples of the third law of flavor in action: by cleverly corralling, separating, or moving molecules, **flavor can be concentrated, extracted, and infused.**

The third law of flavor covers some of the most elemental techniques in cooking: reducing a sauce, making cups of tea or coffee, pressing out juices, deglazing browned-on solids in a pan. These are the techniques you turn to when you think, "I really like the flavor of this thing, but I wish its form were different [thicker, intensified, more liquid, softer]. I want to drizzle this pomegranate. I want to drink this almond. I want this chili to be pourable." And with a little chemistry knowledge, you can do all that, and much more.

How Much of an Ingredient Is Actually, Physically, Flavor?

If you were to take a really powerful microscope to an ingredient with a lot of flavor, like an olive or a lemon, you wouldn't see homogeneous oliveness or lemonness as you zoomed in on it. You'd eventually see a kind of spongy structure made from different types of molecules: a meshy network of solid molecules like cellulose (dietary fiber) or proteins, filled with pockets of liquid, mostly water and some oils or fats. Even at the highest magnifications, you wouldn't actually see many flavor molecules, and those that you would see would be dissolved in the water or oil filling up the solid spongy structure.

Measured by sensory impact, flavor molecules are small but mighty. It only takes a very, very tiny amount of them to make a lot of flavor. Most

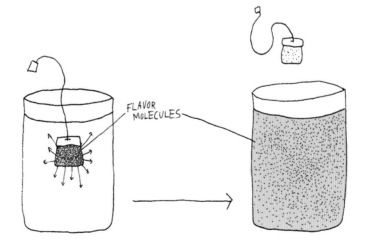

Flavor can be concentrated, extracted, and infused

people can taste sweetness at a little under 1 percent sugar, and that's one of the flavors we're *least* sensitive to. On the most sensitive end of the spectrum, black pepper and green bell pepper both have aroma-based flavors that we're so sensitive to, it would only take a few tablespoons of their molecules to turn an Olympic-sized swimming pool into an intense flavor facsimile of one of them.

Flavors are something ingredients have, not something ingredients ineffably *are*. The vast majority of food's makeup is stuff that isn't flavor—molecules like the aforementioned solid fiber or protein, or liquid water or fat.

From a macro view, this information can seem a little pointless, like expecting to be able to extract a 100 percent cashmere sweater from one knitted from a cashmere-polyester blend.

But since food (and flavor) is molecules, we're free to toss those limits aside and take a molecular view. And in the realm of molecules, chemistry is really, really good at separating different types of substances from one another—substances like, say, solids, water, fats, and dissolved flavor molecules.

Key Rules for the Third Law of Flavor

Lots of cooking techniques boil down to "moving flavor molecules around selectively."

"Moving flavor molecules around selectively" can mean: drying, reducing, deglazing, making infused oils and butters, juicing, brewing tea and coffee, or making dashi or chicken stock, nut and grain milks, or distilled and infused alcohols.

Corralling and moving molecules is easy if you know what they are and how they tend to behave.

The flavors of an ingredient can be concentrated by removing the stuff that has no flavor (fiber, pulp, water).

Flavors can be extracted out of ingredients and infused into something totally different, like fat, water, or alcohol.

Molecules love to move around and mix up with one another so much that all you really have to do to extract flavor from one thing and infuse it into another is to put them in contact for a while.

Like dissolves like: some molecules are more attracted to one another than others.

Oils and fats are fantastic at grabbing on to smell and spicy molecules.

Water is good at grabbing taste molecules and relatively mediocre at grabbing smell molecules.

Chapter 7

Concentrating Flavor

Going back to our zoomed-in microscopic view, let's take a closer look at a few more ingredients. A cut of meat looks like a sponge of proteiny muscle fibers soaked in iron- and amino-acid-rich juice. A lemon resembles a fibrous sponge soaked in acidic lemon juice, surrounded by a waxy sponge soaked in aromatic oils. Even a coffee bean is basically a dry and crackly sponge infused with aromatic and bitter coffee oils.

In fact, most ingredients are basically flavorless sponges, soaked in flavor-molecule-infused juice.

Some of the most versatile techniques for deliciously moving molecules around simply remove some of the stuff that doesn't have any flavor,

holding on to the stuff that does. Like a literal wet sponge, sometimes all you have to do to separate the wet part from the sponge part is just wring it out.

When you squeeze a lemon, you're removing the flavorful juice from the flavorless solids. With a little more force, you can do the same for any fruit or vegetable. Nut, seed, and olive oils are the result of basically the same action on fatty ingredients. And if you flip your focus on removing flavorless water instead of flavorless solids, you get to flavor-concentrated ingredients like reductions, molasses, and dried and sprinklable powders like paprika (made from dehydrated, ground red peppers).

Juicing and Pressing

One of the simplest ways to extract, concentrate, and even purify the flavors of an ingredient is to simply squeeze it hard enough. Literally squeeze it, crush it, juice it—get it to give up its juicy insides.

A few ingredients have big enough pockets of juice (a bigger-holed sponge, if you will) that you can just *wring them out by hand*. Chiefly, citruses,

which actually have two totally separate strata of flavors to squeeze out: sour-sweet juice, from their flesh, and intensely aromatic oils, from their outer layer of skin (the zest or flavedo). A cool flavor investment trick is to peel or zest a citrus before you cut it to squeeze it for juice: either squeeze out the oily flavors from the peel, right away, on or into whatever you're cooking, or freeze them for later. Then, squeeze out the juice to use for a final garnish, a sauce, whatever—you know what to look for, because you read "Sour" (page 48).

Other flavor-spongy, squeezable ingredients require a bit more technological investment to separate flavor from not-flavor. So it's a good thing tool-making is in our evolutionary heritage.

Mostly I'm talking about *blenders* and *juicers*. Fruits and vegetables all have pockets of flavorful liquid in them, but the smaller those pockets are, the more cuts you'll need to make to slice them open. A blender is just a set of knives rigged to slice thousands of times per minute, and a juicer is just a blender that presses the goop it makes against a screen or filter to push the juices out. Basically any fruit (or unstarchy vegetable, like a carrot) that's tasty to eat out of hand makes an amazing juice in a juicer, especially if it's crunchy. Softer fruits like peaches or bananas have a lot of pectin that binds up the juice into a silky puree however you crush or blend them, which is delicious, but adjust your expectations accordingly.

You can, of course, extract juicy flavors to drink on their own, or to add to a soup or braise

Extracting Juice When You Need to Extract a Lot of Juice

If you were processing fruits on an agricultural or industrial scale, you wouldn't use a blender or juicer—you'd crush your fruit. Grapes for wine; apples and pears for cider; cherries, raspberries, and other fruits for brandy and eau de vie: these are all usually loaded into a press, and squeezed out with pressure. Old-style presses use a hand-turned screw to create pressure; newer ones employ a bladder that inflates with air or water to inexorably squeeze the fruit until it's exhausted.

kind of like you would add (tangy, flavorful) wine. But you can also take it a step further and reduce the extracted juice into fruit molasses (page 199), or add sugar and either heat it, to make fruit caramel (page 238), or not, to make juice syrup (page 141).

Try This

If you have a blender or food processor on hand but not a juicer, approach extracting juice like you're making a Mexican-style agua fresca blending the fruit with a little water to loosen it and dilute its pulp a bit.

This works well with juicy and slightly fibrous fruits: honeydews, cantaloupes, watermelons, apples, pineapples, rhubarb, celery, grapes, cherries, raspberries, blueberries.

Combine 3 cups of roughly chopped fruit with ⅔ cup water in a blender or food processor. Carefully pulse the blades until the fruit is a chunky puree, but not totally smooth. Put 2 to 3 layers of cheesecloth in a colander or strainer and set the colander or strainer on top of a receptacle to receive the juice, like a large bowl or quart container. Pour the chunky puree carefully into the cheesecloth, and gather up the edges of the cloth to contain the puree. You can spin the edges shut to apply gentle pressure directly on the puree, or let it sit and drain by itself over a few hours.

To make clear tomato water, pulse very ripe tomatoes into medium-small chunks and forgo the added water. Continue to strain and drain as described above. (Use it in drinks, add it to a salad dressing, or finish pasta with it.)

Or combine fruits (honeydew with green grape, cherry with rhubarb, etc.), add smaller amounts of more silky fruits (like banana, strawberry, or peach), add ginger or other tough spices or herbs, or use whole citruses with an equal volume of water, and add sugar to taste, for a sour-aromatic-bitter lemonade variant.

CONCENTRATING FLAVOR

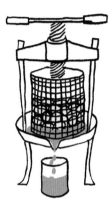

REMOVING FLAVORLESS
SOLIDS

REMOVING FLAVORLESS WATER

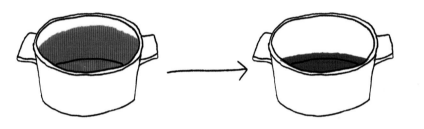

BOILING
REDUCING
FRUIT MOLASSES

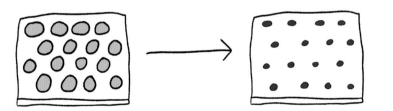

DRYING

Concentrating Flavor by Removing Water

Although ingredients are sponges for flavor, separating their solid parts from their juicy ones is not the only way to extract and intensify those flavors. Like most things that were previously alive, most ingredients tend to contain a lot of water (anywhere from 30 to 85 percent of their total bulk).

Obviously, water itself is flavorless. Removing water from an ingredient is an extremely effective way to concentrate flavors, because the flavor molecules that remain will be, literally, concentrated. Some familiar techniques that remove water include drying fruits to make raisins, sun-dried tomatoes, or paprika; reducing stocks and sauces to evaporate the water by boiling it away, or doing the same to fruit juices to make fruit molasses like pomegranate molasses, grape pekmez, or boiled cider.

Concentrating Flavors with Boiling

Removing water happens fastest with boiling, which obviously requires starting with a liquid. Fill a pan with whatever liquid you'd like to intensify, heat it up until it starts bubbling, and rapid evaporation will do your work for you. You will lose some of the flavor that comes from aroma this way, especially the brighter and lighter top

---

Flavors from Reducing and Boiling

Fruit juice: Pomegranate molasses
Fruit puree: Apple butter, membrillo (quince paste), tomato paste
Plant sap: Maple syrup, sorghum molasses
Alcohol: Wine reduction
Stocks: Double stock, demi-glace, stock reduction

---

notes, since aroma molecules are volatile, too. But you'll keep a lot of them. You'll also keep and *really* intensify the flavors that come from (nonvolatile) taste molecules—sweet, sour, umami, and spicy, as well as bitter, salty, and astringent (think the rough, just-licked-a-paper-towel feeling of a really tannic red wine), which you may want to watch out for. Anything you can taste in a fresh liquid, you'll increase several times over in a reduced one.

Fruit Molasses

Fruit molasses is like looking at a piece of fruit through a powerful magnifying glass. The overall impression of an apple, grape, or pomegranate isn't so clear, but the flavor details, especially tastes, are blown up to an extreme. They're definitely sweet, and deeply, mouthwateringly, toothsettingly sour, all wrapped up in a thick and silky, mouth-coating texture.

In ancient Rome, a culture with a great many Opinions About Sauces, grapes were particularly important for condiments. Boiling down grape juice—removing the water through evaporation and leaving acids, sugars, tannins, and other flavor behind—created saba and defrutum, two different types of grape molasses. (Yes, just one was not enough.) Saba and defrutum were syrupy, fairly sweet, and very sour. Mixed with the fermented fish sauce garum (quite similar to nam pla or colatura fish sauces today), you get oenogarum, a salty-umami-sour-funky sauce that the Romans used like ketchup.

Italians today still make saba or vincotto in mostly the same way, but probably the most famous fruit molasses is made from pomegranates. Pomegranates are pretty intense sweet-sour fruits to begin with, and pomegranate molasses is tart and sweet, and goes on many dishes throughout Iran, Turkey, and the Caucasus (often to cut through the richness of meat stews).

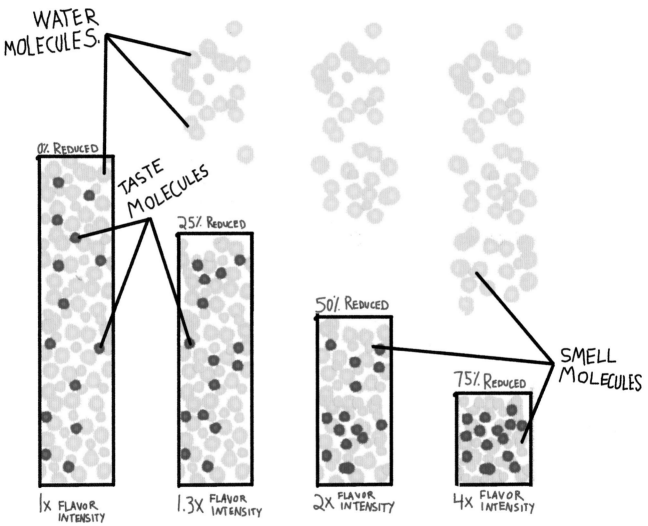

WATER MOLECULES.

TASTE MOLECULES

SMELL MOLECULES

0% REDUCED

25% REDUCED

50% REDUCED

75% REDUCED

1x FLAVOR INTENSITY

1.3x FLAVOR INTENSITY

2x FLAVOR INTENSITY

4x FLAVOR INTENSITY

Besides grapes and pomegranates, you can also find delicious examples of molasses made from figs, plums, mulberries, and apples.

Fruit Molasses

The enemies of fruit molasses are excessive pectin and excessive heat. All fruits have some pectin—which is the glue that holds their cells together—but some hold on to it much better than others. Pectin in a fruit reduction gives it a gel-like structure, and creates a jelly, jam, or fruit cheese instead of a fluid molasses. Heating the juice too hot to speed up the evaporation encourages browning, which can give it a pretty weird flavor; to me, that's not a good trade-off of taste for efficiency.

All you need to make a fruit molasses is a sweet-sour fruit that you can get relatively thin juice from, and patience to slowly keep it at a bare simmer on the stove until about three-quarters of its water has evaporated. Apple cider, grape juice, and pomegranate juice are all traditional. Some other options with potential are cherry juice, tomato water, and not too thick apricot juice. Beet and carrot juice both reduce really well, and while they do get more acidic, like any reduction, they start out so sweet that they make a much sweeter molasses. Juiced squash and sweet potato do, too, but you'll probably have to juice those yourself if you want to try it.

To make a fruit molasses, add **2 cups (475 ml) fruit juice** to a large saucepan. Heat at medium-low until it comes to a simmer, then turn the heat to low so it just barely bubbles. Continue very gently simmering, stirring occasionally to ensure nothing sticks to the pan, until the juice is reduced by three-fourths, down to about ½ cup. Remove from the heat and scrape with a spatula into a plastic or glass container for storage (tightly covered and in the fridge, for up to a month).

Use it in salads (as you would use vinegar), add to marinades and braises, drizzle on ice cream, make a fizzy drink, layer it with tahini or yogurt, lightly sauce vegetables like peppers, cauliflower, squash.

Variation
Beet and Cherry Molasses

Combining already sour cherry juice with super-sweet beet juice creates a molasses that's sour and earthy, sweet and bright, all at once. I've made this with beets I've juiced myself, with packaged beet and cherry juices, and beet-and-fruit juice blends I started seeing in the produce section a few years ago—they all work well.

Follow the Fruit Molasses recipe, starting with **1⅓ cups (315 ml) beet juice** and ⅔ **cup (160 ml) cherry juice** (unsweetened, sour cherry juice if you can find it).

Concentrating Flavor by Drying

Heating at a lower temperature than boiling, for a longer time, lets you remove the water from solid ingredients like fruits (e.g., raspberries, skinned citrus slices, tomatoes, peaches, mango), drying them out and concentrating their remaining flavors. If you stop the removal process before they're fully dry, you'll get delightfully chewy, flavor-concentrated pieces. Hoshigaki, a skinned, semidried, soft, and chewy Japanese persimmon, is an excellent embodiment of this technique. If you continue the drying process until almost all the water is gone and the ingredients are quite brittle, you can grind them into powders—coarse or fine—creating a long-lasting preservation of their flavor that you can dose as a precise, intense sprinkle. (This is commonly done with peppers, for chile flakes or paprika, kimchi, olives, preserved citrus.)

Dried Olive Powder

Drying olives intensifies all the dimensions of their flavor, chief among them funky, salty, and umami. Grinding dried olives into powder creates a dust for adding those flavors to anything, just by sprinkling a little on top. They give dishes a kind of veil of rich olivey flavor, rather than discrete, contrasting bursts, as chopped olives do. Not that either is better—they're just different options, and I like having options.

Try sprinkling this on top of pasta with either a tomatoey sauce or butter, on the mayonnaised side of the bread before you close up a BLT or tuna sandwich, or over steamed broccoli or green beans; or shake it over pizza, spoon into spinach or kale as you cook them with garlic, or sprinkle on any meat you're braising in a Dutch oven.

The volume is arbitrary: you can make as much or as little of this olive powder as you want as long as you spread the olives out enough. It tends to work best with smaller, more intense-tasting pitted olives. Try kalamata, niçoise, arbequina, or black dry- or oil-cured, though it will work with whatever random selection of decent olives you have on hand.

Take **1½ cups (200 g) well-drained olives** and pat dry on paper towels. Set a wire rack into a rimmed, flat baking sheet and spread the olives out evenly over the sheet.

Transfer the sheet *carefully* to the oven, and set the oven to 200°F. Dry out the olives in the oven until they are totally dry and fairly brittle, but not browned. Check on them every 60 to 90 minutes—depending on the air circulation in your oven, the drying time will be in the range of 2 to 8 hours. If you need to, you can shut the oven off with the sheet pan still in it and let it slowly keep drying overnight.

Let the dry olives cool, and coarsely grind into a powder using a food processor (or in a few small batches in a blender). Make sure the food processor is very dry, and keep the resulting powder away from moisture—it will soak it up like a sponge. Store tightly covered and away from light at room temperature, for up to 4 months.

Makes 1 scant cup

Chapter 8

Extracting and Infusing Flavor

If you've ever put a stick of butter back in the fridge and neglected to wrap it securely, or absentmindedly put it in the same compartment as a hunk of cheese or some garlic, you've likely ended up with vaguely cheese-flavored, garlic-flavored, or generically fridge-flavored butter. It's not particularly delicious, but it *is* an illustrative example of the power of extraction. Flavor molecules are ready and willing, if we give them the chance, to be pulled out of one thing—the clove of garlic, the hunk of cheese—and embededd in another.

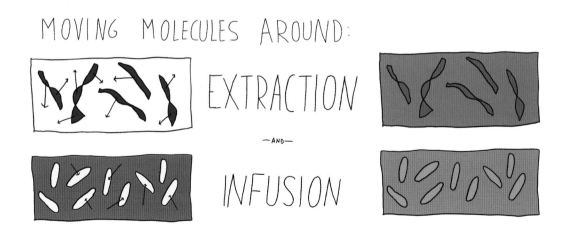

MOVING MOLECULES AROUND:

EXTRACTION

—AND—

INFUSION

There's a limited amount of gastronomic joy to be had from crunching down on coffee beans or chicken bones, but cleverly pushing out their flavors and putting them into water via infusions creates the delicious and rather brilliant inventions of espresso and broth.

Extraction and infusion are two sides of one mechanistic coin, their main difference being perspective. Making a cup of coffee, the hot water extracts aroma, taste, and color molecules from the coffee bean. From the other side, the coffee bean components load into and infuse the hot water. Either way, flavorful molecules are moving around.

Moving Molecules Around, Selectively

Almost any ingredient can be an agent of extraction or infusion, but some combinations work differently than others.

For reasons I find somewhat mysterious, New Englanders generally and Bostonians specifically have both extremely high standards and enormous appetites for ice cream. Like many teenagers in Boston, I had a summer job in one of the many independent ice cream shops sating this demand. I was a little obsessed with the coffee ice cream we made in-house. There were two versions, coffee and white coffee. Coffee ice cream was medium-dark brown (like a cup of coffee with a splash of milk in it) and we produced it by brewing extra-strong coffee, cooling it down, and mixing it into the ice cream base. White coffee was a much paler beige, but had a deeper and more intense coffee flavor and bitterness than regular coffee ice cream; we made it by steeping coffee beans directly into cream, then making that cream into ice cream.

The difference? The water we steeped the coffee in for regular coffee ice cream was able to pull out some of the coffee aroma, but also more of the bitter- and sour-tasting molecules, which are chemically attracted to water. The fat in cream has a far less attractive pull on those taste molecules, so less of those flavors ended up in the cream-steeped white coffee. Where cream (and, specifically, fat) has an advantage is extracting aroma molecules, which are chemically attracted to fats and much less so to water. In an extraction, using chemically different ingredients translates into a different overall flavor profile, because of the chemical affinities of different flavor components.

Like Dissolves Like: A Quick Guide

	Water (polar)	Fats (nonpolar)
Smell molecules (nonpolar)	Weak extraction	Strong extraction
Taste molecules (polar)	Strong extraction	Very weak extraction

Like Dissolves Like: Simple Molecular Rules for Extraction and Infusion

In practice, there are only a few simple rules you need to understand extraction and infusion.

First is that "like dissolves like." Fat molecules and water molecules are as unlike as it gets—fats are what a chemist would call *hydrophobic*, which is why fats and water refuse to mix, and your vinaigrettes tend to split into distinct oily and watery layers.

Smell molecules have a structure that's much more similar to fats than to water, making them slightly hydrophobic. Fats and oils work spectacularly well at extracting smell molecules from ingredients and dissolving them, where water does so much more weakly.

Taste molecules, like salts, sugars, and acids, have structural elements that behave similarly to water, so they're *hydrophilic* and very willing to dissolve in water (and not very soluble in fats).

Making Water Watery and Oil Oily: Polarity

Water molecules are polar, and fats are nonpolar—polar molecules are hydrophilic and mix well with water, and nonpolar molecules are hydrophobic and mix well with fats. All molecules are blanketed with an outer layer of electrons, the same charged particles that build up in dry weather on your hair and rugs and cause staticky frizz and static shocks. In polar molecules, this electron blanket is crumpled and uneven, like somebody who hogs the covers in bed, creating charged spots on the molecule. These charged spots make the molecule act just like a magnet, with a positive pole and a negative pole, sticking to other molecules with these magnetlike properties. And instead of "positively charged in some places and negatively charged in others because of an uneven electron blanket," we just call them *polar* as a shorthand. Fat molecules keep their electron blanket nice and orderly, spread and tucked evenly across themselves. With no uneven, charged spots, they're *nonpolar*. They can still stick to one another, but in a much weaker and undirected way compared with polar molecules. If polar molecules stick together like magnets, nonpolar ones act more like they're covered in Velcro. You can make something with Velcro stick to another thing with Velcro, and a magnet stick to another magnet, but it's much harder to get a magnet and Velcro to stick together.

Giving Smell Molecules a Big Greasy Hug: Extracting and Infusing with Fats

In the rococo heyday of perfume making, workers captured the scents of the most delicate flowers—jasmine, tuberose, gardenia—using a technique called *enfleurage*. High-quality fat was spread in thin layers over glass panes encased in wooden frames, and individual flowers or petals were pressed into the fat to infuse it with their scent. Neatly stacked sets of these enfleurage frames piled up everywhere when flowers were blooming, and when one set of petals was spent, the frames would be unstacked, the flowers peeled off, and a fresh set applied—slowly filling the fat to the brim with dissolved aroma molecules.

Fats are extremely effective at extracting smell molecules from ingredients, regardless of whether we think of those smell molecules as flavors or perfumes. In addition to sharing nonpolar status with smell molecules, fats have a big, floppy surface area for surrounding and mopping up these molecules. They grab on to flavor and stay put.

Fat is where the smell is, so fat is where the flavor is.

Picture this: a hot, perfectly liquid soup, accessorized with drops of parsley oil. A bowl of beautifully crunchy, bright-red, chile-infused oil to drizzle over noodles. A hunk of butter, compounded with charred scallions, spreading flavor all over a piece of bread. The way to get there is infusing fat.

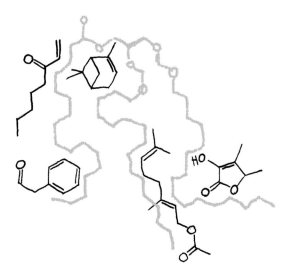

Concentrated, Fatty Extractions

Compound Butter

Brittany is arguably the most butter-crazed part of France—food historian Alan Davidson once put it as that, unlike most of France, Brittany's cheeses were pretty lousy, because they use all their dairy to make cream for butter instead. Brittany is home to the butter maker Jean-Yves Bordier, who, along with three precisely salted levels of excellent plain butter (unsalted, semisalted, and fully salted), makes compound butters with a cult following. Bordier butters might have, kneaded in when semisoft and then chilled into blocks, yuzu, vanilla bean, raspberry, Piment d'Espelette, grains of buckwheat, or one of my favorites, (fresh, oceanic, and addictive) seaweed.

I love butter, so it's probably no surprise that I love it with delicious stuff added to it, too. And because fat exerts a hypnotic pull on aroma molecules, the flavors are less like accessories studded throughout the butter, but actually permeate into the bulk of the butter as well.

Compound butter has a threefold appeal: delicious butter, deliciously infused with flavor, deliciously carrying little concentrated hits of those same flavors.

Among the best-known compound butters are **beurre à la bourguignonne**, butter kneaded with garlic and parsley to get a range of color densities from lightly speckled to fully green, is most traditionally the cap on escargot, melting into and around the snail as it broils. And **Café de Paris butter** is sliced off as chunky rounds to melt onto, and sauce, slices of steak. Whipped into it before forming the log are soft and resinous green herbs, capers and chives, curry powder, alliums, mustard, and anchovies or Worcestershire sauce. But I didn't really grasp the full flavor-infusion potential of the technique until I got hooked on Bordier butter and started branching out into other aromatic ingredients.

Compound butters take advantage of butter's semistable structure: as a water-in-oil emulsion, it has a smooth, deformable texture (what a materials scientist would call "plastic") that holds up even when left out at room temperature to soften. (Melting it all the way is a one-way trip, though: the water-fat structure collapses, and when it solidifies again, it becomes grainy rather than smooth and pliable.) Compound butters especially shine when softened and spread on bread, or applied cold in slices to just-cooked foods, where they will semimelt into a soft sauce.

Cinnamon-Lime Butter

I like this on any kind of grilled meat, or on a fish like monkfish or swordfish.

In a medium to large bowl, combine **2 sticks (about 225 g) softened, best-quality unsalted butter** with **1½ tablespoons (12 g) lime zest**, **2½ teaspoons (6 g) powdered cinnamon**, and **a scant ½ teaspoon (2.5 g) fine sea salt**. Mix together well, then pile on a piece of plastic wrap and roll into a log.

Chill, well wrapped, in the fridge until use. Consume within 3 weeks.

Makes about 1 cup

Cacao Nib–Lemon Butter

This is a salty-sweet dessert on some rich brioche or challah. It's also great on squashes, summer or winter.

In a medium to large bowl, combine **2 sticks (about 225 g) softened, best-quality unsalted butter** (grass-fed and cultured, if you can find it!), **2 tablespoons (20 g) lightly toasted, crushed cocoa nibs, a scant ½ teaspoon (2.5 g) fine sea salt**, and **3 g lemon zest** (just short of 1 medium lemon, zested). Mix together well, then pile on a piece of plastic wrap and roll into a log.

Chill, well wrapped, in the fridge until use. Consume within 3 weeks.

Makes about 1 cup

Burnt Scallion Butter

In this recipe, you don't actually burn the alliums *in* the butter, but it does still make a delicious, cushioning carrier for the sweet-charred-oniony flavors. Great with bread, especially sourdough or Danish rye, or with shrimp.

Take **10 to 12 small to medium scallions**, peel off their outer layer, and trim the dry, tough part of the green ends off. Put them in a single layer on a baking sheet.

Heat a broiler on high for about 10 minutes until it is very hot. Put the sheet with the scallions about 12 inches from the broiler, and broil until the outer layers start to blacken and char, 1 to 5 minutes. Take out the sheet, flip the scallions, and char the other side the same way. Remove the scallions and chop them medium-fine.

In a bowl, combine the chopped, burnt scallions with **2 sticks (about 225 g) softened, best-quality unsalted butter** and **a scant ½ teaspoon (2.5 g) fine sea salt**—or to taste. Mix together well, then pile on a piece of plastic wrap and roll into a log.

Chill, well wrapped, in the fridge until use. Consume within 1 week.

Makes about 1¼ cups

Saucy Oils and Fats with Stuff in Them: Tempering, Chili Crisp, and More

It's the first step in many South Asian recipes as well as the last garnish. *Tempering*, also called *tadka, chhaunk,* or *baghaara*, is the process (and product) of quick-infusing flavorful spices into hot oil or ghee. Even finely ground spices have pockets of aroma that are slow to empty, and this initial dip in fat helps pull them out. The mixture of spices and spice-infused oil can now spread aroma, heat, and color throughout the whole dish faster and with a more encompassing flavor than naked spices could.

Tempering can happen at the beginning of cooking, when some or all of the spices for the dish are simmered in the hot fat before the addition of onions, garlic, ginger, liquid, and vegetables or meat.

As other ingredients are added, they get a coating of this now flavored oil, effectively spreading flavor throughout the whole dish. Since most of these ingredients are mostly water-based, as is whatever makes up the sauce (water, yogurt, tomatoes, juices from the ingredients themselves), they get a bigger and more even dose of flavor from the tempered, flavorful oil than from adding the spices directly to the more watery braise.

A second version of tempering happens after the main dish is cooked through, like a sauce or an additional finish. In a separate pan, you heat up oil or ghee and add a variety of spices. Instead of using this as a base to build the rest of the dish, you now pour the resulting chhaunk or tadka over the finished dish. The hot and flavorful oil sends up a plume of aroma, and floats a layer of flavor over every bite. The spices used here might be simply red chile powder, curry leaves, or cardamom, or a more complex mix of mustard seeds, ginger, cumin, fenugreek, sesame, cashews, and other flavorful and textural elements.

The same infusion and dissolution of smell molecules into fats happens here as happens in infused oils. Leaving the partially extracted ingredients in the oil adds texture and little pockets of flavor—a flavor-infused fatty sauce somewhere between a liquid oil and a relish. It's a technique that works so well, cooks use it all around: chile oils and chile crisps, ginger-scallion oil, salsa macha, and more.

In these recipes, I've played with the concept of saucy, textural fats using some of my favorite flavors and ingredients. Use them to accent flavors that are already in the dish, or contrast them with new ones, especially chile heat, spice aromas, citrus peel, and other flavors that take some drawing out. You can use them wherever a dish feels like it could use a flavorful finish, like spritzing perfume in the air you're about to walk through right before you head out the door.

The Spicy Exception

Unlike smell molecules, most taste molecules don't dissolve well in oil; that's because they are polar, like water. Spicy is the exception. It was already an exception—it's not literally a taste, but rather a form of touch (pain) incorporated into flavor along with (true) taste and smell. Molecules that are spicy, like capsaicin in chiles or piperine in black pepper, are quite nonpolar, so they dissolve really well in fats. They're also pretty big, which is why they can be nonpolar but also not volatile, like a large flightless bird.

Chile-Mint-Sesame Oil

This is delicious with lamb, on hummus, or over grilled eggplant.

In a medium saucepan over medium heat, heat **¾ cup (180 ml) grapeseed oil**. When it's shimmering, but before it's smoking, add **9 tablespoons (80 g) sesame seeds** and **3 tablespoons (20 g) urfa chile flakes,** and fry until the sesame seeds are lightly browned, 1 to 3 minutes. Remove sesame and chile flakes with a slotted spoon and add to a nonreactive metal bowl.

Add **2½ cups (30 g) whole mint leaves** to the hot oil and fry for about 5 seconds, then, using the slotted spoon, immediately remove to the bowl. Turn off the heat, and add **1¾ teaspoons (4 g) cumin,** **1¾ teaspoons (4 g) powdered cinnamon,** and **2 roughly chopped garlic cloves** to the oil. When the garlic is soft but not particularly browned, pour everything from the pan into the bowl.

When the oil cools to very warm, rather than pan-hot, add **¼ cup (60 ml) good olive oil** and **¼ cup (10 g) mint leaves, finely chopped.**

Store in the fridge for up to 2 or 3 days.

Makes about ⅔ cup

Hailing from Veracruz, Mexico, **salsa macha** is much fattier, richer, and less watery than most other salsas. Made with dried chiles (guajillo and ancho chiles, smoky moritas, perhaps small and

Time, Temperature, Surface Area

Extracting flavors and infusing flavors are, at their most basic, just moving molecules around, like you're an aide-de-camp in a war room pushing model ships and infantry units around a giant map. Besides using fat, with its great affinity to smell molecules, as your solvent, there are a few other ways to maximize what you pick up.

First, *time*: moving molecules around relies on their naturally ever wiggling nature, like a young infant who's constantly, aimlessly pumping their limbs. It can take a little while for this random wiggling to push molecules to where you want them (a process called *diffusion*), but the longer you wait it out, the more you'll catch.

Second, *temperature:* heat is, quite literally, the wiggling and friction of molecules, and temperature is just a measure of how fast they're wiggling. More heat, higher temperature, just means molecules are moving a lot faster, so heating things up reduces the amount of time it takes for wiggling diffusion to push things around.

You can think of time and temperature as trade-offs with each other: if you heat up ingredients to capture their flavor, you'll capture it much faster, but you'll need to watch out for things like browning, burning, or oxidation (think: painty-smelling used fryer oil). If you do a cooler extraction, you're handling the molecules a bit more gently, but you'll be waiting longer to capture their flavors.

Rather than trying to steep whole coffee beans to make coffee, we grind them into small pieces first. The reason is also your third tool: *surface area*. A coffee bean's flavor molecules can only infuse into water they can actually touch, so those deeper inside the bean have to wiggle their way to the surface to take part. If you crack the bean in half, what was once the middle of the bean is now an outside surface of each of the halves, and flavor molecules situated there can flow out faster than before. Keep cracking the bean into smaller and smaller pieces, and the distance any flavor molecule has to travel to reach a surface gets smaller and smaller, and the total amount of surface flavor molecules can pass through gets larger and larger—so more flavorful stuff can be extracted, and it can be extracted faster. This works for everything, not just coffee: smash things up a bit (with a mortar and pestle, a grinder, a blender) for faster and more thorough infusion.

spicy chile de árbol) steeped and blended in a lot of oil, with peanuts and sometimes sesame, and an acidic element like vinegar or orange juice, salsa macha often separates into two distinct layers, a flavorful, nutty sludge with deep red-brown oil floating on top. The pasty bottom layer has all the texture, acidity, and salt, and the oil is full of slow-burn spice and nutty-garlicky-chile aromas. It's got a lot in common with Sichuan-style **chili crisp**, which sizzles aromatic dried chiles like erjingtao, Sichuan peppercorn, garlic, and shallots in oil for a layer of crispy bits as well as superaromatic, spicy fat. Chili crisp also makes a versatile base for cooking instead of a garnish, like frying eggs or sweating onions for a braise or sauce. **Hong you**, or Sichuan chili oil, relies on some of the same aromatics plus ginger, star anise, and coriander. Rather than combine every ingredient into the crispy-solid phase, you'd typically infuse everything that's not chilies in very hot oil, then strain it directly over ground dried chilies and buzzy Sichuan peppercorn.

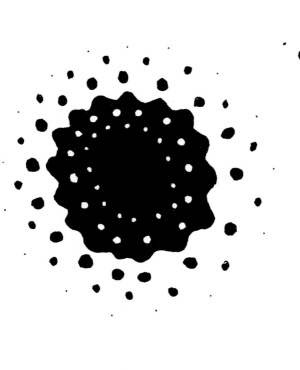
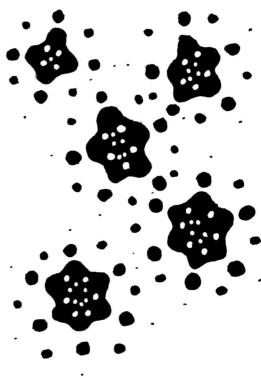

Crispy Morita Chile, Allspice, and Orange Oil

Use this fruity, spiced, smoky chile oil as a crispy condiment as is, or pulse in a food processor to make a more sandy, pastelike condiment. Try it on rice, pork, asparagus, or roasted sweet potato with a little yogurt.

In a medium saucepan over medium heat, heat **1 cup (240 ml) peanut oil**. When it's shimmering, but before it's smoking, add **4 tablespoons (25 g) seeded, crumbled, dry morita chiles**. Fry for 1 to 3 minutes, until the chiles are crisp but not too browned. Remove the chiles with a slotted spoon and add to a heatproof metal bowl.

Next, to the oil, add **1 cup (100 g) sliced shallots**, in two batches. Adjust heat so they bubble and fry constantly. Stir them in the oil as they fry, and cook until dehydrated, golden brown, and crisp. Remove with the slotted spoon and add to the bowl.

Add to the bowl **13 tablespoons (100 g) chopped pecans, 1 tablespoon (8 g) chopped garlic, 8 allspice berries (lightly crushed), and 1 teaspoon (2.5 g) cumin seeds**. Pour the hot oil directly over; everything will sizzle briefly, and then subside. When the oil cools down from extremely hot to just barely touchable, add **3 tablespoons (25 g) orange zest**. Store in the fridge, covered, for up to 2 weeks.

Makes 1½ to 2 cups

On the less fiery end of the flavor spectrum, **ginger-scallion oil** is a fragrant, just slightly piquant, excellent partner to (traditionally) poached or (also good) roast or grilled chicken. Pouring hot oil over minced, salted ginger and scallions immediately infuses flavors through the oil, and wilts and slumps them down to the texture of a jammy confit.

Soft Pickled Chile, Anchovy, Sage, and Crispy Almond Oil

I love this as a soft-spicy-tangy-funky accent to pizza and simple pastas.

Heat **¾ cup (180 ml) grapeseed oil** in a medium saucepan over medium heat. When it's shimmering, but before it's smoking, add **⅓ cup (45 g) roughly chopped almonds** and fry until golden brown, 2 to 5 minutes. Remove from the oil with a slotted spoon.

Then, add **¾ cup (15 g) sage leaves** to the oil and fry until they are crisp, but haven't browned much—1 to 3 minutes. Remove again with the slotted spoon.

Add **2 anchovies** to the hot oil and cook until they start to break down a little. Add **5 chopped cloves garlic** and continue cooking until garlic is soft. Turn the heat to medium-low, add **1 cup plus 2 tablespoons** (225 g) **drained fermented chiles (chopped)**, and cook the chiles, garlic, anchovies, and oil together until the chiles are very soft. Add **2 tablespoons (30 ml) good-quality red wine or sherry vinegar** and let it bubble for about 15 seconds, then remove from the heat.

Pour everything into a nonreactive metal bowl and mix in the fried sage leaves and toasted almonds. Add **¼ cup (60 ml) good-quality olive oil** and stir to combine.

Store in the fridge for up to 2 weeks—use immediately if you want it to stay crisp.

Makes about 1¾ cups

Dill-Scallion-Hazelnut Oil

This is delicious with chicken and dumplings or simple roasted chicken, as well as with cauliflower or new potatoes.

Heat **¾ cup (180 ml) grapeseed oil** in a medium saucepan over medium heat. When it's shimmering, but before it's smoking, add **9 tablespoons (80 g) chopped hazelnuts** and **2½ cups (120 g) chopped scallions** and fry for 1 to 3 minutes, until the nuts have browned a little and the scallions are soft and a little crisp. Turn off the heat and add **2 tablespoons (15 g) finely crushed fennel seeds**.

In a nonreactive metal bowl, combine **3⅓ cups (100 g) chopped dill fronds**, **1½ teaspoons (4 g) chopped garlic**, and **10 to 20 grinds of black pepper**. Pour the still-hot oil, hazelnuts, scallions, and fennel seeds over to wilt the dill and soften the garlic. Let cool until it is warm enough to comfortably touch, then add **¼ cup (60 ml) good olive oil** and **1 tablespoon (15 ml) lemon juice**.

Store tightly covered in the fridge for up to 1 week—use immediately if you want it to stay crisp.

Makes about 2 cups

Moments of Clarity: Infused and Strained Oils

A deeply infused, liquid oil is a pro move: you need to do a little more work extracting and straining it than the simple mixing required for a compound butter or fat. It's basically doing perfume enfleurage, without the framed glass panes. But the product you get out of it is pure, liquid flavor, ready to mix with or drizzle onto anything and release aroma in its wake, without any textural distractions, and very little bitterness or off-tastes. You can infuse nearly any aromatic ingredient into oil, and, like making coffee from gritty and inedible coffee beans, it's especially good for making the flavor of delicious things with difficult-to-eat textures, or tricky techniques, more accessible—like, say, an alternate route to smoke flavors without pulling out the smoker.

Smoke Oil (with Lapsang Souchong Tea)

For someone who has always lived in cramped urban apartments with ineffectual kitchen ventilation, I have an unfortunate enthusiasm for smoky flavors. I developed this recipe while I was living in a studio apartment, the kind where practically speaking, the smoke detector is in the kitchen even if it's in the bedroom. Which is to say, smoking food at home was not on the table.

Enter lapsang souchong tea, which my father drank almost daily during my childhood. Originally from Fujian, China, it is black tea that's dried by being smoked over a pinewood fire. It is very, very smoky, with nice herbal-quincey nuances from the tea itself (from some of the same norisoprenoid molecules in quinces; see page 132). The tannins and astringent-bitter components of tea are not very soluble in oil, so (I hypothesized) an oil extraction would capture mostly just the aroma.

The result was this nicely balanced, very smoky oil with no burning necessary. You can use it to garnish finished dishes, especially raw or cured meats like tartare, or tomato salads. I also like to use it for confit or oil-poaching, which really infuses the food with smoke flavor.

The recipe is completely scalable for any amount of tea or oil you have; I like to do an 8:1 ratio by volume. If you want it to be very concentrated, think like an enfleurage: infuse once, strain out the tea, and then infuse the oil a second time with new tea. And if you're looking for avenues to experiment further, try some other oil you like the flavor of—perhaps a blend of grapeseed and sesame, some sweet and light olive oil, rich and dark-green pumpkin seed oil.

Combine loose **lapsang souchong** tea leaves with **8 times their weight in grapeseed or good-quality canola oil.** Blend in a blender, and let steep in the fridge for 2 days. Strain through cheesecloth or a very clean dish towel (that you don't mind getting very oily). This will keep for a couple of months in the fridge, or indefinitely in the freezer.

Ratios, for reference

For 1 cup of oil, add 2 heaping tablespoons of lapsang souchong tea.

For 750 ml (a regular-sized wine bottle) of oil, add 6 to 7 tablespoons of lapsang souchong tea.

For 1 quart (about 1 liter) of oil, add 8 to 9 heaping tablespoons of lapsang souchong.

Smoke Oil–Poached Fish

Here, both smoked fish and oil-poached fish combine into a succulent, lightly smoky, smoke-detector-friendly center-of-the-plate protein. You're essentially making smoked salmon, without any smoke. Other fish that work well: mackerel, tuna, mussels, and arctic char, as well as white fish like cod and halibut or shrimp.

If you can use an immersion circulator like an Anova (or even a pot, a thermometer, and a sharp eye) to cook the fish in a water bath, it will come out even better and use up less oil.

Besides Smoke Oil this works with any strongly flavored oil: an intense olive oil, a garlic-thyme oil, a ras-el-hanout-infused oil, oil that you've roasted onions in, or even duck or any other animal fat you find yourself with a lot of.

Remove as many **fish fillets or steaks** as you desire, and fit in the pan you're using (I figure 6 to 9 ounces per person for dinner) from the refrigerator about 1 hour before you plan to cook them, so they come to room temperature. Season the outside with **salt** and **pepper** to infuse in while it sits.

Using the Stove and Oven

Survey your cooking pans and identify which one can fit all your fish, with as little overlap, and as little unused space, as possible. The wider it is, the more oil you will have to use to fill it. I suggest an oval Dutch oven, perhaps a deepish pie plate, or even a deep skillet.

Meanwhile, preheat the oven to 170°F (or 200°F if that's the lowest your oven will go). Pour **1 quart (945 ml) Smoke Oil** (page 210) into a not too wide Dutch oven. Heat the whole pot (in the oven, or on a very low burner) to bring the oil to 125°F, measuring with an instant-read thermometer.

Nestle the fish into the warm oil, then transfer to the warm oven for 20 minutes, or until just cooked all the way through, but not tight and opaque. Remove the fish from the oil, which you can strain and save in the freezer, tightly wrapped, for further fish poaching for up to 6 months.

Pat the fish dry on a paper towel, and serve immediately.

Using an Immersion Circulator

Fill a large pot or plastic container with water, and set up the immersion circulator. Set the water bath to 125°F. In vacuum bags to match whatever type of sealer you have, or Ziploc brand freezer bags, lay the fish in one layer and add **1 cup (200 g) Smoke Oil** before sealing.

This will be easy to seal in a chamber vacuum sealer, and harder to seal in a FoodSaver vacuum because of all the liquid oil. If using the Ziploc, zip it closed 95 percent of the way, leaving a fingertip-space of zip open. Push out the air in the bag by displacement: dip the bag into a plugged sink full of water, holding it by the open corner. As you push the bag underwater, the air will be pushed out. Just before the open corner goes under, quickly zip it shut, and you should have a decent zip seal with very little air in the bag.

Cook the bagged fish in the water bath for 35 minutes, or until cooked through but not opaque, then drain and serve as above.

General Infused Oil Steps

- Pick something to infuse, like an herb or spice.
- Weigh it and mix with 2 parts to 5 parts neutral oil.
- Blend the oil and aromatic ingredients.
- Pack into a container and keep away from oxygen.
- Gently warm it up for 1 to 2 hours (see sections below for specific techniques and temperatures), then put it in the fridge or somewhere else dark and cool to continue extracting for 1 to 2 days.
- Strain, my recommendation being a colander or strainer lined with an extremely clean, flat textured kitchen towel over a coffee filter, which clogs easily.

A Thorough Guide to Thoroughly Infused Oils

Infusing oil lets you take anything you like the smell or flavor of (smoky tea and beyond), regardless of texture, and transform it into a smooth, even blanket.

Sometimes the creative constraint of recipes, in their specificity, is the best way to get to know a concept or technique. As a cook, once I get the gist of it, I crave parameters: not just one recipe, but a guide to what I need to pay attention to and tweak to get in the right ballpark with as many kinds of different ingredients as possible. Extracting with oils, in my mind, is the place to match the technique to the flavorful ingredients you have, a set of these parameters that you can adapt to your current mood and materials and figure out what you like through your own experimentation.

STUFF TO OIL + 2X - 8X AS MUCH OIL

GRIND TO INCREASE SURFACE AREA

AND SPEED UP EXTRACTION

STEEP

for a few hours to a few days...

VERY CLEAN CLOTH

TWIST BUT DON'T SQUEEZE

STRAIN W/COLANDER OR LET DRAIN

I wanted this section to feel like paging through a field guide, setting you up with the best starting points for whatever form your favorite flavors start out as. Infusing oils is great for spices, when you don't want a lot of sandy texture, and herbs, because it removes the possibility of water- and enzyme-accelerated oxidation and swampy flavors, freezing their freshest qualities in place. The proof is in the drizzle: a stunningly bright-green color, a limpid clarity, cleanly aromatic and longer-lived (in the fridge or freezer) than a blended herb sauce. I've made, and been served, beautiful oils infused with rose petals, toasted oak chips, lobster shells, or roasted kelp (see page 90).

A broth, a drink, or a scoop of ice cream accented with drops of oil-extracted dill, pine, or fig leaf are very different eating experiences compared with eating those as garnishes. The oil seamlessly incorporates these flavors into a dish rather than being a separate pop or accent—think the difference between garnishing with sesame oil or sesame seeds.

There's more ingredient-specific tips following, but here's the generic mindset I take with oils: I try to pack as much of whatever I'm infusing into the oil, generally ending up with two to five times as much oil as aroma ingredients (by weight). I try to break down whatever I'm infusing to increase the surface area, and agitate the oil around it a bit, to speed up the extraction. I try to keep everything away from oxygen, packing into narrower containers and sealing or covering the top with plastic wrap pressed onto the surface. While it's extracting, I'll warm it up as much as I think it can take for an hour or two. Then, I'll usually give it a day or two to steep, keeping it in the fridge for most of that time.

Soft Green or Fragrant Herb Oils: Blanching and Cool Handling

Herbs are great for maximizing extraction by maximizing surface area because they're so soft and grindable. The softer the herbs are, and the more you blend and moosh their leafy cells, the more their contents will spill out, including enzymes that make grassy (first freshly cut grass, then yesterday's grass) and oxidized flavors. To avoid this (you'll want to avoid it), you can briefly blanch softer herbs in boiling water to inactivate those enzymes.

This approach is best for soft and green herbs like dill or parsley, or tender and fragrant ones like basil, tarragon, mint, shiso, chervil, or lemon verbena. (For more soft green herbs, see page 174; for fragrant herbs, see page 176.) To make more resinous herbs into oil, see Resinous Herbs That Can Take the Heat (recipe follows).

Get a large bowl of **ice water** ready, and heat a large pot of **water** to boiling. Blanch by submerging **the soft green or fragrant herbs** in the boiling water for about 30 seconds (until bright green, but not cooked or wilted at all) then rapidly skim them out and put them directly into the ice water to stop them from cooking. Once cooled off, dry gently on paper towels—you want as little water in the oil as possible.

Puree the herbs with **twice their weight in oil**, transfer to a zip-top bag and push out all the air, or into a lidded container and press a piece of plastic wrap to the surface. Infuse overnight in the fridge before straining into an airtight container. Store for up to 2 weeks in the fridge, or freeze for up to 1 year.

Resinous Herb Oils That Can Take the Heat

Tougher herbs (like rosemary, oregano, sage, thyme, or marjoram) can take some rougher, hotter handling. Heat speeds up extraction, and is a great help where you're not as worried about swampy oxidation flavors.

Puree the **resinous herbs** well with double their weight in **oil**, then transfer to a zip-top bag and push out all the air, or into a lidded container and press a piece of plastic wrap to the surface. Heat the container to 180 to 200°F, using an immersion circulator or a pot of water barely under a simmer over the stove and an instant-read thermometer, for 1 to 2 hours. Then remove the container and let sit in the fridge overnight before straining into an airtight container. Store for up to 2 weeks in the fridge, or freeze for up to 1 year.

Waxy and Aromatic Leaf Oils

The tough, firmly unchewable texture of **waxy and aromatic leaves** (like pine needles, lime leaves, fig leaves, black currant leaves, bay leaves, curry leaves, avocado leaves, peach leaves, patchouli, or even organic, high-quality hay) can hamper access to their sometimes magnificent flavors, ergo, leaf oil. Such as fig leaf oil—I know, kind of sounds like an absurd affectation, but, wow, the layers of herbal, fruity, and creamy flavors it manages to pull out of those leaves compel me, especially as a drizzle garnishing panna cotta or a couple of drops in a gin drink or iced tea (yes, they don't mix, and you do feel the oil, but it's kind of . . . tactile, in a way I like).

Weigh your leaves and blend well with double their weight in **oil** in a powerful blender. Transfer to a thick zip-top bag and push out all the air; or into a lidded container and press plastic wrap to the surface. Heat the container to 140°F/60°C using very hot tap water or a pot on the stove and an instant-read thermometer, for about 1 hour, then infuse overnight in the fridge before straining into an airtight container. Store for up to 2 weeks in the fridge, or freeze for up to 1 year.

Spice Oil

Dense and dry spices (like black pepper, ginger, cumin, caraway, cloves, cinnamon, allspice, cardamom, and pink pepper) can take even more heat than leaves. You can infuse them slowly at room temperature or a bit warmer, but I'm impatient and usually use hot oil. Spice blends like garam masala, ras el hanout, advieh, or shichimi togarashi also work here. A higher ratio of oil to spices is useful, because they're so much drier than herbs they make more of an oily sand than a pourable paste.

Weigh the **spices**. In a medium-sized pot, heat 4 times the spices' weight in **oil** over medium-high heat, until hot but not yet shimmering or smoking. Remove from the heat, add the spices, then cover and steep for 1 to 2 hours on the counter before straining into an airtight container. Store for up to 2 weeks in the fridge, or freeze for up to 1 year.

Allium, Chile, and Mushroom Oils

Onions, shallots, leeks, scallions, garlic, dry or fresh chiles, and mushrooms all have a lot of flavor to give, and can do so in two different ways. You can extract with a little heat, to get soft and relatively clear flavors. Or you can simultaneously cook and extract, for delicious roasted and browned flavors.

Weigh the **allium, chile**, or **mushroom** ingredients. Blend with 2 to 3 times their weight in **oil**, then transfer to a Dutch oven over the stove. Using an instant-read thermometer, heat at 200°F for 1 to 2 hours (for soft, not browned texture); or 250 to 275°F for 4 to 8 hours (for roasted/browned flavors—remove from the oven when it's as dark as you like).

Allow to cool to just warm, then strain into an airtight container. Store for up to 2 weeks in the fridge, or freeze for up to 1 year.

Oils Further Afield: Flowers, Tea, Seaweed, Wood

Oil extracts rich and delicately sweet-smelling molecules from flowers, teas, and seaweeds like red dulse or nori, expressing a completely different side to their flavors than you get from rose water, essential oils, cups of tea, or seaweed as it appears in miso soup or salad. Fragrant-earthy woods like sandalwood or khus (vetiver) are common flavorings for sharbat, a syrup for drinks from the Middle East and South Asia. Considering the amount of dissolved oak flavors you taste if you drink red wine or whiskey, you may already have a taste for sweet-woody flavors. These are admittedly some more out-there recipes, but I'd rather you have a resource for when, say, my bad influence has rubbed off on you enough that when you come across an insanely aromatic chunk of juniper wood or glut of citrus blossoms, your first thought is how to grab on to that flavor and put it in food.

Flower, Tea, or Seaweed Oils

Try infusing rose, elderflower, lavender, jasmine, orange blossom, osmanthus, magnolia, black locust flowers, oolong, lapsang souchong, buttery or floral green teas, nori, or red dulse.

Steep **whole fresh petals unblended,** or **blend seaweeds, teas**, or **dried petals**, with 4 to 5 times their weight in **oil**. Infuse at room temperature for a week, or heat gently to 125°F/50°C, in a Dutch oven over the stove and an instant-read thermometer, for 2 to 3 hours. Strain into an airtight container. Store for up to 2 weeks in the fridge, or freeze for up to 1 year.

Woodsy Oil

For aromatic, food-safe woods like cedar, oak, redwood, juniper wood, cherry wood or bark, black currant wood, sandalwood, mesquite, orrisroot, licorice, vetiver; or tree resins like piney mastic, frankincense, myrrh, vanillaesque benzoin, floral storax.

Use **wood shavings**, or smash the stems of **fresher wood** until they splinter. Combine with 4 to 5 times their weight in **oil** in a heatproof container. Then heat in the oven to 175°F (or if your oven doesn't have a setting that low, then the lowest possible temperature), for 6 to 8 hours. Strain into an airtight container. Store for up to 2 weeks in the fridge, or freeze for up to 1 year.

Extracting and Infusing with Water

Water beats just about anything else in its ability to dissolve other polar molecules: salts, acids, sugars, carbohydrates, free amino acids, and minerals. In other words, it's particularly effective at extracting salty, sweet, acidic, and other taste-based flavors. It's less effective at dissolving nonpolar molecules like aroma compounds, certainly compared with fats. But, fortunately, "less effective" doesn't mean "ineffective," as you know if you've ever enjoyed a flavorful cup (or bowl) of minty herbal tea, aromatic coffee, or deeply chickeny chicken stock.

Technically, polar water and nonpolar things like smell molecules don't like to mix very much. But nature also has an inexorable drive toward disorder, and two totally separate layers of liquid—one polar, one nonpolar—would be Too Orderly. So, water and nonpolar aroma molecules don't completely mix, but they mix a little. But because smell molecules are so potent, just a little mixing can be more than enough. For example, vanillin, the vanilla-smelling-molecule in vanilla, can only dissolve in water up to a concentration of 6 to 7 grams per liter, less than 1 percent. (Compare that with salt, which tops out at 360 grams per liter, or sugar, which water can dissolve double its own weight of.) But, that 7 grams per liter is about *seventy thousand times* higher than the minimum amount of vanillin we can smell in water. "Limited solubility" is far more than enough to deliver lots of flavor, and the same is true for many other aroma molecules and water.

Think of extracting and infusing with water as especially good ways of achieving *strong taste extraction* (sour, sweet, salty, umami, bitter) and *soft aroma extraction* (fruity, herbal, spiced, citrusy).

Infusing Water Quickly

Many things we infuse into water rely on heat for help: coffee, tea, etc. When "cold brewing" versions of these at room temperatures, you usually let them steep for a long time, like overnight. If you want to flavor watery things quickly, try out the third trick, increasing surface area: use larger amounts of the ingredient you want to infuse, and blitz them together for a few minutes under high power in a blender. A higher "brew ratio" of flavorful ingredients to solvent means there's much more flavor that can potentially be pulled, and agitating in the blender increases the contact between them and how fast they mix.

I especially like using juices as solvents for quickly extracting herbal flavors—adding interesting dimensions of flavor while retaining freshness. The result is less sweet and more of a thoughtful beverage than a soda.

Whether cilantro, chiles, lemon thyme, juniper, coriander, cardamom, or another flavor, the goal is less an exhaustive extraction of all flavors, and more a fast and effective one. Quick-infusing flavors by blending is sort of like a very fast cold-brew, but rather than looking for full flavor development (capturing enough of the base notes, getting the right balance of tannins and acidity) from a single ingredient, the juice provides most of the flavor structure already, and the quick-infused aromatics add, well, aromatic dimension.

Apples to Oranges

Apples and oranges actually have nicely complementary flavors, especially orange peel. There's a kind of cold mulled cider element in this, with the added allspice, which goes with both fruits. It works even better if you get some flavorful apples (I'm quite partial to Macoun, for the somewhat easy-to-get, or Cox's Orange Pippin, if you can find them) and juice them yourself. Absolutely leave the skin on, since it contains a lot of flavor molecules that will get extracted into the juice.

Mix **1 quart (945 ml) high-quality apple cider** with **the zest of 2 oranges** and **6 allspice berries** in the carafe of a high-powered blender. Blend together over medium-high speed for 3 to 5 minutes, then strain and chill. Store tightly covered in the fridge for up to 1 week.

Serves 2 to 4

Variations

Cider plays nice with similar amounts of thyme leaves, spearmint, juniper berries, or black pepper plus a sweet spice like allspice or cardamom in place of the botanicals. Try it with lemonade or limeade.

Extremely Green Green Juice

This juice uses the water inside green-flavored cucumber and honeydew to extract even more green flavors from extremely green herbs, which are agitated by the blender to speed up the transfer, and then strained out after their brief but exciting steep.

The flavor profile is, basically, the dream version of what "green juice" can be.

This recipe involves not only infusing the juice, but also using a different type of extraction to get the juice in the first place—see page 195 for more on that.

Wash **2 pounds (900 g) unpeeled cucumbers** and cut them in half lengthwise. Peel and cube **1 small (2-pound/900 g) honeydew melon** to yield **1½ pounds (675 g) honeydew**. Juice the cucumber and honeydew in a juicer (or coarsely pulse in a food processor, then strain in a colander lined with cheesecloth) and mix together.

In a large blender carafe, add the cucumber-honeydew juice and **¾ cup (50 g) each of cilantro sprigs**, **parsley sprigs**, and **dill sprigs**, as well as **half the zest of 1 lime**. Blend on high speed for 3 minutes, then strain the juice off the pureed-up herb bits with a fine strainer. Add the **juice of 1 to 2 limes,** to taste. Serve chilled. This can be stored tightly covered in the fridge for up to 1 or 2 days.

Makes about 3 cups

You're Telling Me You Milked a Nut?

Your local coffee shop probably has, what, three different plant milks available—soy, almond, and oat? Possibly coconut or another nut like macadamia or pistachio. Obviously, these aren't literal "milks": there was no process like milking a cow involved.

Instead, the creamy and opaque stuff in the nuts and grains (mostly fats, proteins, and starches) was extracted with water.

The fats and uncooked starches in nuts, legumes like soy, or grains like oats are basically insoluble in water, as are many proteins. In nut milks, very little of these components really dissolve into the water that extracts them. Technically, what we're making is a suspension: insoluble things microscopically held in water. Milk is also a suspension of proteins and fats, which are what makes it opaque and creamy.

Nut and grain milks go way further back than hippies; medieval Europeans used almond milk instead of cow's milk during various fasts, and they learned the technique from the Arab world, where cooks had been making almond milk for even longer. In ancient Rome, barley water was a famous refresher, and in Mexico and elsewhere in Mesoamerica, corn was blended, cooked or uncooked, into milky drinks.

Peanut Milk

You can substitute nearly any nut or seed for the peanuts (try pumpkin seed, almond, hazelnut, or cashew—watch out for walnut, which has a very tannic skin). You can also even substitute grains like buckwheat, rice, or toasted corn kernels—as long as you make sure not to heat the water too much higher than 150°F (or all the starch will gelatinize and turn it into glue!).

Blend **2 cups (280 g) unsalted, roasted peanuts** with **6 cups (1420 ml) very hot water** in a blender at high speed until they become a uniform, very light brown watery paste.

Line a colander or chinois with a quadruple layer of cheesecloth or a *very* clean kitchen towel you don't mind getting very dirty. Put the strainer-cloth apparatus into a container that can hold it without falling over—a Cambro container or, lacking that, a pasta or stock pot.

Pour the peanut-water paste into the cheesecloth and let the liquid drain into the container for 5 to 10 minutes. Then, pick up all the free edges of the cloth and twist together so all the peanut pulp is squeezed into a ball. Continue twisting to squeeze out as much of the water as possible, leaving behind a dryish peanut paste. The liquid is your peanut milk. It keeps for 3 to 4 days in the fridge in an airtight container.

Makes about 6 cups

Variations

Add an equal volume of demerara sugar to almond milk and heat very gently to dissolve, and you've got a basic orgeat or almond syrup (other renditions add orange flower water and almond extract), which is an essential component to the Mai Tai and other cocktails.

Pumpkin seed, peanut, sesame, and pecan milks all make great syrups—consider using half white sugar and half another flavorful, unrefined sugar (see page 75).

Between Polar Opposites: Vinegar, Alcohol, and Beyond

At the end of a big meal, I love to make like an Italian grandfather and take a glass of amaro as a *digestivo*. Amaro (Italian for "bitter") is both delicious and stomach-settling, but a big part of what makes it so appealing is its many layers of flavors: herbal, spiced, citrusy, and others. Amaro styles are defined less by their specific bitter elements, and more by how these other flavors are balanced and layered: resinous and a bit harsh for Fernet, piney for alpine styles, sweet spices, big orange or rhubarb root notes.

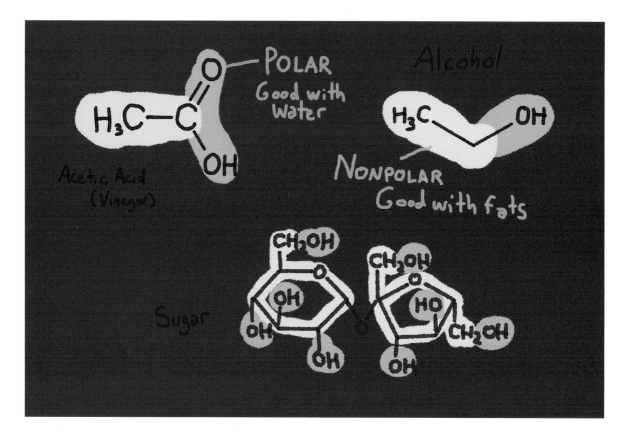

We can thank the extractive abilities of alcohol for all these big flavors. Alcohol, obviously, can mix with water—a typical bottle of hard liquor is still 60 percent water, after all. But it's also got an affinity to nonpolar smell molecules that's closer to fats than to water. Polar vs. nonpolar isn't really a binary; it's more of a spectrum, and alcohol sits somewhere in the middle. The biggest benefit of this hybrid polarity is that alcohol can mix with water, *and* make water better at extracting nonpolar smell molecules.

Playing for Both Teams

Alcohol can do all this because its basic structure is a hybrid: one part of it acts like a polar molecule, another part acts like a nonpolar one. It can act as a kind of mediator between molecules of mismatched polarity. A few other food molecules share this skill—such as acetic acid in vinegar, and sugars—all of which have both polar and nonpolar elements, intermediate polarity, and the power to make water better at pulling out, and holding on to, more flavors than it could on its own.

Let's start simple and sweet, with sugar.

Sweetly Enhancing Extraction with Sugar

With a molecular structure of a nonpolar middle crowned with polar hot spots, sugar bolsters and stabilizes water infusions, holding on to flavor molecules tighter than water alone. This means that adding sugar to your water infusions can make them more flavorful. Many eighteenth- and nineteenth-century formulations for punches call for an oleo-saccharum, dog Latin for "oil-sugar," a unique method for extracting the abundant aroma molecules from lemon or other citrus peels.

Mixing granulated sugar with strips of citrus zest (a little less sugar than zest by volume), massaging them a little, and allowing to sit for a few hours or up to half a day draws out the flavorful peel oil and a bit of water, creating an intensely citrusy syrup that works just as well in individual drinks or other flavoring situations as it does in punches.

Macerating fruits, especially berries or stone fruit, with solid sugar for a few hours does the same thing, but juicier: a syrup pulled out by osmosis (see page 46) and stuffed with flavor by the aroma-grabbing power of the sugar.

Herbed Rich Simple Syrup

When incorporating sugar into water-based extractions, I like to use a rich simple syrup, two-thirds sugar by weight, and pack it with as much flavor as possible. My basic procedure is to boil water, then pour it over weighed-out herbs or flowers and sugar. The heat immediately blanches the botanicals, stopping them from oxidizing, speeds up the infusion, and is usually sufficient to dissolve all the sugar. For plain rich simple syrup, simply omit the herbs. For **around 1 cup (60 g, loosely packed)** herbs, add **3 cups (600 g) sugar** and **1½ cups (350 g) boiling water**. I steep it in the fridge for a day, then strain into an airtight container. Store tightly covered, in the fridge, for up to 3 months.

Makes roughly 2⅔ to 3 cups (600 ml)

Variation

You can also start from whatever amount of herbs you have—add double to triple their volume in sugar, and half as much water as sugar. It's a fantastic method with delicate botanicals like lemon verbena, elderflowers or other wildflowers, basil, shiso, mints, even sturdier thyme branches, dried chiles, or spices. Use to dress or marinate fruits, or add to sodas, coffee, iced tea, or lemonade.

Infusing with Vinegar

For proof of vinegar's ability as a nonpolar solvent, look no further than its many applications in cleaning: most residues, from sticky floors to coffee dregs, stick because of their nonpolar-oily components, which the acetic acid in vinegar handily loosens and carries away.

More deliciously, vinegar can lead us to shrubs (sugar-sweetened vinegars that make delicious sparkling sodas), pickled onions and other vegetables, and surprisingly flavor-packed infused vinegars. My mother makes tarragon vinegar every year, adding a handful of sprigs each to bottles of white wine vinegar, which gives them a soft, richly tarragon-licorice flavor. Other good candidates for this type of extraction are raspberry, citrus, fig, black pepper, mango, bay leaf, ginger, chile, and vanilla.

Mingling of vinegar with flavorful ingredients can create two new ingredients through infusion: soak dried fruit, mushrooms, or rose petals in vinegar, and you'll get both plump, pickled versions of those ingredients, and an extract of their essence into the vinegar itself.

Tarragon-Apricot Vinegar

Use this and the following vinegars in dressing salads or other vegetables, finishing a pan sauce, or making beurre blanc. For a shrub, combine equal parts sugar and infused vinegar, and add a couple of ounces to a glass of seltzer (or cheap and cheerful sparkling wine).

The leftover fruit can be chopped and used as a sweet-tangy relish, folded into cooked barley or farro, or dropped into braising meats.

Combine **1 cup (25 g) fresh tarragon leaves**, **15 dried apricots (about 70 g),** and **2¼ cups (500 ml) white wine vinegar** in a large glass jar. Infuse in a dark, cool spot for 2 to 4 weeks (taste at 1 week and see what you think of the flavor strength). Strain. Store at room temperature, away from light, for up to 6 months.

Makes about 2 cups

"Cherry-Plus" Vinegar

Following the same fruity pattern—but with apricot's cousin, cherry, and its spicier flavors echoed by cinnamon, which shares some of cherry's flavor molecules. It's cherry, but enhanced.

Combine **1 cup plus 1 tablespoon (150 g) dried cherries**, **1¾ teaspoons (4 g) Saigon cinnamon,** and **2¼ cups (500 ml) red wine vinegar** in a large glass jar. Infuse in a dark, cool spot for 2 to 4 weeks (taste after 1 week and see what you think of the flavor strength). Strain. Store the vinegar in an airtight container at room temperature, away from light, for up to 6 months.

Makes about 2 cups

Extracting with Alcohol

As a carrier for captured nonpolar flavor molecules, alcohol is only slightly less effective than fat. And it does things fats just can't do: it's highly drinkable, and alcohol and water mix freely, with the alcohol making the organic stuff much more happy to dissolve and stick around.

Mixtures of alcohol and water—both those with a low percentage of alcohol, like wine, or a higher percentage of alcohol, like liquor—make it possible to cram a bunch more aromatic molecules into water than you ever could with plain water, since they can now stick to the alcohol within.

Medical authorities throughout history have taken advantage of this—getting the medicinal stuff out of an herb or root and into your patient is much easier if they can just drink it, and you can collect a lot of medicinal compounds, more easily, using a water-alcohol mixture than using water alone. Alcohol's not picky about whether molecules are medicinal, flavorful, or both, so a medicinal tincture or wine isn't only a medicinally infused tincture or wine, but also a flavorful one.

Add in the ease of extracting flavor compounds (by accident) while you're extracting medicinal compounds (on purpose) with the innately fun properties of alcohol, and you can see how the line between medicinal and recreational usages of infused wines and tinctures has been innately blurry for a long time.

In the modern world, besides alcoholic tinctures and decoctions actually intended for medicine, we make alcoholic flavor extractions with beer (dry-hopping and barrel-aging), wine (vermouth), and high-proof alcohol (bitters, tinctures, many liqueurs).

Coffee Rum and the Peanut Russian

Coffee and aged rum often end up together in cocktails. The rum reinforces and highlights both the brown and caramelized flavors of the coffee, and the fruity notes, which both beverages already have. Coffee can get bitter quickly when extracted in alcohol (because, like alcohol, those bitter molecules are *also* a mix of polar/nonpolar elements), so this is a quicker infusion than you might use if you were making herbal amaro or vermouth.

Coffee Rum

Source **8 ounces (225 g) fresh, recently roasted coffee beans**, preferably a medium-dark roast. This extraction into alcohol is not very forgiving of staleness.

Grind the beans coarsely, and mix with **one 750-ml bottle dark rum**—Myer's, Goslings, and Pusser's are good choices, as is Plantation dark or O.F.T.D. if you have access to the bottles and the coin.

Steep overnight or for 24 hours maximum and strain. For an even fruitier flavor expression, use a fairly lightly roasted Ethiopian coffee (which tends to have juicy blueberry notes) and a funkier Jamaican rum like Smith and Cross or J Wray and Nephew, or Plantation Pineapple.

Store in the original bottle at room temperature, away from light, for up to

6 months. Makes a scant standard 750 ml bottle

Peanut Russian

Everyone loves a White Russian. But the flavor pattern that the drink expresses—sweet, creamy, roasty—has much more potential than vodka and Kahlúa can fulfill. Here, we use two forms of extraction to make something far more interesting—infusing good coffee into dark rum, and using water extraction to make a creamy milk out of peanuts. It's even vegan! To mix it up, try it with another homemade nut milk like almond, pistachio, or sesame, making sure it's extra-creamy.

Mix **2 ounces (60 ml) Coffee Rum** (above) with **1 ounce (30 ml) Rich Simple Syrup** (page 219).

Put a big piece of ice or 3 to 4 normal-sized ice cubes in a rocks glass or small tumbler.

Top with a float of **2 ounces Peanut Milk (page 217)**.

Serves 1

Boule au Feu Cinnamon Dram

There is a popular sweetened, cinnamon-flavored whiskey that in my opinion ranks among the most unpretentiously delicious industrial food products out there. This is not a "better" version, just a homemade celebration of cinnamon flavor. It's a whiskey-based, complexly sweet-spiced liqueur, which you can sip cold or use in drinks where you'd use another liqueur like Triple Sec or Chartreuse, but when you want a distinctly warm-spiced flavor contribution.

Since the only flavors here are the whiskey and spices, with nothing else to hide behind, you will get a much more consistent result if you weigh your spices rather than using measuring spoons.

In a quart-sized glass jar or similar-sized nonplastic vessel, combine **one 750-ml bottle Rittenhouse rye whiskey** (or other similar whiskey), **45 g Saigon cinnamon, 30 g Ceylon cinnamon, 5 cloves, 15 g fennel seeds, 5 g caraway seeds**, and **3 g cedar or oak shavings**. Cover and store somewhere cool and dark, steeping, for 2 to 3 weeks. Strain the spices out and add **100 g demerara sugar**. Store in the glass jar (or pour it back in the bottle using a funnel) at room temperature, away from light, for up to 6 months.

Makes a standard 750 ml bottle

Vapors and Volatility

Most infusions in cooking involve liquid contact (like steeping a tea) or extraction between solids (like the mingling of ingredients in compound butter). Vapors (or gases—in the frame of reference we're working in, they're chemically the same thing), the third form of matter, can mingle and infuse like a champ, with some weird and exciting side effects.

Distillation, the process we use for making whiskey and other spirits, is a great example. You start out with a relatively low-alcohol mixture, basically a beer or wine. Alcohol molecules and smell molecules already have a tendency toward volatility, which is simply the ability to easily become a vapor (which is how we can smell them in the first place—they have to be gaseous to be able to blow into our nasal cavities!). When you boil a distillation, the vapors it creates are packed full of volatile alcohol molecules and smell molecules, which you then cool and condense down back into a liquid—liquid that is more alcoholic and richer-flavored than what you started with. Beyond taking up moonshining, you can get volatile molecules to diffuse, infuse, and be extracted in their gas form just like they can in their dissolved, liquid, or solid form—but with less accompanying extraction of taste molecules. With a couple of weird but effective tricks, this can mean especially aromatically intense, clean-tasting infusions.

I first came across one of these techniques, which you could think of like a distillation in ultraslow motion, or an extraction that happens in the gaseous headspace above a puddle of booze, in one of Giuliano Bugialli's books on regional Italian foods, in a recipe for a mandarin orange liqueur similar to the Italian liqueur limoncello. But unlike the typical limoncello technique (peeling citrus and mixing those peels directly with strong alcohol for a long soak), you're asked to get a very large jar or crock, put alcohol in the bottom, wrap some mandarins in cheesecloth, and hang them *above* the alcohol. Not touching, leaving room for Jesus. It promised that in several months, through some kind of alchemy, the flavor of the mandarins would be conveyed into the alcohol, at which point you were to remove the fruit and cheesecloth, and dilute and sweeten the liqueur.

I set off to try it myself with three Meyer lemons, cheesecloth, butcher's twine, and Everclear. The lemons were hung o'er the liquor with care, and after a few days the alcohol started turning bright yellow. After a couple of weeks I snuck a taste and found deep, saturated lemoniness, seemingly stronger than biting into a lemon itself.

How? When you seal lemons (or mandarins, or other aroma-rich things) in a jar, their volatile, flavorful smell molecules will naturally diffuse out of them to spread out as a vapor throughout the jar. You get a jar full of scented air. If you add anything else to the jar, it gets bathed in this scented air. If this other thing is alcohol, well, alcohol's whole thing is mixing with and extracting nonpolar molecules, and it doesn't really care whether it gets them from touching a lemon full of smell molecules, or a bunch of air full of smell molecules. In your jar, even if the alcohol isn't touching any lemons, it's extracting the smell molecules out of them, siphoning them over the ether, like a radio picking up an FM signal, no hardwired connection necessary.

The one drawback is that this multistep vaporize-diffuse-dissolve-repeat system is not

very fast, so you need to be patient. But what the technique lacks in speed it makes up for in ease, comprehensiveness, and cheap equipment as a way to perform distillation-like flavor capture without a (usually) illegal still. The method also appears to dissolve very little in the way of bitter and astringent compounds, since these have next to no volatility—so the product isn't muddied by those flavors. I've tried it with things like cacao nibs, grapefruit peels and cardamom, freshly roasted coffee, and juniper berries and coriander in place of lemons or mandarins, and it creates clean, saturated extractions across the board. When I tested it with black pepper, the alcohol picked up some warmth and spiciness as well as, more noticeably, a distinctly perfumed and delicate flavor that's sometimes easy to miss in ground pepper.

Headspace-Infused Meyer Lemon Liqueur

This is a technique that begs to be played with and applied to other aromatic ingredients beyond Meyer lemons. You can try it out with small grapefruits, limes, yuzu, or mandarins; or a few handfuls of pink and black peppercorns, coriander and orange peel, or fennel seeds and allspice, as well as experiment with blending any of the above. The liqueur you end up with after dilution is about 40 to 45 percent alcohol by volume, essentially the strength of gin, with the sweetness of a typical commercial liqueur or limoncello.

To really go wild, you can remove the old lemons after they seem spent, and refresh them with new ones as you carry on the extraction, but this can become even a little *too* aromatic, verging on soapy, if you're not careful about checking the taste frequently (monthly).

Start with a half-gallon, wide-mouthed, sealable glass jar, 2 square feet of muslin or cheesecloth, and thin butcher's twine.

Pour **11½ ounces (325 ml) 190-proof Everclear (**or other very high-proof, neutral alcohol) into the half-gallon jar. It will only fill it about 20 percent of the way and feel a little silly. This means you're doing it right.

Fold the cheesecloth in half into a square, give **3 Meyer lemons** a rinse and dry, and put them in the center of the cheesecloth. Gather the edges of the cheesecloth together loosely, and use the butcher's twine to tie it off into a loose little bundle—basically, just make sure the twine won't slip and the lemons won't fall out. Leave 18-inch tails on both ends of the twine.

Holding the butcher's twine, gently push the bundle 1 lemon at a time into the upper part of the jar, avoiding dropping them into the puddle of alcohol. Pull the strings taut to the outside of the jar, hold them there with one hand, and screw or clamp the lid on with the other. A tight seal on the lid should grab the twine, and prevent it from slipping down. Now the lemon bundle is suspended by the trapped twine, and hanging above the alcohol.

Leave it undisturbed and out of direct sunlight for 2 months, then open and spoon a little out. Dilute 50:50 with water (I don't recommend drinking straight Everclear! It's practically wood stripper at that strength) and taste. If you're happy enough with where the flavor is, stop here, but if you're patient, I recommend letting it go for

2 to 4 more months, because it will keep getting stronger.

When the vapor infusion is finished, carefully remove the lemons and cheesecloth from the jar, and pour the alcohol into a large measuring cup to see how much you have. Some you will have lost to evaporation, wicked through the twine. If the volume you measure is 8½ ounces (250 ml) or greater, add **11½ ounces (325 ml) filtered water** (the same volume as the alcohol you started with), and **5 to 7 ounces (125 to 200 g) white sugar**, to taste, to the infusion, which should get you in the ballpark of 40-45 percent alcohol by volume (or 90 proof). If you ended up with less than 8½ ounces (250 ml) of alcohol, add 9½ ounces (275 ml) water and 4 to 6 ounces (100 to 170 g) sugar. Store in tightly capped glass bottles in the fridge for up to 1 year.

Makes a scant 750-ml batch

Variation

For a more gentle sipper, cocktail ingredient, and cake soak, dilute to 25 percent alcohol by volume by adding **30 ounces (900 ml) water** (or 2.75 times the post-extraction volume of alcohol you measured) and adding **8 to 12 ounces (225 to 350 g) sugar** (20 to 30 percent of the total volume), to taste.

Keeping It Casual: Cooking with an Infusing Mindset

Extraction, infusion, and moving molecules around: simple mechanisms that make for versatile techniques, are all different enough to give you lots of options for incorporating them into your cooking. But they don't happen only when you deliberately set them up to, in molasses-ing a fruit juice or pulling together a spiced oil. Because of the eternally wiggly nature of molecules and

their tendency to haphazardly fill any space they can get to, these processes are happening all the time whenever you put ingredients in contact with one another. One of the reasons that stewed foods taste so good when you reheat them the day after you make them is that all the components have had some time to soak up one another's flavor molecules and meld together.

Further Flavor Extractions "Through the Ether"

Fats and oils are the grand dames of really effective flavor extraction. Throughout history and in different places around the world, this has primarily taken the form of submerging and macerating aromatic ingredients in liquid fat, as well as the early-modern enfleurage process (see page 203) of capturing the odor of jasmine, rose, and other delicate flowers in a thin film of semisolid fat.

An old solvent extraction technique from Persia and Mesopotamia, still in use in some workshops in India for capturing the aroma of jasmine and other specialties, looks a lot like the technique for extracting citrus flavors through the air for limoncello that we just talked about. In this seed-based older sister to enfleurage, the perfumer layers whole sesame seeds or almonds and flower petals in glass or ceramic containers, covers them, and lets the seeds and the petals passively mingle for about a day. The precious volatile molecules slowly diffuse out of the flower petals, into the air around them, and then from the air into the seeds, where they dissolve in their fat and stay put. Since fat can take up flavor molecules like a bottomless reservoir, it can dissolve more flavor molecules than the comparatively minuscule amount in the petals, so periodically the perfumer empties out the jars, discards the spent petals, and relayers the semi-infused seeds or nuts with fresh petals, repeating the process until the seed fats become saturated.

In this form, it's a simple matter of pressing the seeds to squeeze out their now potent oil and using it in perfume.

In modern kitchens, this technique is echoed in thrifty ways of capturing the fleeting and excess scent of luxury materials. Storing whole vanilla beans in jars of sugar before using the beans creates passively infused vanilla sugar, from the aromatic vapor diffusing out of the vanilla bean and around the tiny air gaps between sugar crystals. Likewise, you'll often see truffle stored in jars of rice, to capture the circulating aroma compounds from the truffle in the rice, like pulling water from a stubbornly malfunctioning iPhone. While aroma compounds lost to vaporization could hardly be called waste products, these two methods capitalize on the gonna-lose-it-anyway quality of volatile scent and capture it in much the same way as enfleurage.

So while thinking about discrete techniques is essential for wrapping your mind around these dynamics, they're also a compass you can keep in your (mental) pocket for nudging whatever you're cooking in the direction you want it to go—even if you didn't wake up and say, "Today I am going to do an extraction." Any time you bring ingredients together, there's a way to finesse them to infuse into each other more, or keep them more separate. If there's a flavor you need to liberate from its packaging and physical texture (like big chunks of spices or delicious roasted chicken pan drippings), there's a way to do so via only a minor detour or variation on a recipe.

All you need to do is look for the predictable patterns that create opportunities for extraction that show up in nearly every recipe. I think of it as "casual extraction"—the do-less, infinitely adaptable way to move flavor molecules around.

Sauce: Extracting Butter and Onion Flavors with Tomatoes, Alla Marcella

This simple tomato sauce is in the style of Marcella Hazan, the Italian-born cookbook author who helped popularize an unalloyed and traditional approach to Italian cooking in the United Kingdom and the United States. The sauce got a second wind from online cooking blogs and websites in the 2010s, in part because of the effortless sprezzatura of its preparation. An authentic Italian tomato sauce, but not one that requires the cook to source five kinds of meat and tend for six hours: three ingredients (four if you count salt), done in an hour or less.

It's also a beautiful and minimalist example of flavor infusion: combine tomatoes and butter with a quartered white onion, allow the tomato and butter to extract flavors from each other, and the onion's flavor molecules to infuse into both the tomatoes and butter, then discard the spent onion, having used only its flavor but not its texture.

It's simple enough to make whenever you need it, but I often make extra and freeze it.

In a large saucepan or Dutch oven, combine **2 pounds (4 cups; 900 g) very ripe, very good fresh tomatoes**, crushed and with all their juices (or one 28-ounce can good-quality canned tomatoes), with **12 tablespoons (1½ sticks; 170 g) good-quality butter**, **2 white onions** (peeled and quartered), and **1 teaspoon sea salt (6 g) or kosher salt (3 g)**. Bring to a simmer over medium heat, uncovered, and simmer for about 45 minutes. The sauce will cohere and reduce a bit, but part of its flavor appeal is a freshness that comes from not cooking it for very long. While the total cooking time is flexible, the sauce is at its best when it has cooked just enough to thicken slightly and render the onion translucent, but not long enough to really thicken, turn a deeper shade

of red, and take on a more decisively cooked flavor.

Use a slotted spoon to remove the spent onion pieces and discard. Taste, then adjust salt to your liking.

Eat immediately, or freeze cooled sauce in airtight containers for up to 3 months.

Makes enough to sauce 1.5 to 2 pounds of pasta, or around 3 pounds of pizza dough (as 6 pizzas, below)

Variation
Onion-and-Butter-Infused, Marcella-Sauced Pizza

This sauce makes for a refreshingly light and fresh-tasting pizza, and has a special affinity to burrata or stracciatella cheese, which incorporates strands of high-quality fresh mozzarella with fresh cream in a kind of delicious tangled goo.

Low-Key Infusion and Extractions You're Probably Already Doing

Simmering together tomato, onion, and butter for pasta sauce.

Marinating meat or vegetables.

Cooking in salted water, so salt will infuse into the grains or pasta you're boiling. A flavorful cooking liquid (stock, a mixture of pureed tomatoes and water, adding whole spices or herbs to the cooking water) will infuse even more flavor.

Poaching or steaming chicken or fish with any kind of flavored liquid, tea, or herbs.

Braising substantial greens like kale or collards with a ham hock or other highly flavored meat.

Cooking beans with a bay leaf or other herbs.

Adding a Parmesan rind to simmering soup.

A bouquet garni of fresh herbs, tied into a removable bundle, to enhance simmering stock.

Half of the sauce recipe, a pound of buratta or stracciatella, and 1½ pounds of pizza dough will make 3 decent-sized and delicious pizzas.

Clove and Allspice Pickled Marmalade

I love the flavors of spices and fruit together, but I often find the slightly gritty texture of ground spices a bit of a distraction from the soft and yielding texture of the fruit. A few years ago, I was served some pickled oranges that were the best of both worlds. Arriving alongside several cheeses, they were sweet like a marmalade but runnier, with an addictive vinegary pungency and whole cloves spicing them. The big clove pieces were easy enough to avoid when spooning it from the jar, but their extended contact with the oranges had perfumed and infused the whole thing with spiced warmth. I needed it in my life.

I tried replicating it, first by cooking oranges with wine vinegar to infuse them with acid, and then adding some sugar, which was a try-hard miserable failure and tasted bizarrely overcooked. Next, I tried approaching it like a marmalade, adding spices right at the beginning of cooking (so their flavors would diffuse into the oranges) and neutral-tasting white vinegar at the end, which kept the oranges' fresh flavor intact and had plenty of time to diffuse inward. I included allspice for a more rounded-out spice flavor along with cloves.

Reader, it was perfect. Sometimes I just eat this out of the jar, though it's excellent with cheeses, especially cheddary, blue, or bloomy rind varieties. It is also a great, tangy accompaniment to meats like pork or duck.

The flavor will continue to marry and develop over the first few days, so it's a good idea to make this 3 or more days ahead of time if it's for a specific event.

If you feel like experimenting, start with similar flavors and work outward: with a blood orange or Valencia orange, or Meyer lemon instead of oranges and cardamom instead of allspice and cloves for a more bracing, sour condiment. Play around with spices, especially in the sweet (page 162) or dry and fragrant (page 161) families: cinnamon, sliced ginger, pink pepper, fennel seeds, cardamom, black pepper.

Wash **1 medium-large navel orange**, then cut it crosswise into very thin, ⅛-inch sections (a few can edge toward ¼-inch, but too thick and the texture will be wrong). Cut each section into 4 wedges, the first cut pole to pole, and the second cut perpendicular to this one across the equator—you should wind up with roughly 8½ ounces (250 g).

Put the thinly sliced orange wedges in a large saucepan at least triple the height of the layer they make. Add **8½ ounces (250 g) water** to barely cover the oranges.

Simmer over medium-low heat for 40 to 50 minutes, until the oranges look translucent and floppy and almost all of the water has evaporated. Turn the heat down if it looks like it's boiling or evaporating too fast. The remaining water will be light orange and a bit milky.

Meanwhile, roughly crack **5 allspice berries** and **5 cloves** in a mortar and pestle.

Add **10 ounces (300 g) white sugar** and the crushed allspice and cloves to the oranges and the small amount of water remaining. Gently simmer everything on medium-low heat for about 1 hour, reducing the heat if it starts boiling too fast or sticking. You're both infusing the oranges with sugar and the spice flavors as well as evaporating water to develop a gelled network between the sugar and the orange's pectin and acid. Measure the temperature of the simmering mixture with an instant-read thermometer: you're aiming for 227 to 228°F/109°C, which is slightly higher than we usually go for jam, but we'll be diluting the pickled marmalade in a moment.

When the oranges and sugar are thick and sticky and at the right temperature, remove from the heat and pour into a heatproof, lidded container (or a heatproof container whose mouth can be covered with plastic wrap). Stir in **3 tablespoons plus 1 teaspoon (50 ml) distilled white vinegar**. Immediately cover to avoid evaporating much of the acetic acid away.

Allow to cool to room temperature, then spoon into glass jars with tight-closing lids and store in the refrigerator for up to 6 months.

Deglazing, Extraction, and Pan Sauces. Plus, a Supposedly Dumb Thing I'll Definitely Do Again

You're cooking a piece of meat (which, as we know from page 185, is a network of mostly unflavored proteins soaked in water full of lightly flavored molecules and flavor precursors—yum—that create meaty flavor when you heat them). As you bake, roast, sear, or sauté, the just-add-heat-to-create-meat-flavor juices leak out. As the protein heats and starts tasting meaty, then starts browning, a thin layer of browned-meat juice builds up on the meat and in the pan. (For more on how browning happens and creates flavors, see page 244.) When you're finished, you have a flavorful sear on the meat, and a corresponding cooked-on layer, which a French-trained cook would call a *fond*, stuck to the pan.

Your first thought might be to scrub it off, but if you do wash it down the drain, you'll be missing out on one of the most delicious meaty-flavor substances you can make, as a free by-product of something you're already doing. All you have to do is collect it.

For that, we have a technique called *deglazing*, which is really just making an extraction with a hot, water-based liquid. Deglazing can be as simple as splashing half a cup of wine in the emptied pan and letting it bubble up, dissolve, and extract the cooked-on components as it boils, scraping a little to loosen any stuck parts. (Then, pouring it over your meat, or on any vegetables you eat with it.)

Extraction by deglazing is easy to turn into a pan sauce: use stock, heavy cream, crème fraîche, apple cider or cherry juice, or flavorful vinegar to dissolve the fond. Maybe throw in a pinch of cracked black pepper, cumin, or coriander seeds to infuse your liquid as it bubbles, then fish them out. Gently cook until it's a thickness you find appealing, then remove from the heat and whisk in some mustard or citrus juice, fresh minced herbs, or capers as your heart desires.

Traditional gravy is basically just pan sauce by a different name: fond deglazed and extracted with stock, then thickened with flour. The first time I tried to make gravy without any pan drippings (it'll be fine, I said, there's plenty of flavor in the stock) I was greeted with . . . slightly gelled stock—after that experience, I became an obsessive, an evangelist, for deglazing pan drippings. Sometimes, for flavor insurance on Thanksgiving or another big, roasted-meat meal, I roast a few pounds of chicken drumsticks (or wings, backs, whatever I can get) to make extra food. They are the worst, most dried-out and leathery drumsticks you'll have the displeasure of trying, because when they start getting hot in the oven, I stab them all over with the tip of a knife (just like you're not supposed to do) to make sure there are lots of channels for the juices to run out. I roast them deeply, squeezing and poking them while I turn them. When I'm done, I have roasty bones and meat I can use for stock and, more important, puddles of golden-brown, sticky fond stuck to the pan I roasted them in. Then it's just a matter of pouring in some water, dissolving the fond, and reducing it in a saucepan a little to make sure it isn't too diluted. Scrape it into a small container, and I have about 8 ounces of pure poultry gold, immaculate eighteenth-century osmazome (see page 83), ready to turbocharge gravy (or any savory sauce or braise, pot of beans, etc.) with rich, mouth-filling, roasted-umami flavor.

Making the Most of Infusing the Fatty Ingredients in a Recipe with as Much Flavor as Possible

Creaming butter into cookies, frying pork lardons to start off a boeuf bourguignon, sizzling garlic and onions in oil for a stew or sauté—fats give food an irreplaceable richness and texture, and they also facilitate browning and flavor development during cooking. When they're richly flavored themselves, they spread that flavor around, blanketing every corner of a dish. And since they're so eager to grab on to smell molecules, they're as happy to do so in the middle of any recipe they appear in as they are when you set it up as a special event, like chili oil.

You can take everything you know about infusing fats on their own—whether from a smoke- or herb-infused oil, or a compound butter—and apply it to any situation in which you're cooking with fat.

Any time you're cooking something that involves both a fatty element and aromatic ingredients, you can capture, infuse, and generally spread the flavor out into a cozy blanket by adding the aromatic ingredients to the fat first.

Returning to the Indian culinary technique of tempering (page 205), if you combine any highly aromatic ingredients in a recipe with fatty ones, *before* you combine them with anything else, the fats have a chance to rub elbows with the smell molecules, drawing them out and infusing with them. Adding any more watery ingredients next lets those fats bring flavor with them as they disperse, creating a more even and rich flavor distribution than combining aromatic ingredients with watery ones.

Layered Base to Infuse Cooked Greens

In this recipe, you build and infuse most of the flavors in the initial sautéing step. The chunky, fatty paste that results is sweetly spiced, spicy, brown, nuttily caramelized, and full of salty-umami. Layers upon layers, silkily softly infused throughout its sultry fatty phase with little, salty, sweet sticky bits throughout. Use it for a post-cooking flavor bath for poached or roasted chicken or fish, spoon it over grains or porridge, or, as here, use it as a base to braise greens.

Put **3 tablespoons (50 g) butter** in a skillet or small saucepan and heat it on medium. Keep

Try This

Identifying Extraction Opportunities
When and *how* you combine different ingredients during cooking can take their flavor from good to exquisite: deep, well layered, lingering. Anytime you are cooking, ask yourself, If I'm using fat anyway, can I use a more flavorful version of that fat? Can I put more flavors *into* that fat? Combining fatty and aromatic ingredients first allows the fats to show off their special skills extracting smell molecules.

an eye on it as it melts, foams, and settles into liquid butter over a sprinkled layer of milky solids. When these start turning golden brown, add **¼ cup (60 g) thinly sliced red onion** (about half a medium onion) and cover the pan. The onions will slump and release their internally held moisture, which will stay in the pan and wilt and poach them before they start browning.

When the onions do start to brown a little, add **3 oil-packed anchovies**, finely sliced across and down their length. Let it ride for a few minutes while the anchovy bits start to frizzle, then add **3 finely chopped garlic cloves, 2 teaspoons (4 g) urfa biber flakes, 1 teaspoon (2.5 g) whole fennel seeds**, and **1 teaspoon (2.5 g) cumin seeds**. Cook gently for about 5 minutes, watching carefully so that the garlic doesn't overbrown.

Then add **1 pound (450 g) medium-sliced, stemmed mustard greens, collards, or kale, 1 teaspoon (6 g) of salt**, and **¼ cup (60 ml) water**. Toss, cover, and cook over medium-low heat until the leaves have wilted and softened, about 15 to 20 minutes. Toss again and serve immediately.

Serves 2 to 4 as a side

Capturing and Infusing Flavorful Ingredients into Fats Before Cooking

Making cacio e pepe? Grind the pepper into your butter or olive oil and let it hot-steep a minute before continuing. Making a vinaigrette with herbs? Blend or crush the herb with the oil and let it sit around for 15 to 30 minutes while you're washing lettuce. Making brownies? Add some sweet spice to the butter when you melt it. Nobody wants to read a big block of list-text, so here's a table of times and places to infuse flavorful ingredients into fat you're cooking with—before you add water-heavy ingredients.

What	Aromatic	Fat	When/Where
Cacio e pepe	Black pepper	Butter/olive oil	Melt the butter, add black pepper, then mix with cheese and pasta.
Chili	Chile, cumin, cinnamon, allspice	Oil	Add the spices right into the oil.
Melted butter cookies	Lemon peel, cardamom	Butter	Melt, infuse, cool.
Brownies	Ginger, black pepper, cinnamon	Butter	Melt, infuse, cool.
Pumpkin pie	Ginger, cinnamon, clove, allspice	Cream	Gently heat cream on stove, add spices, infuse for 30 min.
Ice cream	Ginger, tea, coffee beans, mint, basil, lemon verbena, orange peel, pink peppercorn, fennel seeds, cardamom	Cream	Whisk into cream and bring to about 160°F on the stove, hot but below a simmer, then cool and steep for 2 to 8 hours.
Curry	Spices, ginger, garlic	Ghee	Fry with garlic, onions, and ginger, before adding meat, vegetables, or legumes.
Creamed butter cookies	Spices, citrus peel, herbs, lavender, rose, elderflower	Butter	Cream together butter, sugar, and spices/herbs the night before. Let infuse on a cool room-temperature countertop, or in the fridge.
Vinaigrette	Thyme, tarragon, oregano, parsley, dill, red pepper/chile flakes, fennel seeds, cumin, caraway, fennel	Oil	Blend the herbs with the oil and either keep the blender running for a few minutes or let sit for 30 min.
Beans	Bay leaf, rosemary, cumin, chiles, thyme, Mexican oregano, garlic	Oil	Fry or steep them in oil, then add beans and water.
Japanese curry	Coriander, cumin, fenugreek, cardamom, black pepper, fennel seed, cinnamon, clove, star anise, orange peel, turmeric, chili powder, nutmeg, ginger	Oil	Heat the spices in the oil while you're starting the roux.
Any braise	Garlic, bay leaf, thyme, oregano, black pepper, rosemary, citrus peel, fennel seeds, anchovies	Butter/oil	Right when you're heating up the oil—but after you do any searing.

PART 4

The Fourth Law of Flavor

Flavor Can Be Created and Transformed

Key Rules for Creating and Transforming Flavor

When you create new molecules, you create new flavors.

Lots of cooking processes create or alter molecules in ingredients with heat or fermentation.

Heat is a good way to create caramelized, charred, smoky, and roasted flavors in ingredients like sugar, dairy, meats, vegetables, and doughs.

Heat creates flavor by quite literally shaking up molecules, breaking them apart and reacting them together.

Fermentation is a good way to create sour, umami, funky, or sweet flavors in ingredients like vegetables, dairy, grains, and legumes.

Fermentation equals microorganisms: bacteria, yeast, and molds that make new molecules as they ferment ingredients.

Microbes take in one type of molecule in a food—sugar, alcohol, starch, protein—use it for energy, and transform it into a new molecule as a by-product: acids, sugars, free amino acids.

Most fermented foods, like yogurt, kosher pickles, soy sauce, or kimchi, start out as a way to preserve ingredients, and the extra flavors they create are an added bonus.

If you want to have a good time and not a bad one, make sure to match the flavor-creating process you want to use with ingredients that have the right molecular raw materials.

The pinnacle of a medieval alchemist's work was chrysopoeia: the transmutation of matter into gold. It's hard not to feel like that's exactly what I'm doing when I'm caramelizing sugar or browning butter—if I apply just the right set of conditions to simple, innocuous-seeming ingredients, I can conjure up a wealth of deep, complex flavor that there was no visible trace of before. It's uniquely, sometimes even mysteriously, transformative. A poetic metaphor, but happily and more usefully, a molecularly literal one. While I haven't *atomically* changed anything that I put in the pan (which is what you'd need to do to change one element into another), I've taken the comparatively

Flavor can be created, and transformed

sugar × heat = new flavor molecules

bland and boring molecules that are already in sugar and butter, broken them apart, and rearranged them to create new molecules with much more depth and complexity of flavor.

It's a pattern we can find everywhere we turn: jarring up salted cabbage and returning a few weeks later not to salted cabbage but to tangy, pickley sauerkraut, sourness appearing where there was none before. Slowly roasting starchy sweet potatoes into meltingly sugary ones. Pulling a huge depth of umami from the initially tasteless proteins in miso's soybeans or fish sauce's anchovies. Doing these things, we are at the apex of flavor as chemistry: creating flavor at the fundamental level of transforming molecules and creating new ones.

Let's dive in.

Attachment Theory: Breaking and Joining Molecules

Like any other molecule, a flavor molecule is just a bunch of atoms—like carbon, hydrogen, oxygen, nitrogen, or sulfur—held together in a particular arrangement. We call the connections that hold these arrangements *bonds*. Some molecules have very long-lasting bonds and arrangements and stick around for a long time, but none are actually permanent. "Bonding" is less like the bonding power of welding and more like bonds in the form of knots. Some are tighter and more complicated than others, some are harder to tie, but many slip loose pretty easily if you tug or nudge them just right. And with the loose ends created by untying any of these knots, you can easily tie new ones. In other words, join up pieces from old molecules to create new molecules—with new flavors.

Potential Flavors (and Flavor Potential)

Some ingredients have obvious flavors, in the present tense. Herbs, spices, lemons, chiles, salt—they all have a pool of flavor molecules in them, ready to be perceived once you eat them.

Wheat flour, steamed rice and soybeans, a raw chicken? By themselves, in their current state, they have virtually no innate deliciousness. But if you bake the flour into bread, ferment the rice and soybeans, or roast the chicken, the flavor picture shifts: where there was nothing, now you have toasty bread flavors, rich umami-funky miso flavors, and deeply satisfying browned and roasty flavors on offer. These ingredients all had the *potential* for flavor built into their chemical composition; they just needed to be given the right nudge to unlock it and coax it out.

Certain types of molecules that don't have much flavor initially have the potential to turn into ones that do, under the right circumstances. Ingredients that contain these molecules may not currently have a lot of flavor, but they have lots of flavor potential. The list of basic transformations that can transform flavor potential into actual flavor isn't long. But if you know what the options are—what the starting molecules look like, what they have the potential to change into, and the chemical rules and conditions under which you can encourage those transformations to happen—you'll have a much better understanding of what you can do with a given ingredient.

Genres of Flavor Creation

There are essentially two ways of creating new flavor molecules: we can do it purely chemically, using heat, enzymes, and other tools to break and make new bonds, or we can take a biological route, encouraging other creatures (namely, bacteria and other microbes) to do it for us. The first genre of flavor creation covers many of the basic things we consider "cooking": browning, caramelizing, charring, smoking, etc. The second one is the realm of fermentation: what we use to make flavorful things like kosher pickles, kimchi, beer, wine, yogurt, sourdough, and soy sauce.

Starting material	Agent	Transformation	Potential flavor	Examples
Sugar	Heat	Caramelization	Tangy and bitter, cotton candy, caramel, burnt sugar	Caramel
Sugars, carbohydrates, protein	Heat	Burning, charring (pyrolysis)	Charred, burnt, grilled	Charred onions, burnt cinnamon
Wood, carbohydrates	Heat	Combustion/smoking	Smoky	Barbecue, bacon, smoked paprika, tea-smoked duck
Sugar and amino acids	Heat	Maillard reaction	Browned, toasted, seared, roasted, toffee, brown butter	Roasted meat, bread crust, toast, coffee, chocolate, brown butter, "caramelized" onions
Starches	Heat, enzymes/fungi	Amylolysis (breaking starch down into sugars)	Sweet	Malt, sweet potatoes, amazake, koji
Sugars	Lactic acid bacteria	Lacto-fermentation	Sour, pickley, creamy	Sauerkraut, kimchi, yogurt, crème fraîche, sourdough
Alcohol	Acetic acid bacteria	Acetic fermentation (vinegar)	Vinegary, sour, pungent	Wine vinegar, balsamic vinegar, rice vinegar, apple cider vinegar
Proteins	Fungi, some bacteria	Proteolysis (breaking down proteins to amino acids); umami fermentation	Umami and funky	Miso, doenjang, soy sauces, fish sauce, garum

Chapter 9

Creating Flavor with Heat

To a molecule, heat is movement: the more energy a molecule has, the faster it vibrates or tumbles in space, the warmer its temperature is. With enough heat, molecules reach a tipping point, where this motion gets intense enough to start breaking apart at its weakest links. It's like an overexcited horse slipping its hitch knot, or a house falling down in an earthquake. In between hot but intact molecules and catastrophic breakdown, heat energy can alter and rearrange molecules in ways that can drastically change their flavors—slicing up sugars (as caramelization), unlinking a big molecule into its individual pieces (as charring or smoking), or bouncing two different molecules together so hard they entangle, and form something completely new (as the Maillard reaction).

Liquid Gold: Caramelization

The simplest way of creating new molecules with new flavors requires only boring white sugar and a source of heat.

Heat some sugar in a pan and the uniform white crystals eventually collapse and melt into a clear liquid. Keep heating, and you'll get whiffs of cotton candy and sweetshops. The first visual onset of caramelization comes next: straw-colored patches in the clear melted sugar. Continue heating, and these darken to deep gold, and the aroma gets a bit more hefty, suggesting butterscotch and crème brûlée. If you dare to keep going, you can carefully take caramel just to the edge of burnt, turning it deep brown, pleasantly sweet-bitter,

CREATING CARAMEL FLAVOR

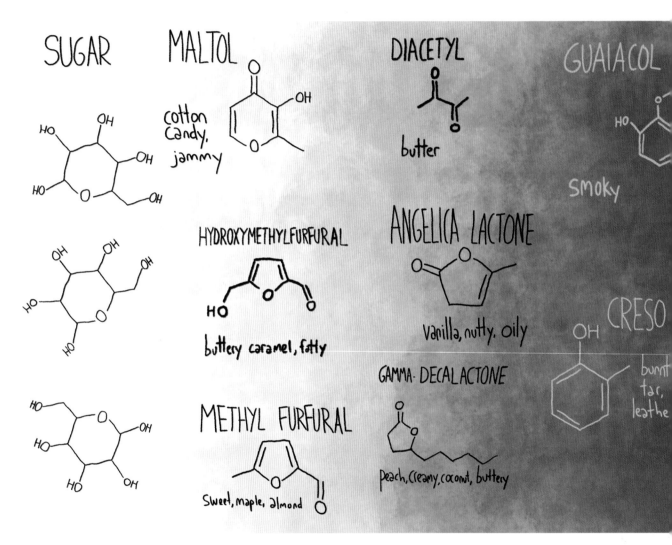

SUGAR

MALTOL
cotton candy, jammy

DIACETYL
butter

GUAIACOL
smoky

HYDROXYMETHYLFURFURAL
buttery caramel, fatty

ANGELICA LACTONE
vanilla, nutty, oily

GAMMA-DECALACTONE
peach, creamy, coconut, buttery

METHYL FURFURAL
sweet, maple, almond

CRESO
burnt tar, leathe

with intense aromas of molasses and nearly charred toffee. (If you just have to go further to see what happens, know that if you keep heating caramel it will eventually fully decompose into a completely black, carbonized mass—and then we're in the domain of burning and charring, which is the next section.)

What you're seeing and smelling is the heat-driven, chemical decomposition (or *pyrolysis*) of sugar.

A sugar molecule can usually be found tied together, head to tail, in a tidy little hexagon or pentagon. Heating it up (shaking it) sends it careening, shearing off little bits of the molecule like a wayward car breaking off its side mirrors, pulling open the knots of the bonds that hold it together.

Breaking any of these bonds leaves behind loose and sticky ends, and they wrap and tangle up with any other loose, sticky, formerly bonded ends they can find. At the beginning of caramelization, this entanglement creates small, light, and volatile aroma molecules, the "sweet" smells you detect before any color even forms. Echoing the original shape of the sucrose (white sugar) molecules they came from, these aroma molecules often have hexagonal or pentagonal elements, like burnt-fruity isomaltol, sweet-creamy gamma-hexalactone, and nutty-vanilla 5-methyl-2-furfural. As you continue to dump energy into your pot of sugar by heating it, the molecules keep sticking and unsticking until they make more complex, heavier, and more burnt aromas, and beefier pigmented molecules called *melanoidins*, which give caramel both its color and its bitterness. A sucrose molecule starts out with twenty-two hydrogen atoms in its structure, and many get knocked off during all this breakdown. They're set loose as hydrogen ions, which are the basic unit of acidity: Caramel gets more acidic and tangier as you heat it.

How Dark Do You Like It?

Maybe you like a slightly bitter, coffeelike flavor and want a surprisingly dark caramel (one of my favorite ice creams is the aptly named Burnt Caramel from Toscanini's in Cambridge, Massachusetts, so dark that it could pass for molasses ice cream). Maybe you like the subtlety and cleanness of a light caramel. There's no right answer, just the right one for you. The best way to figure that out is to caramelize sugar in your kitchen and taste it.

(Be sure to *let the syrup cool before you taste it*.)

I like to put a Post-it note at eye level on the cabinet door nearest to the stove, and keep a pencil nearby to jot things like this down. For even more precision, stick a high-heat-resistant thermometer in the caramel and make a little note about the temperature when it gets to the color and flavor you prefer. Now you know YOUR best caramel temperature.

Here are some approximate guidelines:

Light caramel, 330° to 340°F: Blond to straw brown. Sweet, lightly caramelized, and delicate.

Medium caramel, 350° to 360°F: Golden brown to amber, deeply caramel, taking on toffee notes and more acidity.

Darkest caramel, 375° to 390°F: Dark brown. Deep, coffeelike, and fairly bitter, with an almost smoky flavor.

Caramelized Sugar

Caramelized sugar is not so much its own food as a seasoning, accent, or starting point for something else (like a rich caramel sauce). In that sense, it's similar to the brown (page 73) and unrefined (page 75) sugars detailed in "Sweet"—and it performs well in any of the places where you dissolve those things into something else (added to a braise or beans, or to a syrup or drink).

Freshly made, liquid caramelized sugar will solidify as it cools—so you should pour it into, or onto, wherever you need it to be while it's still hot. That could include the bottom of the container or pan for crème caramel or a fruit upside-down cake. Or try pouring any hot caramelized sugar over thin slices of orange, pear, grapefruit, or apple for a crisp layer that will slowly slump into a delicious sauce. Vietnamese caramel or nước màu is used, often along with salty-umami-funky fish sauce, to sweeten and flavor braised pork and other long-cooked savory dishes.

The enemy of even, glassy caramel is crystallization, which is where the molecules line up into tight and grainy balls rather than crisp and flat caramel. I like to use a little water to help lubricate the sugar molecules while they're melting from solid to liquid. For extra insurance, you can add a little corn syrup or lemon juice, which will impede crystallization by throwing non-sucrose molecules into the mix to interrupt the ranks.

Combine **1 cup (200 g) sugar** with **3 tablespoons (45 ml) water** in a smallish saucepan. Heat over medium heat until the crystals have dissolved, then wash down any crystals on the sides of the pan with a wet pastry brush to dissolve them and prevent them from crystallizing. Ease up on the heat to medium-low if you go past golden brown; it's easy to overshoot and burn the caramel.

Makes 1 scant cup of caramel

Variations
Burnt Honey

Honey can caramelize just like white sugar, and its botanical flavors, tanginess, and lightly fermented complexity makes caramel even more interesting. Bring it to wherever you'd use caramel or caramel syrup, as discussed below, or use as the sweet portion in cocktails, or to glaze or garnish peaches or apples, or with pastry.

Put **½ cup (120 ml) honey** in a saucepan and heat over medium heat. As it heats up, it will thin out in texture. Continue heating until it simmers and stir well. After a few minutes, it will begin to take on a darker color. Cook for 4 to 5 minutes, but pull it when it is still just lightly darkened—you don't want it to literally burn.

Makes ⅓ to ½ cup (approximately 100 ml)

Burnt Sugar Syrup

Make a flavorful syrup to sweeten rum or whiskey cocktails, iced coffee, or lemonade, or to dribble into a salad dressing or sauce to balance sourness with sweetness and toasty depth.

Make Caramelized Sugar as directed, cooking it to your preferred stage (see "How Dark Do You Like It?," page 237) and add **1 cup (240 ml) water**. Stir to dissolve over low heat. Store in the fridge and use within 3 to 4 months.

Makes a scant 1½ cups

Burnt Honey Syrup

Begin making **Burnt Honey** (above). Just as it finishes cooking, add **¼ cup plus 2 tablespoons (90 ml) water**. Stir to dissolve over low heat. Store in the fridge and use within 3 to 4 months.

Fruity Variations on Burnt Sugar Syrups

In a similar vein as the Pineapple Caramel Sauce on page 141, why use water when you could use something more flavorful? Use orange juice, apple cider, grapefruit juice, cherry juice, or pineapple juice in place of the water in one of the syrup recipes, above, for a fruit caramel syrup or burnt honey fruit caramel.

Caramel Sauce

Classic and complexly flavored but uncomplicated. I love caramel sauce over ice cream, with bread pudding, or with anything using chocolate, berries, or fruits in the apple family.

Make Caramelized Sugar as directed, cooking it to your preferred stage (see "How Dark Do You Like It?," page 237). Whisk ½ **cup (120 ml)**

heavy cream into the hot caramel; the caramel will bubble violently and may partially solidify. Cook over medium heat until it smooths out. Remove from the heat and whisk in **3 tablespoons (45 g) unsalted butter** (or, even better, cultured butter). Salt to taste—I suggest using less salt for light to medium caramel and more with dark caramel—perhaps **¼ to ½ teaspoon (1 to 2 g) fine sea salt** for a subtly balanced light to medium caramel or **¾ to 1 teaspoon (3 to 4 g)** for an assertively salted darker caramel. Eat warm, right away.

Makes about 1¼ cups (300 ml)

Turning Up the Heat: Creating Flavors with Burning, Charring, and Smoke

Ever blacken the skin on a pepper before you peeled it off, leaving behind a pleasantly lingering burnt flavor? Or inadvertently leave a casserole under the broiler too long until it chars?

You're essentially taking the thermal breakdown of caramel—as mentioned above, technically a form of pyrolysis—way too far, on purpose (or sometimes by accident, but we won't focus on that).

Pyrolysis: Charring and Burning

With concentrated heat, you can burn and char and create blackened, deep, and somewhat bitter flavors in ingredients, while holding back from actually setting them on fire.

Pyrolysis is destructive creation, intentionally ripping through the material of an ingredient to create burnt flavors within it. Charring ingredients in thin and superficial layers, or on only small areas, makes this complex and intriguing, rather than just carelessly burnt. Mole negro and Phở broth both often start this way: charring onions and garlic, or onions and ginger, on their surfaces before incorporating them into a long-cooked braise or stock. Blackening an onion or an eggplant, or the skin on that aforementioned pepper, adds layers of burnt flavors that stay with them as you continue to cook with the ingredients.

Try This
Burning Vegetable Skins

Sweet bell peppers and eggplants both take *really* well to superficial burning, which is kind of a triple whammy: it gives them some charred flavors, it loosens the skin so it's easier to remove, and it partially or fully cooks the inside. Next time you need cooked peppers or eggplant—to go in a sandwich or pasta, blend into a dip, or eat in a salad or over rice or other grains—put it on a medium-hot grill or under a low broiler, and turn it frequently. Stop just when most of the skin is charred, so you don't burn the whole thing. Eggplants can usually just be scooped out; try putting the hot charred peppers in a zip-top or paper bag for 10 minutes to help steam their skin off before carefully peeling it.

Big Molecules to Small Ones: Woody Burnt Flavors

Tough and woody plant materials are the MVPs of burnt flavors. Whether it's logs, twigs, bark, spices, or rice and tea leaves (for tea-smoked duck), the most familiar charred and smoked flavors come from broken-down, large structural molecules in woody (and wood-adjacent) materials.

Burnt Cinnamon

The idea is not to get a slow and even toast, but to burn the outer layers deeply before the inside has time to roast all the way. This will keep some of the fresh flavor of the cinnamon, along with the burnt notes. Don't try this with powdered cinnamon, which just burns and tastes bitter.

In a dry heavy pan over very high heat, tip in a handful of **cinnamon sticks or large flakes/chips from cinnamon sticks**. Heat on one side until they start lightly smoking and have very dark brown to black patches, then flip them over and burn the other side.

Burnt Cinnamon Syrup

Sweeten lemonade or limeade, a whiskey or rum cocktail, or iced hojicha (Japanese roasted green tea) to taste with this for a charred-spicy depth. Or drizzle a little over cut grapefruit slices or supremes.

Put **3 to 4 sticks (about 15 g) Burnt Cinnamon, 1¼ cups (250 g) sugar,** and **1 cup (225 ml) water** into a medium saucepan and bring to a boil. Turn off the heat and steep for 1 hour. Strain. Store in the fridge for up to 3 to 4 months.

Makes about 1⅔ cups (375 ml)

Woody plants get their tough structure from the molecules hemicellulose, cellulose, and lignin, which are all really big. Like, hundreds to thousands of times the size of a sugar molecule, built up from smaller subunits.

Hemicellulose and cellulose are both very long chains of sugars—kind of like starch, but even bigger. Lignin is particularly fascinating. It's a complex molecule that's not a carbohydrate, protein, mineral, or fat, and it is put together randomly from an assortment of similar molecules (rather than orderly, and from the same molecule, like

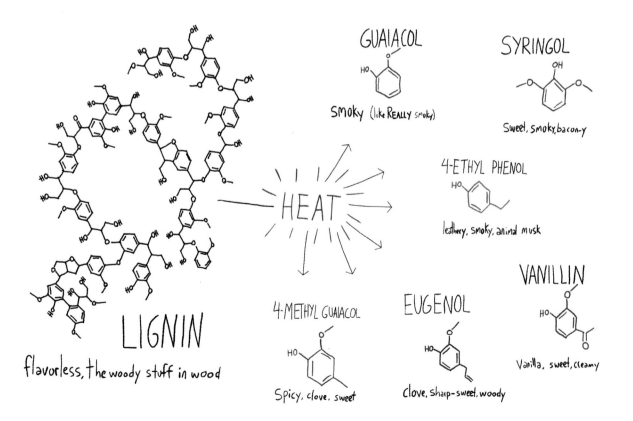

GUAIACOL
Smoky (like REALLY smoky)

SYRINGOL
Sweet, smoky, bacon-y

4-ETHYL PHENOL
leathery, smoky, animal musk

VANILLIN
Vanilla, sweet, creamy

LIGNIN
flavorless, the woody stuff in wood

4-METHYL GUAIACOL
Spicy, clove, sweet

EUGENOL
clove, sharp-sweet, woody

starch or cellulose), which is just a pretty weird thing to see in biochemistry. It's made from some of the same biosynthetic pathways that create phenylpropene aroma molecules like eugenol in cloves (see "Cumin as Chemical Weapon: Herbal and Spiced Molecules," page 153). A chemist would say lignin is a large, highly branched, cross-linked structure of phenolic and methoxybenzene residues—meaning tight and sturdy carbon rings interknitted with oxygen atoms. Lignin makes wood hard and woody instead of leafy and flexible and is indigestible to most organisms except some fungi, which helps protect large plants from rot. (Once plants figured out how to make lignin, there was a stretch of perhaps 100 million years where dead trees piled up everywhere without really rotting. It took fungi figuring out how to make new enzymes to digest the lignin to break down this global woodpile, and put all that carbon biomass back into circulation.)

Burning things made from lignin, cellulose, and hemicellulose (i.e., wood) creates molecules, each with unique ranges of flavor, that echo the structure of these units—a sniff can tell you a story about what the plant was like before you burned it. Lignin transforms into spicy, vanilla, and classic "smoke" flavors, and the celluloses (which are made of sugar, remember) essentially burn-caramelize, making sweet, caramel, and toasty flavors.

In the changing and evolving process of burning, hemicellulose usually starts breaking down first, then cellulose, then lignin, as the temperature gets hotter. In cooperage (barrel making) for wine and whiskey aging, there's usually a step where the almost finished barrel sits over a firepot, pyrolyzing the inside surface of the wood. This is called *toasting*, and different fire temperatures and amounts of time let the cooper dial in a particular flavor profile. Similarly, chemists studying scotch whiskey, which can involve a step of drying barley malt over a peat fire, have found that *which* plants fell into the bog and partially rotted to become the peat can have

lingering, tastable effects on the flavor of the whiskey made from that barley malt. More woody peat versus more leafy peat creates chemically different smoke as it burns, translating into smokier or more malty-caramel flavors, respectively.

Smoke is something new for us to consider—it's a step further than charring. Read on.

Firestarter: From Charring to Flames

Perhaps you're the type to carefully gauge exactly where to hold a marshmallow over a campfire, turning it and pulling it away from the embers as necessary to keep toasting it to an even brown. But maybe you're the pyromaniac who sticks the whole thing in the fire at once, letting it catch and burn into a gooey fireball before you blow it out and enjoy its smoky outer crust.

Setting things on fire means dabbling in combustion, or pyrolysis (burning) plus oxygen (and even more heat). Glowing and flaming are signs that you've heated up wood enough to get it to violently react with oxygen, creating the release of energy that is the hallmark of combustion. Pyrolysis and combustion, charring and fire, are different, complementary routes for creating and applying smoky and pleasantly burnt flavors. They both create flavorful chemical changes with heat, but the addition of oxygen in combustion takes things down some different chemical pathways, creates some different flavors, and makes combustive burning useful for different things than charring (like smoldering oak logs for hours to barbecue a brisket, or cooking fish on a cedar plank thick enough for one side to start smoking and glowing). Combustion is a more destructive and consuming process—it sacrifices one ingredient (like wood, rice and tea leaves, herb branches, or fat) to create smoke, whose flavor molecules then infuse into a second ingredient like paprika, cheeses, or a piece of salmon.

Smoke

Smoke is what you get when you add enough heat and/or oxygen to something (like wood) to get it to start pyrolyzing or combusting but don't take this process all the way to the finish line (which is ashes and carbon dioxide). It's the hallmark of incomplete breakdown. Smoke, kind of like blood, isn't just one substance, it's a lot of substances mixed together (a heterogeneous mixture) by this muddled-up, incomplete breakdown. There are volatile molecules and water vapor that have evaporated out of the heating-up wood, volatile molecules from broken-down cellulose or lignin, gases like carbon monoxide, and tiny carbonized particles, all mixed together into a highly smellable and somewhat acidic haze.

Theoretically, if you burned a piece of wood with perfect efficiency and exactly the amount of oxygen needed to react with precisely every wood molecule, the only products would be carbon dioxide and water vapor. Of course, this doesn't happen in practice, and we intentionally make somewhat oxygen-starved fires specifically for their delicious smoke—for barbecue, smoked salmon, and beyond. Barbecue is a form of *hot-smoking*: create a smoky fire, and then let its smoke sink into tough cuts of meat like brisket or shoulder, as its heat slowly cooks the meat and dissolves its tough connective tissues after many hours. "Hot" is in the name, but the actual cooking happens at much lower temperatures than directly over a grill or live fire. *Cold-smoking* is how you make classic smoked salmon: start with that smoky fire, but let the smoke cool down even more, so it infuses the salmon but doesn't cook away its silky cured texture. Some other cold-smoking techniques use fires made from aromatic ingredients other than wood, as described below.

Both pyrolysis and combustion can create smoke, as anyone (definitely not me!) who's overseared a steak in a small apartment with a sensitive smoke detector can tell you. Smoke generally comes from higher-temperature pyrolysis (since

The Three Types of Burnt, Charred, or Smoky Flavor Creation

1. Charring without combustion: carefully pyrolyze and burn an ingredient to create charred flavors within.
Chemical process: Pyrolysis
Examples: Charred onions, burnt eggplant, charred peppers, lightly scorched doughs, or burnt spices.
2. Perfuming with smoke: use a heat source to burn something flavorful, and put it in proximity with mostly or totally cooked food to collect and extract this flavorful smoke.
Chemical process: Lower-temperature combustion and pyrolysis
Example: Dhungar, tian op, tea-smoked duck, packets of wood chips on the gas grill, "smoking gun" tools, cold-smoked salmon.
3. Cooking with smoke: create a live fire, and put food above it or in its airstream so it collects the woodsmoke but cooks relatively slowly.
Chemical process: Higher-temperature combustion
Examples: Barbecue or hot-smoking.

it takes a lot of energy to break so many big molecules down into volatile and floaty ones) and lower-oxygen combustion.
Pyrolysis: Charring
Combustion: Fire & embers

When you make smoke, you're burning something up, but not completely. There's still plenty of its original stuff in it. If the thing you're burning had any flavors to begin with—aromatic wood, herb branches, spices—some of these flavor molecules survive the burning and end up, intact, in the smoke, adding an incenselike complexity to it. And, sometimes the ingredients we're cooking over a fire essentially smoke

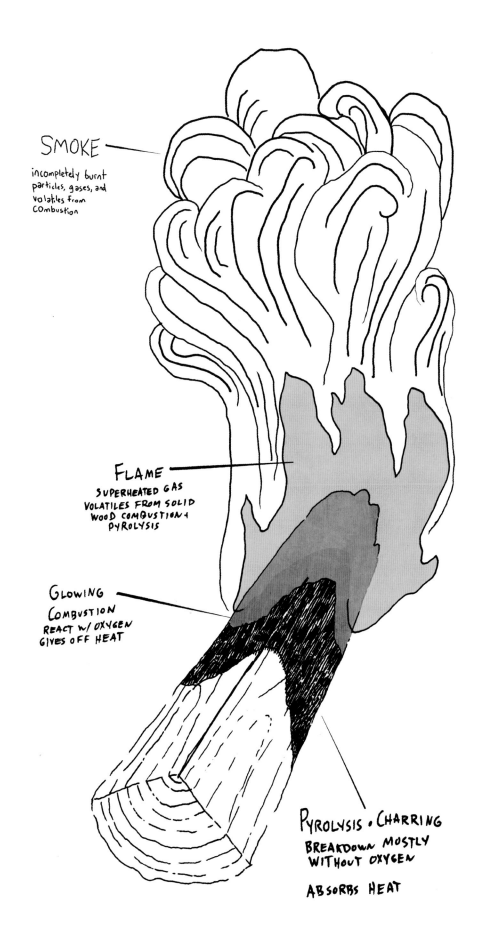

SMOKE
incompletely burnt
particles, gases, and
volatiles from
combustion

FLAME
SUPERHEATED GAS
VOLATILES FROM SOLID
WOOD COMBUSTION +
PYROLYSIS

GLOWING
COMBUSTION
REACT W/ OXYGEN
GIVES OFF HEAT

PYROLYSIS & CHARRING
BREAKDOWN MOSTLY
WITHOUT OXYGEN

ABSORBS HEAT

themselves. Combusting and pyrolyzing fat molecules are essential for classic "grilled meat" flavors, especially from grilling over charcoal, which is basically prepyrolyzed wood and combusts very efficiently, generating less of its own smoke. Fat will melt out of the meat, drop down on the coals, burning and then returning to the meat as smoke. Neat!

The Indian technique of *dhungar* involves perfuming a dish with all of these different origins of smoky flavors as a kind of final seasoning.

Cooks fill a tiny bowl or onion shell with ghee and sometimes spices, float or nestle it in a completed dish, then drop a hot coal in the ghee and seal the whole thing with a lid. The ghee-smoke dissolves in the dish, flavoring it. Thai *tian op* is chemically similar but, instead of butter, uses a special candle with spices mixed into the wax. Light the candle, blow it out, trap it under a bowl together with (usually) a dessert, and the dessert takes on a smoky perfume.

A Whole Lot of Browning: The Maillard Reaction

Complex flavors like caramel are easy to create just by messing around with sugar, a relatively pure, single-molecule substance. The fragmentations and rearrangements that follow create a lot of different, new flavor and color molecules. Charring and smoking, too, generate new flavors from breakdown. But beyond breakdown, one of the most intriguing techniques for creating flavor starts from something a little more complicated.

If you like the aroma of baking bread, the browned sear on a steak, the tawny edges of a roasted cauliflower, the addictive nuttiness of brown butter, or indeed the nuttiness of toasted nuts themselves, you have a complex web of chemical creation called the *Maillard reaction* to thank for them. Named after its discoverer, the French chemist Louis Camille Maillard, this reaction between amino acids (the building blocks of proteins) and sugars is one of the most complex and best-studied for creating flavor.

Starting with a bigger assortment of molecules than just sugar gives you more material to work with and more ways to un-knot and re-assemble things, like adding a new set of Lego to your toy bin. And since so many ingredients have both sugars and amino acids in them, the Maillard

The Maillard Bunch: A Field Guide to the Wide World of Browning

Baked bread

Coffee

Chocolate

Brown butter

Butterscotch and toffee

Caramelized onions (they're called "caramelized," but they're actually Maillardized)

Roast beef

Roast chicken

Roasted vegetables

Malt and beer

Soy sauce, miso, doenjang

Toasted meringue

Toasted marshmallow

Tahdig (Persian rice) and other rice dishes cooked to "scorch" or toast the bottom layer of grains

reaction is behind the flavor of a large number of gently to medium aggressively roasted, browned, baked, and toasted foods. Even a tightly controlled experiment with only one kind of sugar and one kind of amino acid can make hundreds of flavor and color compounds as the reaction proceeds. And since there are more than twenty amino acids and roughly a dozen sugars in foods that can participate, things get wonderfully complicated very quickly.

Maillard Molecules

Because the Maillard reaction involves sugar, many of the carbon-, hydrogen-, and oxygen-containing flavor molecules created by caramelizing sugar are also important for brown and Maillardized flavors. Amino acids, their costars, are the real drivers of complexity and differentiation from simple caramelization. All twenty types of amino acids have at least one nitrogen atom apiece, and two—cysteine and methionine—also have funky sulfur. With more "flavors" of atoms to start with, and more arrangements and shapes to play with, you can rely on the Maillard reaction to come up with lots more new molecules, and more new flavors.

Nitrogen allows for the formation of *pyrazines*, which look a little bit like the pentagonal rings of carbon and oxygen in the furanoids that show up in both caramelization and Maillard, if you swapped oxygen for nitrogen. Those furanoids and other oxygen-containing molecules tend to caramelly sweet, even buttery or fruity; pyrazines and other nitrogen volatiles are more roasted, toasty, nutty, even popcorny. The sulfur molecules derived from cysteine and methionine have even roastier, savory-meaty flavors, sometimes with vegetal or coffeelike elements.

Frequently Asked Questions About the Maillard Reaction

Q: If they both are brown, both involve sugar, and have some similar flavors, how is the Maillard reaction any different from caramelization?

A: An *excellent* question. You can tell them apart, flavor-wise, because the Maillard reaction has many more roasted, toasty, meaty, and coffeelike flavors—from the addition of nitrogen and sulfur from amino acids. You can tell them apart, practically, because the Maillard reaction requires a lot less energy to start—you're reacting and rearranging two molecules, rather than ripping one apart by brute force as happens in caramelization. While caramelization doesn't really get going until you're in the ballpark of 330°F, the Maillard reaction proceeds slowly at around 250°F, and is strongest and fastest from about 280 to 330°F.

Q: What do I absolutely need to have to do the Maillard reaction?

A: The only absolutely key components for the Maillard reaction are some free amino acids, some sugars, and either some decent heat or a lot of patience. White table sugar (sucrose) is actually not very useful for the Maillard reaction—because of how the glucose and fructose units that make it up are assembled, they take up the spots on the molecule that an amino acid would approach to start the reaction off. Fructose and glucose themselves are very able to do the Maillard reaction, and can be found in fruits, corn syrup, honey, agave nectar, and corn sugar (dextrose) or glucose syrup. Other sugars like the maltose in barley malt or malt powder, or the lactose in dairy, are also much better at doing the Maillard reaction than sucrose.

Q: What is nice to have for the Maillard reaction?

A: As long as you've got amino acids and the right sugar, the Maillard reaction will happen. You don't even necessarily need 280°F heat—the Maillard reaction happens slowly, over the course of hours, in simmering broths and in braises, and *very* slowly, over the course of months or years, in aged and brown foods like soy sauce, miso, balsamic vinegar, and even some hams.

MY GLUCOSE BRINGS THE AMINES TO MAILLARD: some Molecules and flavors of the Maillard Reaction

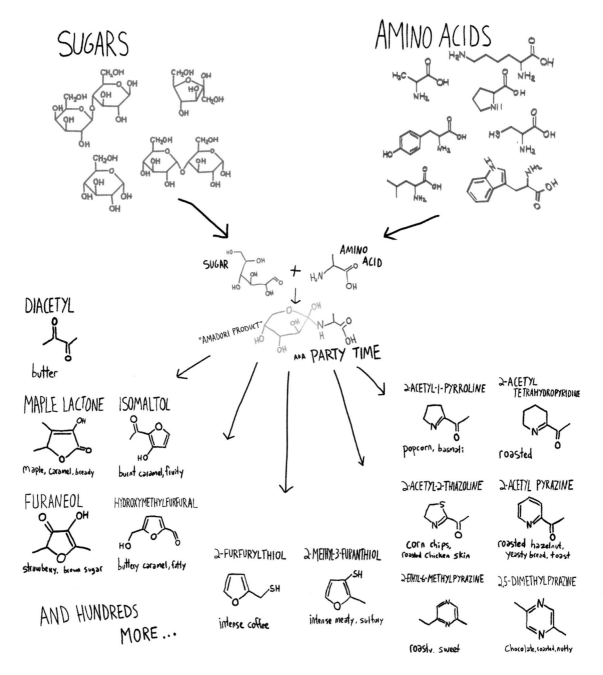

SUGARS

AMINO ACIDS

SUGAR + AMINO ACID

"AMADORI PRODUCT"   ♪♪♪ PARTY TIME

DIACETYL
butter

MAPLE LACTONE
maple, caramel, bready

ISOMALTOL
burnt caramel, fruity

FURANEOL
strawberry, brown sugar

HYDROXYMETHYLFURFURAL
buttery caramel, fatty

2-FURFURYLTHIOL
intense coffee

2-METHYL-3-FURANTHIOL
intense meaty, sulfury

2-ACETYL-1-PYRROLINE
popcorn, basmati

2-ACETYL TETRAHYDROPYRIDINE
roasted

2-ACETYL-2-THIAZOLINE
corn chips, roasted chicken skin

2-ACETYL PYRAZINE
roasted hazelnut, yeasty bread, toast

2-ETHYL-6-METHYLPYRAZINE
roasty, sweet

2,5-DIMETHYLPYRAZINE
chocolate, roasted, nutty

AND HUNDREDS MORE...

But, if you want to maximize your Maillard, here are some things you can do:

- Heat above 250°F (which we've already discussed).
- A dry environment, and drying off the surfaces of your ingredients.
- An alkaline (as in, not acidic) environment.

Q: Why does a dry environment matter?

A: Water is really, really good at absorbing heat. If you add a lot of heat to something with water in it—like a pot over high heat on the stove—the temperature won't rise above the boiling point of water (212°F) until the water has come to a boil, and boiled off. Since the Maillard reaction happens very slowly below 250°F, you won't get much browning while you're waiting for water to finish absorbing heat. Plus, on a molecular level, many of the changes in the Maillard reaction break off oxygens and hydrogens from a molecule and release them as their own water molecule. Trying to make new water molecules when you've already got water molecules right there is kind of like you're a kid asking to go to McDonald's while your mom (aka chemistry) is telling you "We have food at home"—it's loath to make a lot more of what you already have around. Get rid of those existing water molecules, and chemistry is much more willing to do reactions that make more.

Sneaky sources of Maillard-slowing water can include damp or moist ingredients, like meat that might have juices pooled on its surface; when you crowd a pan with pieces of food; and when you use a vessel with too-high walls. Think a skillet stuffed with too many pieces of fish, or a chicken in a really deep roasting tin. Either you generate more moisture than one burner can quickly evaporate, or you trap moisture in and around the food pieces, giving you partially steamed and unevenly browned ingredients. When I want to ensure things brown nicely, I'll make certain their surface is dry: patting with paper towels is good at the last minute. Peking duck, which is all about beautifully browned and crisp skin, uses air-drying to make sure that skin is evenly dried out for best roasting performance. Buy yourself the same insurance by air-drying steaks, roasts, and whole birds in the fridge, uncovered and with plenty of room around them, overnight before you cook them—gadget-friendly cooks, including Marcella Hazan, do it even faster with a hair dryer, right before cooking.

Q: What's the deal with alkalinity?

A: Alkalinity is the opposite of acidity. Acidity is a bunch of free hydrogen ions, and alkalinity is having a significant shortage of these, and the presence instead of ingredients like baking soda, the calcium hydroxide used to cook and nixtamalize corn masa for tortillas, and the sodium hydroxide used to make soap. This matters for the Maillard reaction because of what acidity and alkalinity do to amino acids. The "amino" part of an amino acid contains a nitrogen atom, which has a spare set of electrons it uses to initiate the Maillard reaction with a sugar molecule. If amino acids are in acidic conditions, lots of the nitrogens will grab on to one of the hydrogen ions floating around, rendering it essentially unreactive for Maillard purposes. Remove these hydrogens—make the environment less acidic and more alkaline—and more amino acids will have their nitrogens in "ready" position to react with sugars.

Some cooks use this chemistry directly to their advantage. Traditional pretzels are dipped in an alkaline bath of water mixed with lye (sodium hydroxide) before baking, which turbocharges the dough's amino acids for maximum Maillard reaction during baking, giving the pretzels their iconic deep-brown color and malty-roasty flavor. Alkaline water is added to the dough for moon cakes for the same reason, and pidan or century eggs are made by aging raw eggs in an alkaline brine or wrapping, which slowly browns and gels their whites.

Alkaline ingredients like sodium hydroxide and calcium hydroxide can be dangerously corrosive, and even much tamer alkaline ingredients like baking soda can create bitter, soapy, and fishy off-flavors if you use even a little too much. Some people experiment with adding baking soda to, for example, caramelizing onions to make them browner faster, but mostly what I do with this knowledge of alkalinity and browning is use it to be smart about adding or avoiding acids. If I want something to brown well, I'll keep it away from acidic ingredients until it's well-Maillardized. If I'm going to be cooking something for a very long time, I often bump up the acidity a little—with some vinegar, wine, yogurt whey, fruit juice, whatever fits flavor-wise—to avoid overbrowning.

Q: **What other ingredients have the right stuff (amino acids, sugars, not too acidic, not too wet) for me to strategically Maillard?**
A: **Cocoa beans, coffee, and nuts, seeds, and kernels, generally.** Roasting in a pan, with a stream of hot air (like a coffee roaster does), or watchfully in the oven with frequent stirring will help you create the roasted coffee and cocoa, or the toasted almonds, sesame, hazelnuts, or walnuts of your dreams.

Grain and cereal doughs. A moist oven environment to encourage crust expansion, followed by a hot and dry one will create a just-right-thickness, dark golden brown crust. Pretzel makers use the lye trick I mentioned above, and you could add extra brownable stuff in the form of amino acids to the surface of croissants, pastries, pies, and other doughs with a butter, milk, or egg wash.

Animal proteins, especially mammals and birds. As we learned in "Meaty: Where Does the Flavor of a Steak Come From?" (page 184), a piece of meat is basically a structural sponge soaked in flavor-precursor-laden juice—which includes amino acids and sugars. (The extra amino acids—see page 186—that shellfish produce

make them extra well suited to Maillard reaction browning.) The brown crust of any meat seared in a pan or roasted comes from the juices near the surface, which rapidly lose their water to the heat of the pan and then react in the Maillard reaction. Besides air-drying the surface of meats, dry brining—salting meats uncovered in the fridge—pulls more of the juices from the inside of the meat to the surface through osmosis, with the dissolved amino acids and sugars sticking around there when the water evaporates. Presto, extra stuff to brown.

Butter and dairy. Dairy is full of proteins, free amino acids, and sugars. It's *asking* you to Maillard it.

Maillard in Dairy: The Wonders of Brown Butter

When I was a kid, my parents, following the low-fat standards of the 1990s, didn't cook much with butter. But when I turned sixteen, my former-nurse, health-conscious mother took me to my first fancy French restaurant to mark the occasion. It had starched tablecloths, heavy silver cutlery, and multiple drinking glasses that summer weeks with my etiquette-obsessed grandmother had prepared me for.

What I remember most from that meal was a pool of beurre noisette, or brown butter. Lightened with a little lemon and complexified with a few capers, it garnished a single scallop. This was clearly butter, but intoxicatingly roasted. Like popcorn, but deeper. Like coffee, but sweeter. Like toffee, but more savory. Resisting the urge to lick the plate, I made a mental note to research this substance further. At home, in an old copy of Julia Child's *Mastering the Art of French Cooking*, I found an explanation:

A properly made brown butter sauce has a deliciously nutty smell and taste, but is never black despite the poesy of the title. When you heat butter to the boil, its milk solids begin to darken from golden nutty, *noisette*, to golden brown, *noir*, but you never let it darken to black, burned, and bitter.

I can happily report that further research shows this is the perfect encapsulation of the Maillard reaction, and that there is literally nothing I can think of that isn't improved by it, from cookies to a simple sauce over pasta to pan-cooked vegetables.

A Better Brown Butter, from Heavy Cream

Brown butter, from *cream*?

The flavor of brown butter comes from *milk solids*, which means proteins, sugars, minerals, and free amino acids. Butter has a comparatively small amount of milk solids, because they get washed away during the butter making process, along with the buttermilk. Heavy cream is just butter waiting to happen, ready to have its fat coagulated and its watery, solids-rich buttermilk washed away. If milk solids mean brown butter flavor, then holding on to more milk solids means . . . more brown butter flavor. Your patience and diligence will be rewarded with an abundance of brown butter nuggets, roughly the texture of coarse sand. You'll end up with about 15 percent butter solids and 85 percent butterfat, about five times as much solids as you'd get from starting with whole butter.

You don't have to use exactly 1 pint of cream for this—however, I don't recommend starting with less than 1 cup of cream because you'll lose some to the pan. Since brown butter keeps in the fridge indefinitely, you can make as much as you want. Think of it as a gift to your future self. Besides cookies, simple sauces over pasta, or pan-cooked vegetables, I like to add a little toasty sweet-savory aroma and depth with this in lots of places: a spoonful into mashed potatoes or any kind of meat braise, into fruit I'm cooking

in a pie or crumble, or even in a bitter radicchio salad (page 101).

Put **1 pint (500 ml) heavy cream** in a heavy medium nonreactive saucepan and bring to a mild boil over medium- to medium-high heat, stirring any skin that forms on the top and any solidified cream on the edges back in as it cooks. Stir well on the bottom so that the cream doesn't stick and scorch.

As it boils and water evaporates, the cream will thicken into almost a sauce, then the fat globules will begin to coalesce and split the emulsion, and shortly afterward, the thick, sticky milkwater portion sinks to the bottom and the butterfat rises to the top. Depending on your stove, the size of the burner and the material of your pot, and how spread out the cream is (more surface area means faster evaporation), this should happen in 10 to 20 minutes.

Turn the heat to medium-low and keep cooking. The milk solids will lose enough water to start granulating. Stir well so the granules don't stick together too much and keep heating, watching it like a hawk, until the granules become whichever aroma and shade of brown you prefer: blond, for a very delicate and sweet butter flavor, or, my favorite, several shades darker, a reddish-hazelnut with a more savory flavor. Be careful here, because there's enough residual heat in the pan and the butter to take it right over the line to too brown if you're not paying close attention and leave it to sit. Pour the brown butter into a heatproof glass or ceramic container, cool completely, and store in the fridge, tightly covered, for up to 6 months.

Too long, didn't read? Gently boil heavy cream in a saucepan until it reduces all the way, then splits. Brown the split solids to dark golden. Don't let it burn.

Makes about ¾ cup (200 g)

Toffee Sauce

Toffee is essentially Maillardized butter and sugar, thinned with cream. Table sugar (sucrose) is not very good at doing the Maillard reaction, but it will "invert" or partially break down into glucose and fructose as you heat it, and both of those sugars are great at the Maillard reaction.

Because of the amino acids in the butter and the abundant sugar, this sauce has greater nuttiness and depth than plain caramel, or caramel with butter swirled in—but sweeter, more caramel aromas than brown butter alone. I make it most often to pour over ice cream or bread pudding, but there are few baked or creamy desserts that don't get along with it.

Combine ½ **cup (100 g) sugar** with **8 tablespoons (120 g) unsalted butter** in a medium saucepan. Heat over medium to medium-high heat until the sugar melts into the butter. Continue cooking until the mixture is medium-dark brown. Remove from the heat and whisk in ½ **cup (120 ml) heavy cream** and ½ **teaspoon (2 g) fine sea salt**. Eat immediately or store tightly covered in the fridge for 1 to 2 days before gently reheating.

Makes about 1¼ cups (300 ml)

Chapter 10

Creating Flavors with Fermentation

We've been eating fermented foods for a really, really long time—"before-we-evolved-into-humans" long. Possibly even before our ancestors split from chimpanzees, which happened somewhere in the range of four to seven million years ago, as some genetic evidence suggests. Fermentation can make foods more digestible and easier to extract energy from than their raw forms, as well as extending the window of edibility from a few hours or days to a few weeks or months. There are few things we haven't applied it to— making vegetables or meat easier to keep and eat long after the hunt or harvest; or converting ripe fruit into long-storing alcohol, superperishable liquid dairy into high-protein, nutritionally dense yogurt and cheese, or hard-to-chew grains into bread and beer. Just about every culture in the world has its own set of fermentations, which have been shaped over time by the animals, plants, and seasons of their local landscape. The flavors of these fermented foods are often linchpins of cuisine and cultural identity: fish sauces and pastes, wine, sauerkraut, kosher pickles, kimchi, paocai, soy sauce, cheese, yogurt, beers, ciders, breads, and more.

Fermentation is (*very* technically speaking) a particular biochemical process (metabolizing certain molecules for energy without using oxygen) that some microorganisms, like lactic acid bacteria and bread yeast, are able to do. In cooking and everyday life, it applies much more broadly to processing foods with living microorganisms—like those bacteria and yeast, as well as microbes like acetic acid bacteria, culinary molds, and multi-species blends like the SCOBY (symbiotic colony of bacteria and yeast) used to make kombucha. Did you grow microbes in something, and transform it somehow? You did fermentation.

Like browning or smoking, fermentations create new (and tasty) molecules as part of their

MICROBIAL COLLABORATORS	WHAT THEY TRANSFORM	WHAT THEY CREATE
YEAST *Saccharomyces cerevisiae*	**SUGAR**	**ETHYL ALCOHOL** (alcohol) **CARBON DIOXIDE** $O=C=O$ (fizz)
ACETIC ACID BACTERIA *Acetobacter pasteurianus* *Gluconobacter oxydans*	**ETHYL ALCOHOL** **OXYGEN** O_2	**ACETIC ACID** (vinegar) Sour + pungent
LACTIC ACID BACTERIA (L.A.B.) *Pediococcus acidilactici* *Lactiplantibacillus plantarum* *Lactococcus lactis* *Leuconostoc mesenteroides*	**SUGAR** **SALT** NaCl Sometimes (vegetables, fruit)	**LACTIC ACID** Cleanly sour
MOLDS *Aspergillus oryzae* *Aspergillus sojae* *Rhizopus oligosporus*	**STARCH** **PROTEIN**	**SUGARS** sweet **AMINO ACIDS** umami + ENZYMES
MOLD ENZYMES **YEASTS** **ACETIC ACID BACTERIA** **LACTIC ACID BACTERIA**	**SUGAR** **STARCH** **PROTEIN** **SALT** NaCl (Soybeans · Salt · Molded Starter)	**LACTIC ACID** **AMINO ACIDS** **ETHYL ALCOHOL** **ACETIC ACID** (+ Many other ferment-y flavors)

What Flavors Can You Create with Fermentation?

Things you can ferment	What gets fermented	Who does the fermentation	What do you make?	What flavors do they create?
Malted grains, fruits	Sugar	Yeast	Beer, wines, cider, sake, and other alcoholic drinks	Alcohol and fizzy carbon dioxide, fruity flavors, winey flavors
Wine, beer, other alcoholic beverages	Ethanol (alcohol)	Acetic acid bacteria	Vinegars, kombucha (which starts with yeast turning sugars into alcohol)	Tangy and pungent acetic acid
Fruits, vegetables, dairy, grains	Sugar	Lactic acid bacteria	Kosher pickles, sauerkraut, kimchi, paocai, and other lacto-fermented pickles; yogurt, crème fraîche, cheese, kefir, and other lacto-fermented dairy; sourdough breads, kvass, idli, and other lacto-fermented breads and grain drinks	Clean and tangy lactic acid, buttery, creamy, and pickley flavors
Rice and other grains, soybeans	Starch and proteins	*Aspergillus oryzae, Rhizopus* fungi, *Bacillus* bacteria	Koji, meju, nuruk, qu, and other fermentation starters	Sweet, fermentable sugars and umami-y free amino acids, nutty and funky flavors
Koji, meju, or other fermentation starters, plus soybeans or other legumes	Sugars	Lactic acid bacteria and special salt-tolerant yeasts	Miso, soy sauce, doenjang, doubanjiang and douchi/salted black beans	Tangy lactic acid, malty and brown flavors, funky flavors
Soybeans	Proteins	*Bacillus* bacteria; *Rhizopus* fungi	Natto, cheonggukjang, tempeh, and other alkaline (as in, not acid-producing) fermentations	Savory, cheesy-funky, flavors

transformation. Sour acids and umami-flavored free amino acids are some of the most predominant. Like plants, microbes produce smaller amounts of lots of other flavors as part of their secondary metabolism: fruity and winey flavors; bready and yeasty ones; floral, nutty, creamy, and pickley ones.

It's possible to take a very directed approach to the transformation and flavor creation that is fermentation: creating a sterile environment with a sterilized ingredient, selecting a single, purified microbe strain to transform it, and tightly controlling that transformation. Like making things on an assembly line, you give up some variation and personality in exchange for no-surprise predictability.

Personally, I approach most fermentations more like a collaboration between myself and a collective of microorganisms, with the goal of creating flavor. Our environments—our hands, the surfaces around us, and especially the outsides of our ingredients—have ready-to-ferment microbes all over them, and each of them—a lactic acid bacteria, a yeast, a mold—is a skilled artisan that specializes in a particular technique and medium. They can create collaboratively—if each has the right environment, the right tools, and the right materials. They can even keep out intrusive spoilage microbes, essentially self-regulating their process, as long as I provide those just-right conditions and food sources. In practice, this looks like adding a certain amount of salt to a fermentation, keeping out or adding air, and making sure that whatever I'm trying to ferment has the right stuff in it—sugars, alcohol, starches—for the fermentation I'm trying to do. Their process might be a little idiosyncratic or hard to follow, or involve odd requests—but they're masters at what they do, and if I respect the process and let them do their thing, the flavor results can be spectacular.

Making Vinegar: Creating Pungently Sour Flavors out of Alcohol

Most cuisines make some kind of vinegar. It might be a grape vinegar, a grainy malt or rice vinegar, balsamic vinegar, black rice vinegar, akasu (the vinegar favored for seasoning sushi rice), or a vinegar from banana, coconut, guava, or sugarcane. Vinegar is such a widespread product because it's so simple to make that it essentially makes itself—if you leave an alcoholic drink out for a while.

The prime movers in vinegar making are acetic acid bacteria (or AAB, for the acronym lovers among us). AAB make vinegar by eating up alcohol (specifically ethyl alcohol, also called ethanol) and oxygen, and creating acetic acid—so much so that they're named for it. A few things make them highly effective at their jobs. For one, they can tolerate being in an alcohol-saturated environment more than most other microbes. Second, the acetic acid they create has a double whammy of possessing both acidity (which drives away most microbes) and its own innate toxicity to whatever microbes stick around. Third, acetic acid bacteria are everywhere—on surfaces, on the feet of fruit flies, floating in the air, etc.—so AAB reaching an open container of alcohol is more a matter of *when* it will happen rather than *if* it will happen.

In short, **alcohol + acetic bacteria + oxygen = vinegar.** All of this makes acetic bacteria the bane of a wine, beer, or mead maker's existence, but the happy, easygoing friend of anyone who wants to make vinegar.

Microbes All Around Us

This fact sometimes shocks people when they learn it, but most things, like tabletops, fruits, cabbage leaves, cow udders, or your hands, have bacteria and other microorganisms all over them. This includes clean surfaces, and basically anything you didn't very recently sterilize with high-proof alcohol or bleach. Most of these microorganisms are relatively harmless to us and can even be beneficial.

Left to their own devices, they get onto things like fruits, vegetables, or pails of milk, consume what they can, and create fermented products in the process. "Being fermented" is essentially the natural end-state for organic material; when it happens unintentionally, we tend to call it "rotting." But if something is inevitable, like this blanket of microorganisms and the things they do, we can plan for it, and enjoy the results as yogurt and cheese instead of spoiled milk, sauerkraut instead of spoiled cabbage, vinegar instead of spoiled wine.

Any Wine Vinegar

Commercial wine vinegar isn't usually made out of the nice stuff, even historically, because it makes more economic sense to sell whatever wine you can as wine (for a better profit margin), then use whatever's left to make vinegar. Any wine you drink at home is of much higher quality than the wine that goes into any wine vinegar you can easily buy. So if you like very dark, robust red wine, or full-bodied, buttery, white wines, or slightly sweet rosé or orange pet-nats, some of those special qualities will come out if you make that wine into vinegar.

Since vinegar kind of makes itself, this is less a strict recipe than a strategy to apply. The general rule of thumb for making a vinegar with a familiar level of acidity (about 4 to 7 percent) is to start in the ballpark of 8 percent alcohol. Most wines are much higher than this, 11 to 15 percent. I find that the results are a little more predictable and palate-friendly if I dilute the wine slightly.

The starting dose of acetic acid bacteria comes from using unpasteurized vinegar. You can use a commercial unpasteurized vinegar, like Bragg apple cider vinegar, or a wine vinegar you have made previously.

Covering the jar with a dish towel or other cloth lid lets air in, which is necessary for the acetic bacteria to do their work, while keeping out dust and fruit flies, which are less desirable additions.

Clean a half-gallon glass jar with soap and hot water and let dry. This recipe will only half-fill the jar, but it's useful to have that headspace for air to diffuse into the vinegar. Pour **one 750-ml bottle (about 3⅓ cups) wine of your choice** into the glass jar, and add **2 cups (475 ml) filtered or good-quality tap water**. (If you are using a lower alcohol wine, say 11 percent, add a little less water, 1¼ cups or 300 ml).

Add **1¼ cups (300 ml) unpasteurized vinegar** (such as Bragg apple cider vinegar or homemade vinegar), to the jar. Stir well to aerate, then cover with a clean dish towel, tied tightly around the mouth of the jar with a piece of sturdy twine to keep out dust and fruit flies.

Put the jar in a warm place away from light. Start tasting after 2 weeks, with a clean spoon. It should convert to a puckery vinegar in 2 to 3 months. If it starts smelling like nail polish remover, it's just going through an awkward teenage phase, and should mellow as it continues fermenting. If you notice mold, you should toss it. "Vinegar eels" are invertebrates that can live in vinegar, and they may possibly show up to the party—they're kind of gross-looking, but not harmful, and FDA guidance for vinegar makers is just to filter them out of the finished product. Your kitchen, your choice.

Once the vinegar is fully fermented, put it into bottles or other glass containers, and fill them as full as possible—at this point, more air exposure is not helpful. Store in these tightly capped bottles and use right away, or age in the dark for a few months to a year to further mellow it. A good idea: save a cup of it to inoculate the next batch of vinegar.

Makes about 5 to 6 cups (1 to 1.5 liters) vinegar, depending on evaporation

Whatever alcoholic beverage people make a lot of in a place is usually what becomes their local vinegar, defining the contours around the sour flavors of that place's cuisine. Rice wines beget rice vinegars, apple ciders beget cider vinegars, red wine begets wine vinegar. They all have acetic acid in common, and therefore its sourness, as well as its unique aromatic flavor qualities: nose-clearingly zippy, refreshingly pungent.

Beyond that, the flavors of vinegar are funkier, slightly muted versions of their former selves. Fruity things keep some of their fruity notes, malty things retain malty flavors, and sweet things hang on to their sweet ones. Picking out something whose flavors speak to you, then making it into vinegar, lets you tailor your own flavor preferences to a reliable pungent you can use for dressing salads or vegetables, finishing soup, or punching up a pan sauce.

Porter Vinegar

Commodity malt vinegar is obligatory at our house on fish and chips (my husband grew up in England, where it's the traditional seasoning). I've been delighted by the malty-sour-nutty-sweet flavor profile that results from a malt vinegar made with flavorful beers, especially ones with roasted malts and residual sweetness, which give a balsamiclike balance to its sourness.

I recommend starting with beers that are not very hoppy, as their bitterness and slightly dank hop character can be tricky to finesse.

High-gravity (sweet and alcoholic) porters and milk stouts have depth from their dark-roasted malts but are usually surprisingly delicate on the hop front. Beers like Ballast Point Victory at Sea, Stone Coffee Milk Stout, Evil Twin Even More Jesus, or North Coast Brother Thelonious (technically a Belgian-style strong ale, not a porter) all fit the bill nicely.

Clean a half-gallon glass jar with soap and hot water and let dry. This recipe will half-fill the jar, but it's useful to have that headspace for air to diffuse into the vinegar.

Pour **two 12-ounce (350 ml) bottles of high-gravity porter beer or milk stout** with a bit of sweetness (see above) into the glass jar. Most of these heavier styles are 10 to 12 percent alcohol by volume, so a little diluting is in order. Stir the beer until it goes flat; you should have about 24 ounces (700 ml). Add **¾ cup plus 1 tablespoon (200 ml) filtered or high-quality tap water** and **¾ cup (165 ml) unpasteurized vinegar** (such as Bragg apple cider vinegar or homemade vinegar).

Pour the mix into the glass jar, and cover with a clean cloth tied around the mouth of the jar with string. Keep in a warm place away from light. You can start tasting it with a clean spoon after 2 weeks. It should convert fully to puckery vinegar in 2 to 3 months. Store in a tightly capped bottle away from light; it will stay good for a year or longer.

Makes about 3 to 4 cups (¾ to 1 liter), depending on evaporation

Vinegar is the natural end-state, or the undead afterlife, of alcoholic fermented beverages. But if you want to get a little cheeky about it (and I usually do), all you *really* need to make vinegar is a little bit of alcohol, a little bit of acetic bacteria, oxygen, and some time. You could start with an alcoholic drink, but you can also skip the logistics of figuring out "What booze do they make that has exactly the flavor I want in it?" and just add alcohol to a liquid you already like to reach 8 percent alcohol by volume.

This kind of thinking about vinegar was one of the big things that pulled me onto the strange, is-it-being-a-scientist, is-it-being-a-cook, why-not-both path I'm on today. In the early 2010s, the great Southern-experimental restaurant Mc-Crady's was experimenting with cultural signifiers by serving dishes sauced with Mountain Dew vinegar. When I showed up at the doorstep of the Nordic Food Lab in Copenhagen around the same time with a vague idea of writing a research paper there, I learned how to make vinegars they'd developed from celery juice, pine needle tea, and elderflower syrup in exchange for acting as a kind of pocket chemistry reference. I got two dissertation chapters out of it and, more important, some cool techniques to share. It's an extremely useful example of looking at a product—vinegar—seeing a pattern in how it's made—alcohol and other flavorful stuff—and using that pattern to improvise in interesting ways.

Moxie (or Anything-You-Want) Vinegar

This balsamic-inspired, but highly irreverent, vinegar builds on the herbal, citrusy, slightly bitter flavor of Moxie Cola and turns it into a surprisingly tasty and complex vinegar. Moxie is a Maine classic that, growing up in New England, I was aware of but didn't come to appreciate until I was an adult. Now I love it: it's basically Amaro Nonino, if it was a soda instead of an alcoholic liqueur.

The key with adding alcohol to something juicy to make vinegar is hitting the right proportion based on proof, or ABV (alcohol by volume). If you want to dilute the flavors the least, look for 190-proof, 95 percent ABV Everclear. As it says on the label, it's 95 percent alcohol, so you won't need much of it to reach 8 percent. Other alcohols are usually 40 percent ABV to 75 percent ABV

(80 to 150 proof), with the remaining 60 percent to 25 percent being water. You'll need to add more of them to get the same amount of alcohol, so I like to use an alcohol with a lot of flavor to make up for the greater dilution.

This recipe makes about 1 liter of vinegar, so reaching 8 percent is relatively simple arithmetic. You could use any one of these—whatever you pick, go by the ABV on the label for dosing:

Everclear: 95% ABV; 90ml/3 ounces
Overproof rum (such as J Wray and Nephew): 63 to 69% ABV; 140 ml/4.75 ounces
"Navy strength" rum (such as Smith and Cross): 54 to 57% ABV; 170 ml/5.75 ounces
Bonded rye or bourbon whiskey: 50% ABV; 190 ml/6.5 ounces
Typical-strength whiskey or brandy: 40% ABV; 250 ml/8.5 ounces

You can mix almost any liquid, alcohol, and vinegar starter with these proportions and end up with vinegar. I've had it with new-Nordic bases like celery juice, pine needles steeped into tea, carrot juice, juiced fennel, diluted elderflower syrup. . . .

Clean a half-gallon glass jar with soap and hot water and let dry. This recipe will half-fill the jar, with the headspace providing air to diffuse into the vinegar.

In a large bowl, stir **26 ounces (3¼ cups; 775 ml) Moxie Cola** (from 3 cans or two 20-ounce bottles, with some left over) until it is flat. Add **7.5 ounces (225 ml) unpasteurized vinegar** (such as Bragg apple cider vinegar or homemade vinegar) and **3 ounces (90 ml) Everclear**, or a flavorful lower-proof alcohol to reach 8 percent (see the suggested amounts above).

Pour the mixture into the glass jar and cover with a clean cloth tied around the mouth of the jar with string. Keep in a warm place away from light. You can start tasting it with a clean spoon around week two. It should convert fully in 2 to 3 months. Store in a tightly capped bottle away from light and it will stay good for at least 1 year.

Makes about 3½ to 4 cups (1 liter)

Gin and Blackberry Vinegar

This vinegar is inspired by the bramble cocktail, which uses gin, lemon, and the blackberry liqueur crème de mûre, where the spice and juniper notes bring out a really complex side to the blackberry.

I like using it to season more blackberries or other berries, or bitter lettuces in a salad.

This admittedly slightly extravagant vinegar is best when blackberries are in season and it's easy to get a lot of really ripe ones. If blackberries are not in good shape, you can substitute other very ripe fruits to your liking and switch the alcohol to match. Peach and bourbon, perhaps, or melon and tequila or Sun Gold tomato and rye whiskey. For less juicy fruits, juice in a juicer instead of crushing.

Because of the solids in this recipe, I recommend keeping a watchful eye on it, especially early on, for mold or wrinkly surface yeast. Discard if it gets moldy, and skim off any wrinkly white surface yeast. The recipe can also be scaled down if you can only get 1 pound of fruit (but I wouldn't advise trying to make less than that). If the berries are not yielding much juice when you crush them, try adding 1 cup (240 ml) water to loosen them, along with 3 tablespoons (45 ml) extra gin and 3 tablespoons (45 ml) extra unpasteurized vinegar for every cup of water you add.

Clean a half-gallon glass jar with soap and hot water and let dry. This recipe will half-fill the jar, but it's useful to have that headspace for air to diffuse into the vinegar.

In a large bowl, carefully crush **2 pounds (900 g) extremely ripe blackberries**, as soft as possible, by pressing on them with a potato masher using even pressure—you're looking to just rupture them, not mash them into a puree. Keep the skins and pulp in. Add **¾ cup (175 ml) good-quality gin of around 45 percent ABV** (Tanqueray, Bombay Sapphire, or your favorite) and **⅔ cup (165 ml) unpasteurized vinegar** (such as Bragg apple cider vinegar or homemade vinegar) as a starter.

Pour the mixture into the glass jar, and cover with a clean cloth tied around the mouth of the jar with string. Keep in a warm place away from light. Every other day during the first week, use a clean spoon to give it a stir, so the top surface doesn't dry out. You can start tasting it late in week two (with a clean spoon). It should convert fully in 2 to 3 months. Gently strain out any solids, using a colander or sieve as necessary, before using. Store in a tightly capped bottle away from light and it will stay good for at least 1 year.

Makes about 2½ to 3 cups (600 to 700 ml)

Lactic Fermentation: Creating Tangy Flavors from Trace Sugars

Pickles, kimchi, lacto-fermented summer sausage, and yogurt or cheese can last months or sometimes years if carefully stored, while a naked cucumber, a lone cabbage, some raw pork, or a glass of milk will long since have rotted away. In fact, lactic fermentation is responsible for a lot of pickled and soured heavy hitters: yogurt, sourdough, pickles, and kimchi, among many others. If it's tangy and its not vinegar, rhubarb, or sour fruit, it's probably gone through lactic fermentation.

In lactic fermentations (sometimes the process is called lacto-fermentation, but they're the same), lactic acid bacteria (LAB) consume sugars and create lactic acid. Since lactic fermentations are limited only by their need for sugar, they can happen with any ingredients that are even a little sweet—vegetables, fruits, dairy, grains and doughs, even some meats. Lactic acid, like acetic acid, is sour; but it doesn't have a pungent smell—it's closer to plain, clean acidity.

Rather brilliantly, LAB have evolved so that the primary by-product their metabolisms produce, lactic acid, is toxic to most other microorganisms, including spoilage-enacting ones, but safe for humans to consume—even delicious. Many LAB can also tolerate a level of salt that kills other microbes, so salting and souring developed as a double layer of protection in many foods, like pickles and other fermented vegetables.

Crème Fraîche: Creating Cleanly Sour Flavors in Dairy

Making crème fraîche is an easy way to dip a toe in the waters of flavor creation via lactic fermentation that doesn't require any special equipment or holding it at a super-specific temperature. I think it's also the most delicious and rewarding lactic fermentation to do with dairy—the slow, overnight fermentation makes it even more delicious than the store-bought version, and you can stop it (or let it go) until it becomes exactly as acidic as you want.

Food producers use pure LAB cultures for making crème fraîche. But the simplest way to make it at home is to start with a small amount of already lacto-fermented dairy: store-bought crème fraîche, in this case, but I've also used buttermilk and kefir to delicious effect.

This crème fraîche recursion, making new crème fraîche from old crème fraîche, which itself was made from an earlier batch of existing crème fraîche, is a time-honored method called *backslopping*, well known and loved in traditional

beer making, vinegar, and "seeding" new batches of miso. Whenever you make a successful fermentation (like the crème fraîche in the recipe below, or vinegar), hold on to some of it and use the microbial community in it to kick off the next one.

Crème Fraîche

To make crème fraîche, start out with crème fraîche. The next time you have some on hand, save 2 tablespoons of it for this recipe. The lactic acid bacteria in the original crème fraîche will provide a new supply of lactose to consume, digest, and convert to lactic acid, and the increase in acidity will gently denature some of the dairy proteins, thickening the cream. At a warm room temperature (80°F/25°C), the process will take roughly overnight or maybe a couple of hours more; at 65°F/18°C, I've needed to let the crème fraîche ferment for about 40 hours. If getting crème fraîche is inconvenient, buttermilk and kefir both work great as starters.

Wash a large glass jar and lid well with hot water and soap, then air-dry. With a clean long-handled spoon, add **2 tablespoons (30 g) store-bought crème fraîche** to the jar. Pour ½ **cup (120 ml) heavy cream** into the jar and, using the spoon, mix together, breaking up the crème fraîche and incorporating and diluting it into the cream. Add an additional 1½ **cups (360 ml) heavy cream,** stir well, and loosely cover the jar with the lid. Put the jar of cream in a slightly warm part of your kitchen, away from the light.

In the morning, use a clean spoon to taste the crème fraîche. It should be tangy, like sour cream, and thickened. If it hasn't reached this stage yet, re-cover it and let ferment for 8 hours more. If the cream seizes and separates into a chunky curd and yellowy whey, and the aroma is aggressively cheesy, it's gone too far—discard and don't eat it. Store in a covered container in the fridge for up to 1 week.

Makes 2 cups (450 ml)

Things to Do with Crème Fraîche

Eat it with potato chips and a few snipped chives (and caviar, if you can).

Deglaze a pan of chicken drippings with it and a little broth.

Make a pasta sauce by cooking 2 chopped shallots and 3 tablespoons of capers in 1 tablespoon of butter, then deglazing with 8 ounces crème fraîche.

Mash potatoes with crème fraîche and butter.

Dollop it over poached cooled salmon and sprinkle with chopped dill.

Mix equal parts crème fraîche and grainy mustard and drizzle it over roasted brussels sprouts or roasted cooled carrots.

Whip crème fraîche into whipped cream for a richer, faintly tangy topping.

Overwhip the whipped cream until it gets granular and condenses into fat globules, strain off the liquid, and now you have cultured butter (eat it quickly, it's more perishable than store-bought) and delicious tangy buttermilk.

Pickles and Beyond: Creating Sour Flavors in Fruits and Vegetables

Sour, pickley, lip-smacking deliciousness is why lacto-fermented foods are such a fixture in so many cuisines. It's also this tangy promise that has drawn a lot of cooks to it more recently, developing general rules to apply lactic fermentation techniques to new ingredients.

Sturdy vegetables and low-sugar fruits are the mainstays of classic lactic fermentation—like cabbage, mustard greens, turnips, peppers, cucumbers, unripe plums, eggplants, and green tomatoes. All these take on fermented sourness easily and keep well. I think about lactic fermentation as a kind of flavor-creating seasoning, something to add sourness and complexity in any ingredients with a little sugar.

Before proceeding further, make sure you carefully review the safety information in the box.

Brine, Water, Pressure

The simplest salty lactic fermentations just involve water-heavy vegetable matter, salt, and light squeezing. **Sauerkraut** is a classic example: shred cabbage, add some salt, and massage, squeeze, or even lightly pound, which allows the salt to get at the water-swollen cells inside the cabbage through the cut edges. The salt draws out the water via osmosis, wilting the leaves and creating a brine. The brine covers the cabbage and provides a medium for lactic acid bacteria to circulate, sip water, multiply, and ferment in as well as a salty safeguard against spoilage bacteria and an air barrier against molds.

Other fermented products like **kosher pickles, radish dongchimi (water kimchi),** and Sichuan cabbage or mustard green **paocai** involve a separate water brine, made by dissolving salt in water and pouring it over the ingredient, sometimes with extra seasonings like garlic and coriander, or chiles and ginger. Generally, these pickles tend to ferment faster, and develop a slightly less intense funky-sour flavor than "self-brined" fermentations, because of the increased water activity and diluted sugar content. Because less water is pulled out of the ingredients, water-brined fermentations tend to be a little crisper, which works nicely with crisp vegetables like radishes, cabbage, and green beans. Beyond textural considerations, I (very broadly) tend to use brines with more intense-tasting vegetables, and just salt for less-intense ones.

Still other lactic fermentations, especially for certain styles of tsukemono (Japanese pickles), use pressure or air-drying to expel more water from

With any fermentations involving salt or molds like *Aspergillus* or koji (which we'll get to later—see page 272), I work exclusively by weight, not by volume. Weight is the most accurate and easy way to ensure safety, which is largely determined by salt concentration. For this reason, I don't generally write fermentation recipes using volumes for anything except water (which, usefully, is completely interchangeable between grams and milliliters).

Skilled fermenters can work by volume or even by eye and get reliable and safe results. Since I am not there in your kitchen to teach you personally, I can only responsibly recommend working by weights, preferably grams and kilograms. A cheap-but-decent digital scale that measures in 1-gram increments will suffice, and can be acquired online for around twenty dollars if you can't find one in a brick-and-mortar store near you.

This brings up some general safety considerations. Fermentations have been used for a long time to safely preserve foods, but there is always some risk of spoilage. When you ferment, it is your responsibility to make sure you're doing so safely, both for your own safety and especially the safety of others who might be eating your fermented goodies. Use good judgment and look out for signs of spoilage (below). If you are *ever* unsure if something is safe, err on the side of caution and don't consume or serve it. By using any of the recipes in this book, you acknowledge that the ultimate responsibility for food safety belongs to you. Reach out to your local state university's food science or agriculture extension office, or the USDA's National Center for Home Food Preservation (nchfp.uga.edu) if you have doubts or need further guidelines.

Always wash your hands and your containers with soap and hot water, and rinse off any produce, which should be fresh and high-quality.

Your nose and your eyes are the best gauge for safety when fermenting: look for mold, and smell for really unpleasant smells. Many fermentations develop a deposit of harmless, filmy white yeast on their surfaces—you can scrape or spoon this off. Mold looks decidedly fuzzy, white, green, black, or sometimes orange. If you catch mold when there's just one or two small spots, it's usually all right to remove it by spooning it and *the product around it for at least 1 inch in all directions* out, with a very clean spoon. Smooth out the surface and sprinkle on a little more salt. Greater amounts of mold mean a fermentation should be discarded. Brine fermentations will turn cloudy from (good) bacterial growth—this is not the same as mold. The best defense against mold is to keep fermenting vegetables completely submerged in brine.

Usually fermentations will taste and smell pretty weird for the first 3 to 10 days, including a mild sulfury smell (especially anything involving cabbage, turnips, or garlic). This doesn't necessarily or usually mean it's gone off, and you should wait until it's been going for around 2 weeks before making a judgment call.

Brine can thicken during the early part of fermentation, but it usually thins out again. If it doesn't, smell/taste check (smell first!) to see if it is very off somehow.

Smells such as a *strong* sulfury smell, dirty diaper, rotten food, and anything arrestingly off-putting are a sign that you should throw out and not taste your fermentation.

When in doubt, throw it out. Your health is not worth risking for one food experiment.

PACKING AND WEIGHTING FERMENTATIONS

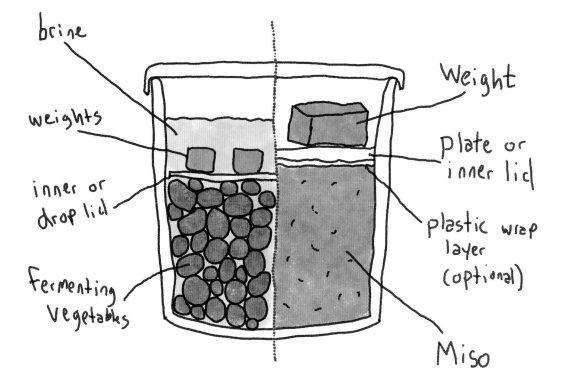

brine

weights

inner or drop lid

Fermenting Vegetables

Weight

plate or inner lid

plastic wrap layer (optional)

Miso

LACTIC FERMENTATION

MISO FERMENTATION

FERMENTATION CONTAINERS

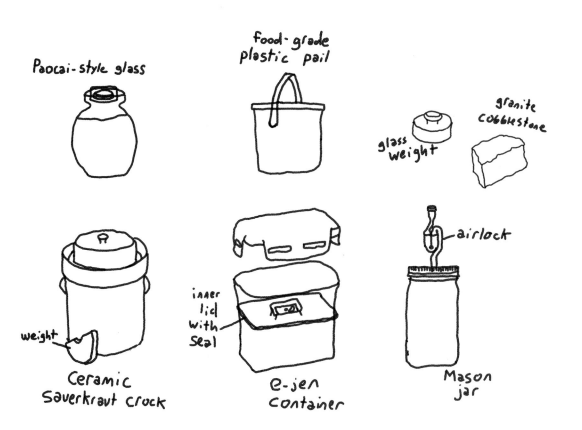

Paocai-style glass

food-grade plastic pail

glass weight

granite cobblestone

Ceramic Sauerkraut crock

weight

inner lid with seal

e-jen Container

airlock

Mason jar

A Note About Containers and Equipment

Keeping fermenting vegetables submerged in their brine is your best insurance for good flavor development, and against spoilage. The brine is salty and acidic, protecting the fermentation from most spoiling microbes. It's also a barrier against oxygen, which halts both oxidation reactions (which can lead to browning, softening, and blackening) and mold growth (which you *really* don't want).

Many different styles of container can be pressed into service for lactic fermentation. The most effective ones have a design strategy to keep everything submerged in brine—often flat weights or drop lids that sit on top of the fermenting vegetables, under the waterline of the brine, in addition to an outer lid covering the whole container. Traditional ceramic crocks and pots often come with ceramic weights that fit inside, and it is easy to find glass weights that fit wide-mouth canning jars online. I personally don't recommend using clear glass apothecary jars, mostly because

big glass containers are an iffy combination of heavy and fragile, and the shape of the lid makes it easy to chip when you lift or lower it and then you've got near-invisible glass chips in your food. I've had a lot of success, on a restaurant production scale, with Korean-made containers called *e-jen*, which have an inner drop lid with a gasket that helps keep it submerged, and a loose-fitting top lid (look for them online).

The loose-fitting top lid is important because lactic fermentations generate carbon dioxide gas, which needs some way to escape the container. Other designs deal with this with a one-way valve or airlock, letting the carbon dioxide out without letting outside air in. You can spend anywhere from almost no money to hundreds of dollars on a fermenting vessel; the cheapest way is usually a wide-mouth glass canning jar, with a double-bagged zip-top bag filled with water or brine sitting on top of the vegetables to weight them down. (See illustration, page 264, for the options.)

---

the ingredients before or during fermentation, to get firmer textures and sometimes sourer flavors. For the Kyoto-regional specialty called *suguki*, for example, a variety of turnip is weighted down, traditionally in wooden vats called *tenbin* that use a kind of cantilevered arm with stones attached to it. The finished pickle is dense and almost meaty, quite unlike a fresh or cooked turnip. Umeboshi, salt-pickled ume fruit (which are stone fruit like plums and apricots), are usually weighted while they are fermenting, then semi-dried after they become sour, so their texture becomes intense and jammy.

Brine-Fermented Chiles

In contrast to more heavy-flavored fermented Aromatic Fermented Pepper Paste (page 111), using a brine to ferment chiles keeps their flavor lighter, brighter, and zippier. I like to use mostly mild but aromatic aji dulce or Caribbean seasoning peppers, plus a couple of habaneros, for this. Since this recipe will work with most chiles (as long as they're small enough to fit in your jar), use the most full-flavored ones you can find.

Select a fermentation vessel, lid, and weight (see "A Note About Containers and Equipment," above) that can hold ½ gallon or 2 liters. Wash them with soap and hot water.

Make a 5 percent salt brine by combining **½ gallon (1.9 liters) water** and **95 g uniodized salt**, such as kosher salt or sea salt, in a large pot on the stove. Heat over medium-low heat until the salt dissolves. Let cool to room temperature.

Wash, stem, and seed **1 pound (450 g) fresh chiles** and put them into your fermentation vessel. Pour the brine over to fill the vessel up to 1 to 2 inches below the opening; you won't use all of it. Put the vessel on a baking sheet or large plate to catch any spillover brine. Weight down the chiles using your weight to keep them submerged in brine, and cover the vessel using the lid—making sure, whatever lid you choose, that there's some way for gas to escape.

Let ferment in a room-temperature or slightly cool spot, away from the light. As it ferments, it will start to bubble and the brine will turn cloudy—these are good signs of bacterial activity. Check every day for yeast bloom or mold and skim any off when you see it, then keep fermenting. If you see mold beyond a few initial specks, you should discard the fermentation (see "A Note on Recipes, Units, Weight, and Safety," page 262, for more information).

Around day 7 to 10, or when any bubbling is finished and the brine is cloudy, use a clean spoon to fish a chile out and taste it. It should be a bit softened and quite tangy. If it still tastes fizzy or yeasty, or isn't as sour as you like, let the jar ferment further, and check every other day until the chiles are roundly tangy.

Store tightly covered in the refrigerator, submerged in brine, for up to 6 months.

Makes about 1 pound fermented chiles and 3 to 6 cups brine

Brined Chile Hot Sauce

The chile brine is nearly as delicious as the chiles themselves, and I recommend saving it to season vegetables, seafood, and marinades with sour-salty gentle heat even after you finish off a jar of fermented chiles. What I look for in a hot sauce is aroma, tangy sourness, and good depth of flavor in addition to some heat—this light and fluid, brine-heavy recipe is a great way to get there.

Combine **¾ cup plus 1 tablespoon (200 ml) chile brine** (from the preceding recipe), **2 cups (300 g) Brine-Fermented Chiles, drained** (page 265), and **2 tablespoons (30 ml) white vinegar** in a blender jar and blend well. Add more vinegar to taste if necessary. Store in a tightly capped container in the fridge for 4 to 6 months.

Makes about 2 cups hot sauce

Orangey Brined Chile Styled like Yuzu Kosho

Yuzu kosho is a paste-y Japanese hot sauce, made by salting and pureeing chiles with salt and large amounts of zest from yuzu, an intensely aromatic citrus. It's spicy, it's salty, it's incredibly citrusy-flavorful and refreshing. A little dab livens up raw or grilled seafood, fried tempura, even broths or grilled beef.

This take on it started with a large quantity of already fermented chiles and the thought "What if I made this as fresh and citrusy as possible?" It's also fantastic over asparagus or broccoli rabe cooked with a little bit of char on them, to season mayonnaise (to your taste) for a chicken salad sandwich or BLT, or added bit by bit to a dressing for arugula or crisp romaine lettuce.

Combine **¾ cup (120 g) drained Brine-Fermented Chiles** (page 265), **1 tablespoon plus 1 teaspoon (10 g) grated orange zest**, **¼ cup plus 1 tablespoon (70 ml) freshly squeezed orange juice**, and **1 tablespoon (15 ml) lemon juice** in a blender or food processor.

Pulse to combine and roughly grind but stop before the mixture liquefies. It should have a pasty texture, and some liquid will seep out. Taste and adjust the lemon juice (to make it sourer) and orange juice (to make it sweeter or balance the sour/salty). Keep in the fridge for up to 6 months in a tightly covered container with as little headspace as possible.

Makes 1 scant cup

Two Different Lactic Fermentation Strategies Demonstrated on Eggplant

As mentioned in "Brine, Water, Pressure" (page 261), simple changes to a lactic fermentation, like how much water is present or how much weight holds the vegetables down, can have deliciously different effects on flavor and texture. The next two recipes explore this by fermenting the same vegetable, eggplant. One is water-brined and juicy, the other is pressed and meaty-textured, something that works especially well on the spongy structure of eggplant as well as vegetables like turnips or daikon.

Brine-Fermented Eggplant with Savory Spices

This makes a sour, savory, slightly earthy-spicy, low-key snackable eggplant pickle using a water brine. I eat it right out of the jar, or have it on the side of stewed lentils or chickpeas, meat and rice, or grilled chicken.

Try to find smaller eggplants, which are generally more flavorful and less spongy than large ones. Choose specimens that seem heavy for their size, with the thinnest skin and no blemishes or soft spots. If you have thicker-skinned ones, peeling them is a good idea, because the skin can stay tough. You can adjust the quantity of eggplants up or down; just be sure to make enough brine to cover them, and keep the salt in the brine at 5 percent. If black garlic hasn't made it to where you are, it's widely available online, or you can omit it.

Select a fermentation vessel, lid, and weight (see "A Note About Containers and Equipment," page 265) that can hold ½ gallon or 2 liters. Wash them with soap and hot water.

Make a 5 percent salt brine by combining **1½ quarts (1400 ml) water** with **70 g uniodized salt**, such as kosher salt or sea salt, in a large pot on the stove. Heat over medium-low heat until the salt dissolves. Let cool to room temperature.

Wash **2 to 3 small to medium eggplants (8 to 10 ounces/240 to 300 g)** and cut into roughly 1-inch (2.5-cm) cubes. Put the eggplant in the container along with **1 minced garlic clove, 2 teaspoons (7 g) black garlic powder or 2 whole cloves black garlic, 1 heaping tablespoon (4 g) lightly crushed coriander seeds, 2 teaspoons (5 g) whole cumin,** and **½ teaspoon (1 g) Turkish urfa biber flakes or another rich, medium-hot dried flaked or ground pepper such as guajillo or ancho.**

Put the vessel on a baking sheet or large plate to catch any spillover brine. Weight down the chiles using your weight to keep them submerged in brine, and cover the vessel using the lid—making sure, whatever lid you choose, that there's some way for gas to escape.

Let ferment in a room-temperature or slightly cool spot, away from the light. As it ferments, it will start to bubble and the brine will turn cloudy—these are good signs of bacterial activity. Check every day for yeast bloom or mold and skim any off when you see it, then keep fermenting. If you see mold beyond a few initial specks, you should discard the fermentation (see "A Note on Recipes, Units, Weight, and Safety," page 262, for more information). As the eggplant ferments, the brine will stain purplish brown from the skin—this is normal.

Once the bubbling seems to be slowing down (which, depending on the temperature, sugar content of the eggplants, and local bacterial flora, can take 3 to 7 days, or even a bit longer), fish out a piece of eggplant with a clean spoon and taste it: it should still have some texture and be pleasantly sour-salty all the way through. Let the fermentation go for a few more days if it's not there yet or if you're unsure. The eggplant can be stored in the fridge (with some brine covering it, just like a jar of pickles) in a tightly sealed container for 2 to 3 months.

Makes about ½ to ¾ pound (2 to 2.5 cups)

Variation
Branching Out with Lactic Fermentation and Vegetables

This water-brined recipe has the same proportions as the water-brined chiles above, with some extra flavor added to the brine. You can think of it as a template for trying out lactic fermentation on a variety of vegetables—green beans, green or white asparagus, okra, sections of raw corn on the cob, carrots (sliced lengthwise), brussels sprouts (cut into halves or quarters), napa or other green cabbage (cut lengthwise through the core into halves or smaller pieces to fit your jar), cauliflower florets, slightly underripe cherry tomatoes, radishes, small turnips, or ramp bulbs are all a great place to start. An easy place to experiment and find what you like is with flavorful additives—adding to taste or around a couple of teaspoons per ½-gallon jar, garlic, citrus zest, dry and fragrant or earthy spices (see page 161), spice mixtures (see page 163), and herbs like thyme, mint, dill, tarragon, shiso, or cilantro. Other excellent additions include small amounts of fruit, which would tend to get very yeasty when lacto-fermented on their own: cubed apple or pear, blackberries or cranberries, dried cherries, chunks of lemon, Meyer lemon, or orange (figure a small handful or roughly 15 grams of spices or herbs into a ½-gallon jar as a good proportion to start).

Eggplant "Gabagool"

Several years ago I was in Kyoto, Japan, learning hands-on techniques for kyotsukemono (traditional, local-to-Kyoto pickles), courtesy of Murakami Jyu Honten, a family company that has been producing them for generations. One of my favorites was suguki, a turnip pickle that is fermented under pressure until it becomes almost meaty in texture. That meatiness made me think about using a similar technique with seasonings that take some flavor inspiration from salumi. Gabagool (the Philly–Jersey–New York way of pronouncing the whole-muscle cured pork cappocollo, also called coppa), at its best, is seasoned with garlic, spices, and smoke—as is this. I like to slice this into thin strips and snack on it; it's also great in pasta sauces or finely chopped as a little garnish to a charcuterie plate.

In this fermentation, you will start fermenting sliced and salted eggplant under weights in zip-top bags, then season and move the pressed slices to a fermentation container to finish fermenting. For weights, you can use very clean rocks or cobblestones—run them through the dishwasher or give them a good scrub. Twenty-eight-ounce tomato cans, multipound boxes of kosher salt, a large cast-iron Dutch oven, or bricks will also work for the first pressing step. For the second fermentation step, depending on your container's shape, wrap the underside of your weight in plastic wrap, or nest a second container into the first, then put the weights inside the top container.

The recipe calls for a few eggplants, but you can scale it up to work with larger amounts.

Slicing, Salting, and Pressing the Eggplant

Wash **2 to 3 small to medium eggplants,** cut the ends off, and cut into ¾-inch-thick (2 cm) slices, as uniform in thickness as possible. Weigh all the slices together—you should have 8 to 9 ounces (225 to 250 g). Put the eggplant slices in a glass, stainless-steel, or plastic bowl and evenly sprinkle them with **4 percent of their weight in uniodized salt,** such as kosher salt or sea salt (for 250 grams eggplant, this is 10 grams salt). Toss the slices with clean hands to cover them as evenly as possible with the salt; it will make only a patchy layer.

Keeping as much of the salt or any leaked juices on them as possible, put the eggplant slices in a single even layer in a zip-top freezer bag, adding another even layer of salted slices on top. Use more bags as needed to fit all the slices in two layers. Push as much air out of the bag(s) as possible, keeping the slices in their even layers. To fully remove any remaining air, fill a sink or stockpot with water. Crack the zipper just a bit at one corner. Lower the bottom of the bag into the water and as you continue to submerge it, the water will press in around the bag and push out the rest of the air through the opening. Close the zipper before the whole thing goes underwater. You can also layer the eggplants in vacuum-sealer bags and seal them that way.

Put the bag(s) of eggplant slices on a flat baking sheet, or a wire rack on top of a rimmed baking sheet. Place another baking sheet on top of the bags and put 5 to 10 pounds of weight on it (see headnote). Let sit, stacked and weighted, at room temperature for 2 days to compress.

Seasoning and Second Fermentation

Select a fermentation vessel, lid, and weight (see "A Note About Containers and Equipment," page 265) that can hold ½ gallon or 2 liters. You will need a "hat"—a drop lid, a small plate, a yogurt container lid, or several clean cabbage or collard greens leaves—to help press the eggplant. Wash the container and all accessory parts—lid, weights, etc.—with soap and hot water.

After 2 days, the slices should be very flattened and have given up some water to make a brine. Decant off the brine into a large bowl, and mix it with **1 tablespoon (10 g) smoked paprika**, **2 minced cloves garlic**, **1 teaspoon**

(3 g) crushed fennel seeds, and **½ teaspoon (1 g) smoked/morita chile** (or another flavorful dried chile like guajillo or peperoncino).

Add the flattened eggplant slices to the bowl and toss with the spice-brine mixture.

Layer the coated slices evenly in your fermentation vessel, and place the vessel's interior lid or other "hat" on them to hold them down, then place a 3- to 5-pound weight on top. Cover the whole thing with a loose-fitting lid, or a clean dish towel tied with string to keep out dust, insects, and mold spores.

Ferment the seasoned eggplant slices under the weight for 4 to 6 days, checking on the fermentation pot every day to make sure the slices are still under brine and to remove any yeast blooms or spots of mold that may form.

At this point, use a clean spoon or fork to remove a slice to taste. It should be salty and sour; if it tastes too weak or is still a little fizzy, repack it and keep fermenting it for a few more days.

When it's finished, the eggplant should be quite dense, spicy, and tangy-salty. To serve, cut into thin (¼- inch/1 cm or thinner) slices with a sharp knife. Store the eggplant covered in its paste in a sealed container in the fridge for up to 3 months.

Makes 6 to 7 ounces

(Almost) Anything Can Be a Brine

Let's get metaphysical about brine for a sec. "Brine" can have a primary connotation of "salty water," but we can also think of it as "the stuff that covers and holds fermenting things away from air and in a certain salt concentration." In practice, that doesn't necessarily have to mean "water" or "juices." Some of the most delicious fermentation techniques for Japanese tsukemono (fermented pickled vegetables) make thickly coating "brines" from sake lees (yielding kasuzuke and narazuke), rice bran (nukazuke), or miso (misozuke). Instead of vegetables floating in a jar, they look like vegetables buried in a bed of flavorful paste. Whatever's interesting in the brine has a chance to infuse into the pickles, and they can even encourage specific, tasty microbes to grow better than they would in a liquid brine, making each style a unique tool for creating flavors.

Garlic Fermented in Miso (Garlic Misozuke)

Misozuke uses salty, funky, umami-packed miso as a "brine" to ferment things in. Fermenting garlic like this is one of my favorite techniques, and it creates two ingredients in one fermentation: a roasty-tasting, garlic-infused miso and a rich, salty, umami-infused, pickley garlic. I slice the garlic thinly and eat on its own or as a nibble with cheese, incorporate it into mayonnaise or a vinaigrette, or use it to flavor sautéed vegetables or roasted meats. The garlic-perfumed miso is great incorporated into a marinade, glaze, or dressing, especially with a dash of rice vinegar as a condiment for vegetables or grilled meat.

I recommend a medium-intense commercially made miso for this: red (aka) miso, yellow (shinshu) miso, or barley (mugi) miso are good bets.

The weight of salt and of sugar you add, 10 grams, is ⅓ of an ounce, but please make your life easier and measure in grams.

Select a fermentation vessel and lid (see "A Note About Containers and Equipment," page 265) that can hold at least 1 quart or 1 liter. Wash them with soap and hot water.

Using very clean bowls and utensils, mix **1 pound (450 g) miso** with **10 g uniodized salt**, such as kosher salt or sea salt, and **10 g brown sugar**. In the fermentation vessel, spread a ¼-inch-thick (1 cm) layer of the miso mixture. Start placing **peeled garlic cloves from 3 to 4 firm heads of garlic** on the surface of the miso in a single layer so they are close but not touching. Using a spoon or thin spatula, spread another layer of miso mixture over the garlic, covering them and filling the spaces in between the cloves. Add another layer of garlic cloves and another layer of miso, repeating until you have just enough miso to cover the top layer of cloves with a ½-inch (1.5 cm) layer. Use a piece of plastic wrap to push down gently on top of the layers, to push out any air pockets. Leave the plastic wrap in place, pressed onto the surface of the miso, and loosely cover the container with the lid, closing it loosely or leaving it partially ajar so gas can escape.

Ferment the miso and garlic together for 1 month at room temperature. Use a clean spoon to fish out a piece of garlic to test; it should be medium brown and look rather marinated. If the garlic still looks too white or solid, let it ferment for another month. The garlic misozuke should smell like fermented soy and lightly roasted garlic; *if you sense any rotten or off-flavors or the garlic gets overly slimy or grows green or black mold*, you should throw the batch out and start again.

The garlic and miso can be stored together, tightly covered, in the refrigerator for up to 1 year.

Makes about 1 pound (2 cups) of garlicky miso and about 5 ounces (150 g) pickled garlic cloves

Creating Umami: Fermenting with Fungi

For a basic taste, umami can be surprisingly elusive in the wild. A few raw ingredients contain a lot of it—seaweeds like kombu, tomatoes, shiitake mushrooms, shellfish. To branch out further, we have to create it ourselves, which means finding a way to break up protein molecules to get at the amino acids within.

The tried-and-true strategy is to locate an enzyme to do the work for us, and then give it time to work. Enzymes are highly specific protein tools, shaped to do one particular job or action. If we're looking for enzymes to break down proteins, we're (by definition) looking for a protease. And just like recruiting a specially skilled lock picker or getaway driver for a (highly realistic) cinematic heist, you just have to know where to look.

Living cells pretty much all produce proteases for the occasions they need to recycle a protein into a different protein. When you salt a pig's leg and age it to make ham, or set a youthful wheel of Parmesan on its shelf to age, you're giving these generalist proteases time to do their work.

Specific plants like nettles, cardoons, pineapples, figs, and papaya make higher-than-normal amounts of proteases, which will absolutely ruin your (proteinaceous, gelatin-based) molded Jell-O constructions if you attempt to include them untreated.

Digestive tracts usually have a lot of proteases in them, for animals to consume dietary protein and break it up for their own use. And in a slightly gross, but brutally efficient stroke of human genius, we use them as hybrid ingredient-tools for things like curdling cheese (originally done with the proteases produced by baby cows) and making fish sauces and pastes like (Thai) nam pla, (Vietnamese) nước mắm, (Filipino) bagoong, or (ancient Roman) garum—all of which involve blending fish or crustaceans, with their guts and

digestive juices and a lot of salt, to essentially digest themselves.

Superspecialist microbes like the molds *Aspergillus oryzae*, *Aspergillus sojae*, *Penicillium roquefortii*, and *Rhizopus oligosporus* and the bacteria *Bacillus subtilis* are the Cadillacs of protein-chewers, the I-don't-get-out-of-bed-for-less-than-ten-thousand-dollars-a-day visionaries of protease production. If you want to get as much protease as possible, generate umami quickly, or glutamatize protein-rich plants cleanly, these are who you call up.

Microbes That Break Down Proteins

Working with most major fermentation microbes (lactic acid bacteria, yeast, acetic acid bacteria) is a bit like using your dairy-farm-neighbor's manure to fertilize your vegetable farm. You luck out that

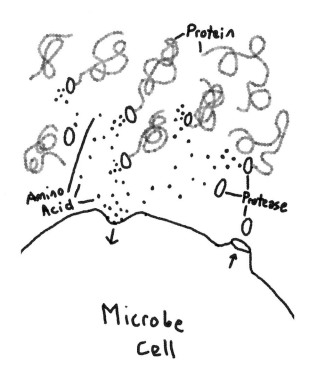

something you want—the molecules lactic acid, alcohol, acetic acid—happens to be a by-product of some other kind of regular business. These molecules happen to be useful for the microbes, too—they're good at warding off other types of microorganisms we consider spoilage microbes, like molds, which would otherwise compete to consume whatever you're trying to ferment. But they're a waste product, not the main event for your microbes.

Creating umami means fermenting for free glutamate, which is a bit more like breaking into a brick factory and stealing a few bricks to finish a wall. Microbes that are really good at breaking down proteins into amino acids do so as an intermediate step—where the next step is using those amino acids to build their own proteins. The microbe strains we use to create umami produce way more free amino acids than they actually need, which is what creates their umami flavors in the first place; however, they're produced in the first place for their own use.

Originally, things like miso and fish sauce were probably not made with the primary intention of flavor creation. Rather, the flavors we create are a happy side effect of wanting to make protein-rich foods last a long time. Salting them and fermenting them were really effective ways of keeping food from going bad, and the flavor creation was an added bonus that eventually became the primary goal.

Building on this preservation tradition, many fermented ingredients with rich stores of free glutamate are deliberately, painstakingly constructed monuments of umami, setting up the right strains of microbes, the right sources of protein, the right balance of water and salt to ferment and slowly break down those proteins over weeks, months, even years. Over that time, many other biological and chemical reactions are churning, slowly creating other flavor molecules as accessories to umami: toasty-caramel flavors, funky ones, cheesy ones, nutty, even fruity or floral.

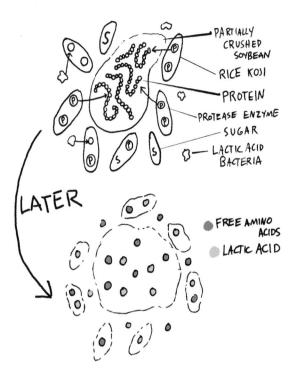

Miso, which starts life as soybeans, (usually) rice, and salt and ends up a richly aromatic and deeply flavored paste, is an excellent example. To get there can take anywhere from three weeks to two years of fermentation, in two distinct phases. In the first phase, you cook rice (occasionally barley or soybeans), and inoculate it with spores of *Aspergillus oryzae* mold, then let the mold grow in a warm, humid room for two days. This relatively brief window is like putting money in an index fund: an initial, relatively minor investment in flavor that will build over time like compound interest. *Aspergillus* makes its own aroma compounds—it can be distinctly chestnutty, floral, even fruity, with some mushroomy notes—but we're really interested in its enzymes. Along with the now-familiar, umami-creating proteases, it also makes amylases, which specialize in breaking down starches into their smaller sugar subunits.

The next step is to take the mold-inoculated rice, called *koji*, and mix it with soybeans, salt, and a little water. Molds need oxygen to grow, so, the *Aspergillus* cells in the koji no longer directly participate in the action. But their enzymes are still there, like working machinery left behind in an abandoned factory. These amylases and proteases get to slow work on the carbohydrates and proteins in the soybeans. Added to the sugars they extracted from the rice's starch molecules, the stage is now set for a second, totally separate fermentation phase. For the next few weeks or months, former-mold enzymes churn through protein and starch, and bacteria and yeast consume previously inaccessible sugar molecules and create new, flavorful compounds.

The ratios of koji to soy, the amount of moisture, and the amount of salt are a relatively simple set of levers that miso makers can use to set up a surprisingly wide range of flavors. The oldest, darkest misos, like deep-savory, deep-brown hatcho miso from Nagoya, have the most salt, and the least rice (hatcho miso actually has no rice—the *Aspergillus* is grown on soybeans to create the koji). On the other extreme, sweet white saikyo miso from Kyoto uses up to twice as much rice koji as soybeans, with less salt and a brief fermentation time (a few weeks), keeping many of its sugars unfermented and giving it a sweet, floral-chestnut flavor. In between you can find slightly older and more savory white misos, yellow misos, and older red misos.

While I've researched and learned miso making in Japan, I first learned how to do it in Denmark. My friend Lars Williams, then head of R&D at Noma, asked the question: What role does soy play in miso? What function does it serve? If it's largely as a source of plant protein, is there something else that can fill that role? The highest-protein local legume turned out to be Danish yellow peas, which, when combined with koji made from Danish barley, birthed yellow pea miso, or "pea-so," a funky, umami, tangy condiment that Noma continues to use today, and which put the idea of combining canonical fermentation techniques with novel, local ingredients on many cooks' radars. Making your own

miso is a way to discover umami and shape it to your tastes, or use ingredients you might not find fermented elsewhere to create slightly unpredictable, but delicious flavors to use.

Pumpkin Seed Miso

I love the already-savory, buttery quality of pumpkin seeds, and of the many things I've used in place of soybeans in miso, they're one I like best. Like soybeans, they've got a ton of protein in them (up to 30 percent of their weight), and using them for miso kind of echoes their sometime appearance in ogiri, a West African umami condiment of aged fermented oilseeds.

Based on the proportions for a relatively sweet white miso, this is a pretty quick method for getting to usable finished product, and the relatively young age keeps a kind of bright, fruity, fresh-nutty quality. I use dried koji and am a big fan of the pineapple notes in Miyako koji, which is pretty easy to find online in the United States.

As I mentioned earlier, I always measure by weight, in grams, when I'm doing fermentations, so I can be accurate with overall ratios and especially salt, which is key for safety.

Select a fermentation vessel, lid, and weight (see "A Note About Containers and Equipment," page 265) that can hold ½ gallon or a hair under 2 liters. Wash them with soap and hot water.

Crumble **400 g dried koji** into a blender or food processor, and pulse to grind it coarsely. Remove to a large mixing bowl, then pulse **525 g raw, good-quality pumpkin seeds** in the blender until most of them are reduced to a coarse meal and about one-fourth of them are whole or slightly chunky. Add to the koji in the bowl and mix together by hand. (You may want to wear latex or nitrile gloves, but you certainly don't have to if your hands are clean.)

Measure out **400 ml filtered or good-quality tap water** into something you can pour with. Add the water gradually to the koji and pumpkin seeds, mixing it in as you go, stopping when the miso will hold a ball without crumbling when you squeeze it but isn't so wet that it squishes between your fingers. For the weights as written, add **70 g uniodized salt**, such as kosher salt or sea salt, to the miso and mix in thoroughly by hand. If you've used a lot more or less than 400 ml water, stop and measure the weight of the miso mass. Whatever it weighs, calculate 5 percent of it (or divide by 20—the math works out the same); this is the amount of salt you need to add.

Carefully pack the salted miso into your fermentation vessel, making sure there aren't any air pockets inside or under the miso. Press a piece of plastic wrap to the surface of the miso, then add on your container's inner lid, or a small plate or plastic disc cut-to-size from the lid of your container. Put your weight—it should be 2 to 3 pounds (a container of small rocks, a brick, or some other object)—on top of the plate.

Ferment the miso in a warm spot (90°F/32°C is ideal) for 2½ to 3 weeks, or a bit longer if the room is cooler. Stop fermenting when the miso is succulent and light brown, with fruity-caramel and slightly funky aromas. Keep in the fridge, with plastic wrap or parchment paper pressed to the surface, for up to 6 months or freeze flat in zip-top bags.

Makes about 2.5 pounds (5 cups) miso, depending on evaporation

Sweet White Rice and Soy Miso

I like to use this version when it's young and still rather sweet, but you can age it for up to 3 months if you want to deepen the flavor further. If you feel like experimenting, you could substitute cooked chickpeas (or other cooked legumes, like fava beans or good-quality pinto beans) for the cooked soybeans. Not to state the obvious, but high-quality dried beans (such as Rancho Gordo) that you cook yourself will make a miso

---

Where Did Fermenting Soy into Miso Come From?

Fermenting soy to make richly umami pastes began with soy-based jiang in China sometime before 200 B.C.E. During the early Tang dynasty (seventh century C.E.) jiang production was exported out of China and into neighboring nations, carried with the spread of Buddhism. In Korea, it became doenjang, and in Japan, it became miso. For a thousand or so years prior to fermenting miso in Japan, there were many indigenous techniques for fermenting similarly umami hishio, based on fish or meat such as crab, deer, carp, or squid. (These used either a microbial starter or digestive enzymes, like the rich and inky, squid-based ishiri that's still made in Japan's Noto Peninsula.)

---

about an order of magnitude more flavorful than mass-produced dried or canned beans.

Proceed as for pumpkin seed miso, using **400 g dried koji** and substituting **800 g of soaked, cooked by boiling until soft, cooled soybeans** for the pumpkin seeds, reserving the water you cooked them in. Use the **bean cooking liquid** in place of the tap water to hydrate the miso as necessary. It should hold a ball without crumbling when you squeeze it, but isn't so wet that it squishes between your fingers. Because the soybeans already have lots of water in them, you'll probably need only about 200 g of cooking liquid. Add **70 g salt** if you followed the above weights exactly, including the cooking liquid; otherwise, weigh and add 5 percent (or one-twentieth) of the total weight of the miso in salt. Put the weighted miso assembly in a relatively warm, dark place. Leave undisturbed for 3 weeks, then begin checking it: it should be forming a more coherent mass, turning a light brown, and giving up a small amount of sticky golden liquid that has risen to the top. Taste it: it should be getting a little tangy,

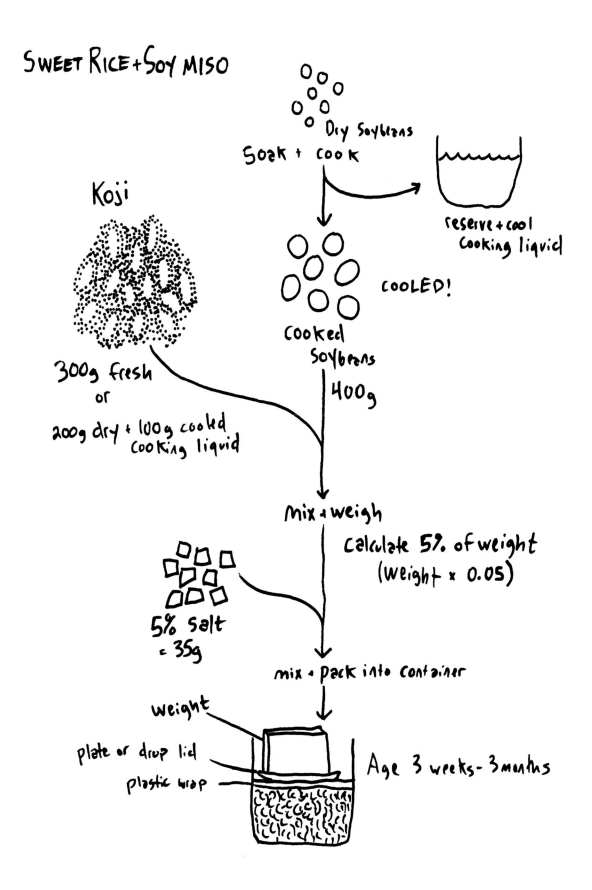

SWEET RICE + SOY MISO

Dry Soybeans

Soak + cook

reserve + cool
Cooking liquid

COOLED!

Cooked Soybeans

400g

Koji

300g fresh
or
200g dry + 100g cooked
Cooking liquid

Mix + weigh

Calculate 5% of weight
(weight × 0.05)

5% salt
= 35g

mix + pack into Container

weight
plate or drop lid
plastic wrap

Age 3 weeks - 3 months

with umami flavors and caramel undertones. If it's not there yet, let it go another week and taste again. Repeat if necessary.

Store in the fridge, with plastic wrap or parchment paper pressed to the surface, for up to 6 months or freeze flat in zip-top bags.

Makes about 2½ pounds (5 cups) miso, depending on evaporation

Meyer Lemon Miso

Between pickled-vegetable-packed namémiso and hishio, ginger and sansho in kinzanji miso, and yuzu miso from Kyoto, Japanese miso makers have developed tons of styles that include extra-flavorful ingredients alongside koji and soy.

Adding Meyer lemon zest to the miso at the outset of fermentation, especially for a sweeter white miso, will make it develop sweet, perfumed, almost cakelike aromas as it progresses. Adding aromatics early leads to a different result than mixing them in just before using the miso: more blended, certainly, but also heightened in flavor.

Alternatively, yuzu zest is also an amazing inclusion. Yuzu kosho, mixed in at the same stage, will create a spicy-citrus-floral quality in the miso. Grated fresh turmeric in the same proportions beautifully enhances the earthy notes of the miso.

Follow the instructions in Sweet White Rice and Soy Miso (page 274) or Pumpkin Seed Miso (page 273). After you have added the liquid to get the texture right, add the **grated zest of 4 Meyer lemons** (roughly 8 to 12 grams). Then, starting with mixing in the salt, continue to finish the recipe.

Incubating Fermentations

With all fermentations, the biggest arbiter of fermentation speed (and flavor development) is temperature. At cooler temperatures, microbes are more sluggish and the fermentation proceeds slowly. Kind of like cold-blooded lizards, they

Experiment: Perfumed and Aromatic Miso

Misozuke (see Garlic Fermented in Miso, page 270) uses miso as a medium to both ferment and infuse garlic with its flavor. In return, the miso pickling bed gets very garlicky. I was thinking about this when my friend and chef Angela Dimayuga, for whom I'd made Meyer lemon and turmeric miso, told me she was cooking a dinner to celebrate the launch of a new perfume. I have a bunch of perfume aromatics at home—resins, woods, other things like that. They smell great, but the texture isn't nice to eat—they're not chewable. So, with red sandalwood, orrisroot, and frankincense (separately), I made Sweet White Rice and Soy Miso (page 274). Rather than mix in the aromatics, I layered them in kind of like misozuke, with cheesecloth to help keep them contained. Layering right at the stage where you're packing the miso crock is a great time to incorporate more potent, less mixable aromatics. The miso juices and microbes can seep in and extract flavors, perfuming the whole thing as it ferments. And each of the misos was, indeed, perfumed, but in a deliciously edible way.

speed up as they get warmer, and are very active at summery temperatures above 75 to 80°F. Some fermentation styles count on seasonal temperature changes: many misos and soy sauces are started over the winter to slowly ramp up, and then ferment for up to 12 or 18 months, going faster in the summer, and slower in the autumn and the following winter.

There's no real substitute for waiting a few months (or that full year-plus for richer styles of miso) to reach full flavor development, but you still can get something pretty damn good if you speed the process up with a little heat. A very warm environment of 95°F/35°C can take the aging time of a saltier miso down to 2 to 3 months

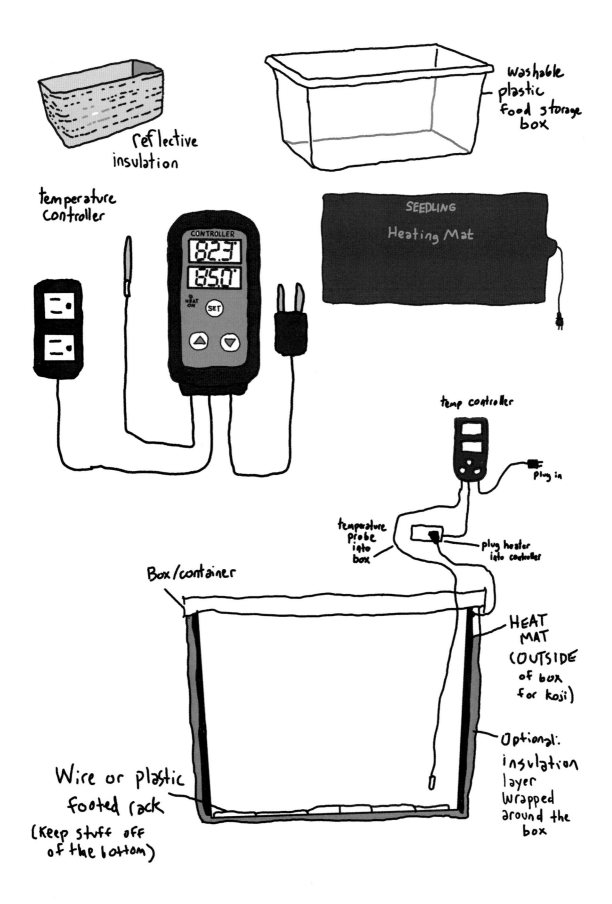

reflective insulation

washable plastic food storage box

temperature controller

CONTROLLER
82.3°
85.0°
HEAT ON
SET

SEEDLING
Heating Mat

temp controller

plug in

temperature probe into box

plug heater into controller

Box/container

HEAT MAT (OUTSIDE of box for koji)

Optional: insulation layer wrapped around the box

Wire or plastic footed rack
(keep stuff off of the bottom)

rather than a year, or for a sweeter miso, to a few weeks instead of a few months.

To keep a space that warm without steaming everybody out of the apartment (or prep kitchen), I like to make fermentation boxes—essentially, incubators. I've made these from the scale of a shipping container all the way down to a cardboard box sized to hold 6 bottles of wine. The two essential parts, besides the housing, are a temperature controller (essentially a plug-and-play thermostat) and a low-powered heat source to plug into it. You set a temperature on the controller, which is attached to a thermometer you put inside the box, and the controller turns the heater on when it senses temperatures below the set point, and shuts it off if the temperature goes above the set point. The most accessible equipment is designed for reptile enthusiasts and indoor gardeners; controllers made by Inkbird usually run about thirty-five dollars online, and seedling heat mats—basically small, flexible, waterproof electric blankets—perhaps ten to fifteen dollars apiece.

For miso, you can set up a simple incubator by wrapping a heat mat around the fermentation vessel, plugging it into the controller, and putting it all in a clean cardboard box large enough to hold the container. If I'm making multiple batches, or making koji (pages 279–84), I like a 17- to 22-gallon plastic food storage box (such as those made by Cambro), which are easy to find online, especially from restaurant supply websites (which are usually happy to sell to home cooks). They're fine getting wet, and can be cleaned out easily. You can attach one or two heat mats to the bottom or sides of the container with a bit of duct tape (or, my preference, double-sided VHB tape) on each corner. If I know I'm going to ferment in it for a while, I buy a roll of reflective insulation (essentially, bubble wrap with foil on both sides) from the hardware store and tape it around the outside of the box. It keeps the heat in more efficiently, and you can slide it off or untape it if and when you need to clean the container.

History Break: Enzyme-Rich Mold and Bacterial Starters

The original, mold-cultured rice or wheat dough fermentation starter was called *qu* or *chhu*. What we call qu now are the group of fermentation starters used in China for rice wine and liquor, soy sauce, and soy paste. Qu spread from China into other parts of East Asia, probably alongside Buddhism during the Tang dynasty.

Qu was, as far as we can tell, originally used as a wine starter. The molds involved tend to produce amylases, which break down starches into sugars. Growing them on grains or doughs creates fermentable sugars, and the potential for alcohol, where starch otherwise offers no incentive to yeasts. From its use making beverages, qu became a starter for vegetable and meat pickles and relishes called *jiang*. Soybeans got involved later in the form of soy jiang or doujiang. Mixed in with this evolution, people also started fermenting mold-cultured soybeans or soy nuggets on their own, becoming doubanjiang.

Qu has many descendants today. The microbes involved are often our old friend *Aspergillus oryzae* and other *Aspergillus* species, other fungi like *Rhizopus*, as well as *Bacillus subtilis*, not a mold at all but rather a heat-tolerant, alkaline bacteria that makes a lot of protease enzymes.

In Korea, soy sauce (ganjang) and paste (doenjang) start with **meju**, blocks of cooked soybeans naturally inoculated with molds and bacteria by wrapping the blocks in straw or twine and letting them hang in a warm indoor spot. Rice-based **nuruk**, with similar microbes, is used for rice wines and liquors. In Japan, *Aspergillus oryzae* is bred and produced as a totally domesticated mold, and the mold's purified spores, called *koji-kin*, are used as a starter for koji. Besides miso, koji is the starting point for soy sauce, sake, distilled shochu, briefly fermented seasonings like shio-koji, and the sweet rice drink amazake.

It's a little bit of a project, and many people may not want to go to the hassle and expense, but I include it for the curious and for the few of you whose interest in miso and other fermentations has been piqued enough to want more.

Sweetly Breaking Down Starch to Sugar: Amazake

Japanese-style amazake is a milky, thick, rich, sweet drink—and it's completely plant-based. I've enjoyed it cold out of the fridge and hot after a winter afternoon making miso in an unheated workshop. It starts a bit like miso, using koji to break down a second ingredient. You just skip the salt, and add the koji to rice, so the amylases create sweet sugars. Once the amazake is sweet, you stop the incubation before any secondary, yeasty-alcoholic fermentation happens.

Squash Amazake

Amazake usually uses rice as a source of starch, but that's not your only option: amylolytic starters like koji can break down basically any source of starch, rice or otherwise. Here, it's a starchy, delicious, roasted squash.

Preheat the oven to 400°F and roast **about 1 pound (500 g) kabocha squash**, cut into small cubes, until cooked through and just a little brown, 25 to 30 minutes. Cool the squash.

Lightly mash 300 g of the roasted squash with **500 ml very hot, but not boiling, water** (about 160°F/70°C). Mix in **200 g rice koji**. The texture should be soupy; if it's too thick, add more hot water. Take the temperature of the mix—it should be 140°F/60°C. You want to hold it at this temperature for 6 to 12 hours, ideally at least 8.

To keep it warm during the starch breakdown, this is a great opportunity to build your own simple fermentation box (see page 278). If you have

a rice cooker or instant pot, it may have a setting for sweet porridge or yogurt; if so, simply put the amazake mix in the pot of the cooker and warm it as directed above.

After 8 hours or when the amazake is sweet, you can puree the resulting orangey squash-rice amazake to a smooth consistency, and either drink as is (hot or cold), diluted with water, or strain like Peanut Milk (page 217) for a very liquid drink. Serve immediately.

Makes about 4 cups (1 liter)

Variation
Black Rice Amazake

Follow the instructions for squash amazake, but substitute an equal amount of cooked black forbidden rice for the squash. The resulting amazake will be deep reddish purple and a bit aromatic-fruity. Serve immediately.

Making Sour Flavors, and Umami, by Lacto-fermenting Rice Koji: Shio-Koji

Literally "salt koji," shio-koji can be used as its own flavoring or sauce, and as a flavor creation agent—it has active amylase and protease enzymes from the koji, which keep working if you use shio-koji to marinate vegetables or meat. You can often buy it premade from Japanese grocery stores, but it's pretty easy to make yourself. Think of it as like a very wet miso, made with koji and no legumes, or like a sourdough starter that happens to be salted and uses koji. Simply puree koji, water, and salt together, and let them incubate and ferment together briefly. Protease and amylase enzymes create umami-tasting free amino acids and sweet sugars, and lactic acid bacteria ferment some of those sugars, making the shio-koji tangy and buttery.

Using Shio-Koji

Mix 1 tablespoon shio-koji, 2 tablespoons butter, and 2 tablespoons minced preserved lemon into a sauce for 4 ounces of dried pasta.

Used in a marinade, the active protease enzymes in shio-koji can happily break down some of the proteins in meats. This creates new umami richness, as well as physical tenderization. Briefer marination seasons the meat, slightly firms it, and keeps it juicy, like brining does.

Marinate chicken with 2 to 3 tablespoons of shio-koji per pound, overnight or for up to 3 days in the fridge, prior to breading and frying, roasting, grilling, or broiling. Wipe excess koji off before cooking to prevent excess burning from the sugar.

Marinate fish fillets or steaks in a light coating of shio-koji for 30 minutes before cooking, which will firm them up deliciously, as well as flavoring them.

For even more umami, rub shio-koji all over a steak and marinate for 2 to 3 hours in the fridge before cooking (wiping off the excess); do the same for a larger piece of meat you plan on roasting, but marinate overnight or up to 2 days.

Shio-Koji

In a very clean container (a mason jar or even a Tupperware container works well), mix together **200 g fresh or dry koji**, **270 g water**, and **45 g sea salt**. Stir well to combine and dissolve the salt. (Note: if using dry or firm granular koji, you can pulse the mixture quickly in a blender a few times to make sure the koji is well combined with the water.) The texture should be thick and puddingy, but looser than a firm paste. Add a bit more water if necessary to thin the texture, but make sure to add a little extra salt as well (you're aiming for 9 to 10 percent salt, which you will hit using the numbers above). Cover loosely with the lid and put the container in a warm, dark area to ferment. Stir with a very clean spoon every 1 to 2 days. It will start to smell funky and sweet, bubble a little, and loosen up even more in texture. (As always, if you notice really unpleasant smells or mold, throw out the batch.) Ferment until clean and tangy and no longer bubbly, about 5 to 10 days depending on the temperature of the room. Store in the fridge for up to 6 months in a tightly covered, fully filled container.

Makes about 1 pound (2 cups)

Rice or Barley Koji

You absolutely don't have to attempt this recipe. Honestly, if you just want to make miso or amazake, almost all of the home fermenters I know in Japan use dried or otherwise store-bought koji and don't bother making their own, and you can do the same.

If you *would* like to try a bit of a challenge and learn something new, though, making and using your own koji is an incredibly rewarding way to experience building complex flavor from start to finish. Fermentation-created umami, done just how you like it; or an exploration of local or favorite ingredients. Once you get the hang with barley, there are tons of fascinating and unexpected flavor creations you can explore with other grains, with a little finessing: Carolina Gold or forbidden rice, pearled barley or pearled wheat grano, even corn grits or buckwheat. Carolina Gold rice and black-eyed pea miso? Cracked heirloom corn and ayocote morado bean miso? Buckwheat amazake? The sky's the limit.

This recipe, which is a project, will take about 48 hours to complete: cooking and cooling the grains, sprinkling them with spore to inoculate them, then two phases of fermentation—one where the koji tends to get too cold on its own, followed by one where it can overheat itself. The technique of starting the fermentation with the rice loosely bundled, then spreading it out, was recommended as a way to get higher consistency in small batches by one of the biochemists I visited with at Higuchi Matsunosuke Shoten, an Osaka-based koji starter producer.

Like any fermentation, making koji is a collaboration between yourself and microbes. Part of what makes these processes interesting is getting the hang of them through experience—and determining for yourself what you can do to nudge the fermentation and flavor profile one way or another. While you're in the learning stage, your first batch or two of koji may fail. The most common reasons for failure are overheating, too much water, not enough air, and drying out. For more discussion on these, see the box on page 284.

Aspergillus oryzae spores (also called tane koji or koji-kin) can be mail-ordered domestically from GEM Cultures, and you can get imported Japanese brands (Higuchi Moyashi, Bio'c, and Akita Konno are excellent ones) online from websites like Kawashimaya, Modernist Pantry, Kalustyan's, and even eBay. For the first fermentation, you will need a large food-safe plastic bag, like a turkey brining bag or a 5-gallon bucket liner. You will also need perforated trays, such as 2-inch-deep perforated half-sized hotel pans from any restaurant supply store, and a waterproof box, cooler, or other container (see page 264) that can hold the perforated trays of koji as they ferment. If you have solid, wooden trays—in which case, you probably have some idea of how to make koji—those will work in place of perforated pans. To line the plastic bag and trays, you'll need large, very clean kitchen or dish towels or quadruple-layered cheesecloth, large enough to line the bottom of your trays and come at least partway up the sides. You'll likely use two or three for the initial fermentation, and two per tray after that. I recommend buying an inexpensive 8- or 10-pack of side towels for the purpose, and washing them in hot water in the washing machine before using. They should have a smooth surface, not terry cloth or some other fuzzy texture.

Make sure grains are in prolonged direct contact only with cloth or other porous and absorbent materials while fermenting—sitting in direct contact with metal or plastic will encourage water pooling or inhibit airflow, both of which will damage the koji by encouraging bacterial, rather than mold, growth.

Making koji is more of a formula than an exact-batch recipe. You can make it with as much rice or barley as you want. I'd recommend 1 pound (450 g) dry grains to start, which cooks up to about 2 pounds (900 g), so you have enough to work with to get the hang of it. This is one of the few fermentation recipes where I rarely measure the components, especially the spores—I learned to do it by eye, and each brand of spores

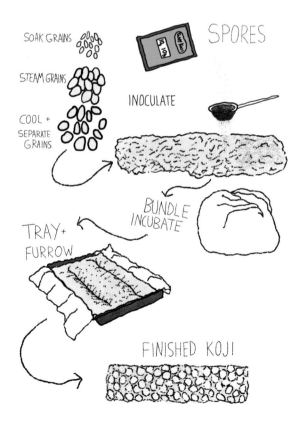

SOAK GRAINS
STEAM GRAINS
COOL + SEPARATE GRAINS
SPORES
INOCULATE
BUNDLE INCUBATE
TRAY + FURROW
FINISHED KOJI

has different recommended inoculation rates. Typically, it's around 0.2 g to 1 g of koji spores per kg (2 pounds) of steamed rice. To make the spores easier to handle, you can bulk them out with pasteurized cornstarch or rice starch—heat the starch in a pan over medium heat on the stove until it's very hot (but not smoking or burnt), let cool to room temperature, and mix the spores from the packet you buy with about 4 times their weight in heat-treated starch. (Then use 5 times the amount recommended on the packaging.) GEM Cultures sells small, 15-gram packets of spores that have already been bulked out with starch.

Day 1: Prepare the Rice

Start by soaking **at least 1 pound (450 g) pearled barley or medium-grained rice** in cold water for at least 4 hours, preferably overnight. In the morning, rinse the grains in more cold water to release any residual surface starch.

Steam the rice or barley in a rice cooker or on the stovetop until cooked through, but not mushy.

The cooked grains are now essentially sterile—we'll be intentionally adding microbes to it, which could include contaminants if our hands or equipment aren't clean. Wash your hands well and put on disposable gloves (doubling up on gloves can help with the heat). Separate and fluff the grains of barley or rice and spread them out on clean baking sheets or very clean towels. As they cool, lightly stir and toss the grains several times to help separate them and prevent them from clumping. Let cool until just barely warm.

Day 1: Inoculate the Rice

Check the temperature of the rice with a clean thermometer; it should be 85 to 99° F (30 to 37°C), just at body temperature or slightly lower.

Make sure:

- All the grains are fully cooked, but not mushy or slimy. Remove any blown-out or over-cooked grains.
- All the grains are separate and not sticking together. Gently crumble apart any clumps.
- Water hasn't condensed or pooled underneath the grains. (If it has, dab it off with a clean paper towel.)

Spread the grains out evenly 1 to 2 inches deep. You may need several baking sheets or you can spread them on a triple layer of plastic wrap on the counter.

With fresh disposable gloves, get the **koji spore starter** ready in a small container or cup, using the dosage amount recommended on the package (or 2 to 3 times as much, if you're just starting out, to get the hang of it).

Hold a fine-mesh tea strainer 2 to 3 inches above the surface of the grains. Pour a small amount of the spores carefully from the cup into the tea strainer, and sift the spores slowly and methodically over the surface grains in a Zamboni-like pattern for full coverage. Only transfer about one-fifth to one-sixth of the spores to the tea strainer at any one time, so they don't fall out too fast.

Use one-third of the spores to dust all the grains in this first pass. After you've added the first dose of spores, put the spores and strainer down carefully where they won't get knocked over or blown on.

Carefully mix the grains you've just inoculated with spores to evenly distribute them throughout the pile. Use a gentle scoop-and-turn motion to gradually mix and flip over the whole pile in palmfuls at a time. With clean gloved hands, pay special attention to the grains on the edges and in the corners, scooping them up, dropping them in the middle, and evening the pile out into an even layer.

Add the second dose of spores in the same way as the first and mix the same way. If you've gotten the hang of adding the spores slowly and have enough left for a third pass, do a third pass.

When all the spore powder has been added, give the grains another thorough mix including the sides and corners. Repeat several times. Thorough mixing ensures that the spores are well distributed throughout the grains and that the grains are separated and aerated.

Day 1: Ferment in a Bundle

Soak 2 to 3 very clean kitchen towels or cheesecloth in hot water and wring them out very well. You want them to be just slightly damp. Gather the inoculated rice or barley into a pile on top of one of the towels and use the others to wrap it loosely so it's totally covered by cloth. Transfer the bundle to a large plastic bag (see headnote—line the bag with the towels and then add the inoculated grain if moving as one bundle proves difficult).

If using a fermenting box (see page 278), thread a thermometer probe through the cloth bundle into the middle of the grain mass. Close the bag loosely and put the bundle in the fermenting box with the temperature set to 91°F/33°C. If going without a fermentation box, keep the bundle warm and check the temperature every few hours with a clean thermometer.

Watch the temperature of the fermentation. If it edges up to 99°F/37°C, give it a loose stir with clean hands to help incorporate air and dissipate some of the heat.

Avoid letting the koji get hotter than 105°F/40°C. These temperatures will start killing the mold and encourage the growth of *Bacillus* species of bacteria, which will ruin the fermentation.

After 12 hours, open the bag to gently stir the koji bundle. You probably won't see any mold growth. If the cloth has dried out, spray a little water on it to dampen, but not so much that it pools or soaks it.

Day 2: Ferment in Trays

After 24 hours, carefully open the bundle; you will probably begin to see some mold growth and clumping of the grains. Mazel tov! With clean hands, gently crumble the grain clumps into individual grains. Carefully stir the mass of grains by hand, with a scooping and flipping motion, so the grains that were in the outer and inner parts of the bundle are well mixed.

Gather and clean perforated hotel pans—this much koji will fill 2 to 3 half-sized hotel pans (see headnote). Soak two clean, large dish towels or quadruple-layered cheesecloth, each large enough to cover the tray, for each tray or hotel pan you're using, and wring out very well. Lay one cloth to cover the bottom of each pan and pour the semifermented grains into the pans to a depth of about 1 to 1½ inches (2.5 to 3.5 cm). Make two lengthwise furrows in the grains with your hands and cover with the second soaked towel. Put the thermometer probe into one of the pans and load the pans into the fermentation box—if stacking vertically, rotate each pan by 90 degrees from the last one, so they're staggered. Crack the lid slightly to let heat escape and air to circulate.

Keep an eye on the temperature; it tends to rise rapidly around hour 36, which is also when you will start seeing significant mold growth, which will bind the grains into a single cake or mass. Now is a good time to pull the trays out, let them cool a little bit, and rearrange them when you put them back in, so the ones that were on the top of the stack are now at the bottom.

Again, *avoid letting the koji get hotter than 105°F/40°C.* These temperatures will start killing the mold and encourage the growth of *Bacillus* species of bacteria, which will ruin the fermentation. If the temperature starts to rise this high, take the trays out of the box for about 20 minutes, then put them back in and leave the lid cracked slightly open.

Day 3: Harvest

At about hour 48, the mold should have fully grown into a fluffy mat that binds all the grains into a fairly solid block. The aroma should be pleasantly fruity and floral with a delicate mushroomy undertone. If the mold looks patchy or underdeveloped, let it go for another 6 hours. Avoid letting the mold grow until it looks very fuzzy. At this point, the mold is putting its energy into creating spores, so the koji quality will be reduced.

Use immediately for miso (pages 273–76), Shio-Koji (page 280), or amazake (page 279), or refrigerate in single layers wrapped in parchment paper, then plastic wrap or plastic containers, to cool for up to 3 days, or freeze for up to 6 months, well wrapped in a layer of parchment paper, then plastic wrap.

Makes about 2 pounds (900 g)

Watch Out for These Koji Problems

Overheating: By roughly the second day of the koji fermentation, the mold is so active that it generates its own heat. Keeping it wrapped up or insulated during this time without airing it out or mixing it to shed some of this heat can shoot the temperature upward well past 100°F/40°C, which can kill off the mold cells and allow heat-tolerant alkaline bacteria to take over. If it smells eggy, or like fruit that's starting to rot, or looks slimy, you have bacterial contamination and should throw out the koji and start over, cleaning all your equipment very well.

Too much water: If the steamed rice or barley gets too wet or sits in its own pooled moisture or takes on too much condensation, the mold cells, which need oxygen, can essentially drown. As with overheating, alkaline bacteria can then take over.

Not enough air: With insufficient air circulation—for example, if you wrap the fermenting koji too tightly, put it in a too-small container, or lack ventilation on the bottom of your trays—you'll suffocate the mold.

Drying out: The heat generated by the koji, plus airflow (ambient or heated airflow, since you need some external heat at the beginning), can dry the koji out and result in poor mold growth.

Selected Bibliography

General Resources

Dunn, Rob, and Monica Sanchez. *Delicious: The Evolution of Flavor and How It Made Us Human*. Princeton, N.J.: Princeton University Press, 2021.

Maarse, Henk. *Volatile Compounds in Foods and Beverages*. New York: Marcel Dekker, 1991.

McGee, Harold. *Nose Dive: A Field Guide to the World's Smells*. London: John Murray, 2020.

———. *On Food and Cooking: The Science and Lore of the Kitchen*. New York: Scribner, 2004.

Mouritsen, Ole, and Klavs Styrbæk. *Umami: Unlocking the Secrets of the Fifth Taste*. Translated by Mariela Johansen. New York: Columbia University Press, 2015.

Patterson, Daniel, and Mandy Aftel. *Aroma: The Magic of Essential Oils in Foods and Fragrance*. New York: Artisan, 2004.

———. *The Art of Flavor: Practices and Principles for Creating Delicious Food*. New York: Riverhead, 2017.

Shepherd, Gordon. *Neurogastronomy: How the Brain Creates Flavor and Why It Matters*. New York: Columbia University Press, 2013.

Shurtleff, William, and Akiko Aoyagi. *The Book of Miso: Savory Fermented Soy Seasoning*. Lafayette, Calif.: CreateSpace, 2018.

Steinkraus, Keith. *Handbook of Indigenous Fermented Foods, Revised and Expanded*. 2nd ed. Boca Raton, Fla.: CRC Press, 2018.

———, ed. *Industrialization of Indigenous Fermented Foods, Revised and Expanded*. 2nd ed. Boca Raton, Fla.: CRC Press, 2004.

Stevenson, Richard. *The Psychology of Flavour*. Oxford: Oxford University Press, 2009.

Part 1

Axel, Richard. "Scents and Sensibility: A Molecular Logic of Olfactory Perception (Nobel Lecture)." *Angewandte Chemie International Edition* 44, no. 38 (2005): 6110–27.

Buck, Linda, and Richard Axel. "A Novel Multigene Family May Encode Odorant Receptors: A Molecular Basis for Odor Recognition." *Cell* 65, no. 1 (April 5, 1991): 175–87. https://doi.org/10.1016/0092-8674(91)90418-x.

Bushdid, C., M. O. Magnasco, L. B. Vosshall, and A. Keller. "Humans Can Discriminate More than 1 Trillion Olfactory Stimuli." *Science* 343, no. 6177 (March 21, 2014): 1370–72. https://doi.org/10.1126/science.1249168.

Chaudhari, N., and S. D. Roper. "The Cell Biology of Taste." *Journal of Cell Biology* 190, no. 3 (2010): 285–96. doi:10.1083/jcb.201003144.

Chen, X., M. Gabitto, Y. Peng, N. J. Ryba, and C. S. Zuker. "A Gustotopic Map of Taste Qualities in the Mammalian Brain." *Science* 333, no. 6047 (2011): 1262–66.

Goff, S. A., and Harry Klee. "Plant Volatile Compounds: Sensory Cues for Health and Nutritional Value?" *Science* 311, no. 5762 (February 10, 2006): 815–19. https://doi.org/10.1126/science.1112614.

McGann, John P. "Poor Human Olfaction Is a Nineteenth-Century Myth." *Science* 356, no. 6338 (May 12, 2017). https://doi.org/10.1126/science.aam7263.

Rowe, Timothy B., and Gordon M. Shepherd. "Role of Ortho-Retronasal Olfaction in Mammalian Cortical Evolution: Olfaction and Cortical Evolution." *Journal of Comparative Neurology* 524, no. 3 (February 15, 2016): 471–95. https://doi.org/10.1002/cne.23802.

Saive, Anne-Lise, Jean-Pierre Royet, and Jane Plailly. "A Review on the Neural Bases of Episodic Odor Memory: From Laboratory-Based to Autobiographical Approaches." *Frontiers in Behavioral Neuroscience* 8 (July 7, 2014). https://doi.org/10.3389/fnbeh.2014.00240.

Shepherd, Gordon M. "The Human Sense of Smell: Are We Better than We Think?" *PLOS Biology* 2, no. 5 (May 11, 2004): e146. https://doi.org/10.1371/journal.pbio.0020146.

———. "Smell Images and the Flavour System in the Human Brain." *Nature* 444, no. 7117 (November 2006): 316–21. https://doi.org/10.1038/nature05405.

Yeshurun, Yaara, and Noam Sobel. "An Odor Is Not Worth a Thousand Words: From Multidimensional

Odors to Unidimensional Odor Objects." *Annual Review of Psychology* 61, no. 1 (January 2010): 219–41. https://doi.org/10.1146/annurev.psych.60.110707.163639.

Zhang, Yifeng, Mark A. Hoon, Jayaram Chandrashekar, Ken L. Mueller, Boaz Cook, Dianqing Wu, Charles S. Zuker, and Nicholas J. P. Ryba. "Coding of Sweet, Bitter, and Umami Tastes." *Cell* 112, no. 3 (February 2003): 293–301. https://doi.org/10.1016/S0092-8674(03)00071-0.

Part 2

Boelens, Mans, and Ronald Boelens. "Classification of Perfumes and Fragrances." *Perfumer & Flavorist* 26 (2001): 10.

De Pelsmaeker, Sara, Gil De Clercq, Xavier Gellynck, and Joachim J. Schouteten. "Development of a Sensory Wheel and Lexicon for Chocolate." *Food Research International* 116 (February 1, 2019): 1183–91. https://doi.org/10.1016/j.foodres.2018.09.063.

James, Andrew. "How Robert Parker's 90+ and Ann Noble's Aroma Wheel Changed the Discourse of Wine Tasting Notes." *ILCEA: Revue de l'Institut Des Langues et Cultures d'Europe, Amérique, Afrique, Asie et Australie*, no. 31 (March 1, 2018). https://doi.org/10.4000/ilcea.4681.

Noble, A. C., R. A. Arnold, J. Buechsenstein, E. J. Leach, J. O. Schmidt, and P. M. Stern. "Modification of a Standardized System of Wine Aroma Terminology." *American Journal of Enology and Viticulture* 38, no. 2 (January 1, 1987): 143–46.

Noble, A. C., R. A. Arnold, B. M. Masuda, and S. D. Pecore. "Progress Towards a Standardized System of Wine Aroma Terminology." *American Journal of Enology and Viticulture* 35, no. 2 (January 1, 1984): 107–109.

Urdapilleta, I., A. Giboreau, C. Manetta, O. Houix, and J. F. Richard. "The Mental Context for the Description of Odors: A Semantic Space." *European Review of Applied Psychology* 56, no. 4 (December 1, 2006): 261–71. https://doi.org/10.1016/j.erap.2005.09.013.

Salty

Breslin, Paul A. S. "Interactions Among Salty, Sour and Bitter Compounds." *Trends in Food Science and Technology* 7, no. 12 (December 1, 1996): 390–99. https://doi.org/10.1016/S0924-2244(96)10039-X.

Breslin, P. A. S., and G. K. Beauchamp. "Salt Enhances Flavour by Suppressing Bitterness." *Nature* 387, no. 6633 (June 1997): 563. https://doi.org/10.1038/42388.

Drake, S. L., and M. A. Drake. "Comparison of Salty Taste and Time Intensity of Sea and Land Salts from Around the World." *Journal of Sensory Studies*

26, no. 1 (2011): 25–34. https://doi.org/10.1111/j.1745-459X.2010.00317.x.

Kasahara, Yoichi, Masataka Narukawa, Yoshiro Ishimaru, Shinji Kanda, Chie Umatani, Yasunori Takayama, Makoto Tominaga, et al. "TMC4 Is a Novel Chloride Channel Involved in High-Concentration Salt Taste Sensation." *Journal of Physiological Sciences* 71, no. 1 (August 25, 2021): 23. https://doi.org/10.1186/s12576-021-00807-z.

Lebert, André, and Jean-Dominique Daudin. "Modelling the Distribution of aw, pH and Ions in Marinated Beef Meat." *Meat Science* 97, no. 3 (July 1, 2014): 347–57. https://doi.org/10.1016/j.meatsci.2013.10.017.

Volpato, G., E. M. Z. Michielin, S. R. S. Ferreira, and J. C. C. Petrus. "Kinetics of the Diffusion of Sodium Chloride in Chicken Breast (Pectoralis Major) During Curing." *Journal of Food Engineering* 79, no. 3 (April 1, 2007): 779–85. https://doi.org/10.1016/j.jfoodeng.2006.02.043.

Sour

Al-Kadamany, E., I. Toufeili, M. Khattar, Y. Abou-Jawdeh, S. Harakeh, and T. Haddad. "Determination of Shelf Life of Concentrated Yogurt (Labneh) Produced by In-Bag Straining of Set Yogurt Using Hazard Analysis." *Journal of Dairy Science* 85, no. 5 (May 1, 2002): 1023–30. https://doi.org/10.3168/jds.S0022-0302(02)74162-3.

Amato, Katherine R., Elizabeth K. Mallott, Paula D'Almeida Maia, and Maria Luisa Savo Sardaro. "Predigestion as an Evolutionary Impetus for Human Use of Fermented Food." *Current Anthropology* 62, no. S24 (October 1, 2021): S207–19. https://doi.org/10.1086/715238.

Bernalte, M. J., E. Sabio, M. T. Hernandez, and C. Gervasini. "Influence of Storage Delay on Quality of 'Van' Sweet Cherry." *Postharvest Biology and Technology* 28, no. 2 (2003): 303–12.

Fereidoonfar, Hossein, Hossein Salehi-Arjmand, Ali Khadivi, Morteza Akramian, and Leila Safdari. "Chemical Variation and Antioxidant Capacity of Sumac (Rhus Coriaria L.)." *Industrial Crops and Products* 139 (November 1, 2019): 111518. https://doi.org/10.1016/j.indcrop.2019.111518.

Frank, Hannah E. R., Katie Amato, Michelle Trautwein, Paula Maia, Emily R. Liman, Lauren M. Nichols, Kurt Schwenk, Paul A. S. Breslin, and Robert R. Dunn. "The Evolution of Sour Taste." *Proceedings of the Royal Society B: Biological Sciences* 289, no. 1968 (February 9, 2022): 20211918. https://doi.org/10.1098/rspb.2021.1918.

Ganzevles, Paul G. J., and Jan H. A. Kroeze. "The Sour Taste of Acids. The Hydrogen Ion and the Undissociated Acid as Sour Agents." *Chemical*

Senses 12, no. 4 (December 1, 1987): 563–76. https://doi.org/10.1093/chemse/12.4.563.

Gharezi, Maedeh, Neena Joshi, and Elnaz Sadeghian. "Effect of Post Harvest Treatment on Stored Cherry Tomatoes." *Journal of Nutrition and Food Sciences* 2, no. 8 (2012). https://doi.org/10.4172/2155-9600.1000157.

Goli, T., P. Bohuon, J. Ricci, G. Trystram, and A. Collignan. "Mass Transfer Dynamics During the Acidic Marination of Turkey Meat." *Journal of Food Engineering* 104, no. 1 (May 1, 2011): 161–68. https://doi.org/10.1016/j.jfoodeng.2010.12.010.

Kneifel, W., Doris Jaros, and F. Erhard. "Microflora and Acidification Properties of Yogurt and Yogurt-Related Products Fermented with Commercially Available Starter Cultures." *International Journal of Food Microbiology* 18, no. 3 (May 1, 1993): 179–89. https://doi.org/10.1016/0168-1605(93)90043-G.

Levy, Y., A. Bar-Akiva, and Y. Vaadia. "Influence of Irrigation and Environmental Factors on Grapefruit Acidity." *Journal of the American Society for Horticultural Science* 103 (1978): 73–76.

Lorente, José, Salud Vegara, Nuria Martí, Albert Ibarz, Luís Coll, Julio Hernández, Manuel Valero, and Domingo Saura. "Chemical Guide Parameters for Spanish Lemon (Citrus Limon [L.] Burm.) Juices." *Food Chemistry* 162 (November 2014): 186–91. https://doi.org/10.1016/j.foodchem.2014.04.042.

Mayuoni-Kirshinbaum, Lina, and Ron Porat. "The Flavor of Pomegranate Fruit: A Review." *Journal of the Science of Food and Agriculture* 94, no. 1 (2014): 21–27. https://doi.org/10.1002/jsfa.6311.

Nikfardjam, Martin S. Pour. "General and Polyphenolic Composition of Unripe Grape Juice (Verjus/Verjuice) from Various Producers." *Mitteulingen Klosterneuburg* 58 (2008): 28–31.

Obenland, David, Salvatore Campisi-Pinto, and Mary Lu Arpaia. "Determinants of Sensory Acceptability in Grapefruit." *Scientia Horticulturae* 231 (January 27, 2018): 151–57. https://doi.org/10.1016/j.scienta.2017.12.026.

Öncül, Nilgün, and Şeniz Karabiyikli. "Factors Affecting the Quality Attributes of Unripe Grape Functional Food Products." *Journal of Food Biochemistry* 39, no. 6 (2015): 689–95. https://doi.org/10.1111/jfbc.12175.

Pérez-Díaz, I. M., Fred Breidt, R. W. Buescher, F. N. Arroyo-López, R. Jiménez-Díaz, A. Garrido-Fernández, J. Bautista-Gallego, S.-S. Yoon, and S. D. Johanningsmeire. "51. Fermented and Acidified Vegetables." In *Compendium of Methods for the Microbiological Examination of Foods*, 4th ed., 531–32. Washington, D.C.: American Public Health Association, 2013. https://doi.org/10.2105/MBEF.0222.056.

Peters, Anna, Petra Krumbholz, Elisabeth Jäger, Anna Heintz-Buschart, Mehmet Volkan Çakir, Sven Rothemund, Alexander Gaudl, Uta Ceglarek, Torsten Schöneberg, and Claudia Stäubert. "Metabolites of Lactic Acid Bacteria Present in Fermented Foods Are Highly Potent Agonists of Human Hydroxycarboxylic Acid Receptor 3." *PLOS Genetics* 15, no. 5 (May 23, 2019): e1008145. https://doi.org/10.1371/journal.pgen.1008145.

Plane, Robert A., Leonard R. Mattick, and LaVerne D. Weirs. "An Acidity Index for the Taste of Wines." *American Journal of Enology and Viticulture* 31, no. 3 (January 1, 1980): 265–68.

Salji, Joseph P., and Anwar A. Ismail. "Effect of Initial Acidity of Plain Yogurt on Acidity Changes During Refrigerated Storage." *Journal of Food Science* 48, no. 1 (January 1983): 258–59. https://doi.org/10.1111/j.1365-2621.1983.tb14839.x.

Siddiq, M., A. Iezzoni, A. Khan, P. Breen, A. M. Sebolt, K. D. Dolan, and R. Ravi. "Characterization of New Tart Cherry (Prunus Cerasus L.): Selections Based on Fruit Quality, Total Anthocyanins, and Antioxidant Capacity." *International Journal of Food Properties* 14, no. 2 (February 28, 2011): 471–80. https://doi.org/10.1080/10942910903277697.

Skryplonek, K., David Gomes, Jorge Viegas, Carlos Pereira, and Marta Henriques. "Lactose-Free Frozen Yogurt: Production and Characteristics." *Acta Scientiarum Polonorum Technologia Alimentaria* 16, no. 2 (June 30, 2017): 171–79. https://doi.org/10.17306/J.AFS.0478.

Sowalsky, Richard A., and Ann C. Noble. "Comparison of the Effects of Concentration, pH and Anion Species on Astringency and Sourness of Organic Acids." *Chemical Senses* 23, no. 3 (June 1, 1998): 343–49. https://doi.org/10.1093/chemse/23.3.343.

Stahl, Ann Brower, R. I. M. Dunbar, Katherine Homewood, Fumiko Ikawa-Smith, Adriaan Kortlandt, W. C. McGrew, Katharine Milton, et al. "Hominid Dietary Selection Before Fire [and Comments and Reply]." *Current Anthropology* 25, no. 2 (April 1984): 151–68. https://doi.org/10.1086/203106.

Teerachaichayut, Sontisuk, and Huong Thanh Ho. "Non-Destructive Prediction of Total Soluble Solids, Titratable Acidity and Maturity Index of Limes by Near Infrared Hyperspectral Imaging." *Postharvest Biology and Technology* 133 (November 2017): 20–25. https://doi.org/10.1016/j.postharvbio.2017.07.005.

Tsantili, E., K. Konstantinidis, P. E. Athanasopoulos, and C. Pontikis. "Effects of Postharvest Calcium Treatments on Respiration and Quality Attributes in Lemon Fruit During Storage." *Journal of Horticultural Science and Biotechnology* 77, no. 4 (January 2002): 479–84. https://doi.org/10.1080/14620316.2002.11511526.

Turkmen, F. Ucan, H. A. Mercimek Takci, H. Saglam, and N. Sekeroglu. "Investigation of Some Quality Parameters of Pomegranate, Sumac and Unripe Grape Sour Products from Kilis Markets." *Quality Assurance and Safety of Crops and Foods* 11, no. 1 (February 2019): 61–71. https://doi.org/10.3920/QAS2018.1293.

Ubbaonu, C. N., N. C. Onuegbu, E. O. I. Banigo, and A. Uzoma. "Physico-Chemical Changes in Velvet Tamarind (Dialium Guineense Wild) During Fruit Development and Ripening." *Nigerian Food Journal* 23, no. 1 (2005): 133–38. https://doi.org/10.4314/nifoj.v23i1.33609.

Wang, Yu-Tao, Shao-Wen Huang, Rong-Le Liu, and Ji-Yun Jin. "Effects of Nitrogen Application on Flavor Compounds of Cherry Tomato Fruits." *Journal of Plant Nutrition and Soil Science* 170, no. 4 (August 2007): 461–68. https://doi.org/10.1002/jpln.200700011.

Sweet

Feng, Ping, and HuaBin Zhao. "Complex Evolutionary History of the Vertebrate Sweet/Umami Taste Receptor Genes." *Chinese Science Bulletin* 58, no. 18 (June 1, 2013): 2198–2204. https://doi.org/10.1007/s11434-013-5811-5.

Mohos, Ferenc A. "Appendix 1: Data on Engineering Properties of Materials Used and Made by the Confectionery Industry." In *Confectionery and Chocolate Engineering: Principles and Applications*, 555–78. Chichester, Eng: John Wiley, 2010. https://doi.org/10.1002/9781444320527.app1.

Moskowitz, Howard R. "The Sweetness and Pleasantness of Sugars." *American Journal of Psychology* 84, no. 3 (1971): 387–405. https://doi.org/10.2307/1420470.

Stone, Herbert, and Shirley M. Oliver. "Measurement of the Relative Sweetness of Selected Sweeteners and Sweetener Mixtures." *Journal of Food Science* 34, no. 2 (1969): 215–22. https://doi.org/10.1111/j.1365-2621.1969.tb00922.x.

Yamaguchi, Shizuko, Tomoko Yoshikawa, Shingo Ikeda, and Tsunehiko Ninomiya. "Studies on the Taste of Some Sweet Substances." *Agricultural and Biological Chemistry* 34, no. 2 (February 1, 1970): 181–97. https://doi.org/10.1080/00021369.1970.10859599.

Umami

Chaudhari, Nirupa, Ana Marie Landin, and Stephen D. Roper. "A Metabotropic Glutamate Receptor Variant Functions as a Taste Receptor." *Nature Neuroscience* 3, no. 2 (February 2000): 113–19. https://doi.org/10.1038/72053.

Curtis, Robert I. "Umami and the Foods of Classical Antiquity." *American Journal of Clinical Nutrition* 90, no. 3 (September 1, 2009): 712S–18S. https://doi.org/10.3945/ajcn.2009.27462C.

Ikeda, Kikunae. "New Seasonings." *Chemical Senses* 27 (2002): 847–49.

Keast, Russell S. J., and Paul A. S. Breslin. "An Overview of Binary Taste–Taste Interactions." *Food Quality and Preference* 14, no. 2 (March 1, 2003): 111–24. https://doi.org/10.1016/S0950-3293(02)00110-6.

Kurihara, Kenzo. "Glutamate: From Discovery as a Food Flavor to Role as a Basic Taste (Umami)." *American Journal of Clinical Nutrition* 90, no. 3 (September 1, 2009): 719S–22S. https://doi.org/10.3945/ajcn.2009.27462D.

Maga, J. A. "Umami Flavour of Meat." In *Flavor of Meat and Meat Products*, edited by Fereidoon Shahidi, 98–115. Boston: Springer US, 1994. https://doi.org/10.1007/978-1-4615-2177-8_6.

Maga, Joseph A., and Shizuko Yamaguchi. "Flavor Potentiators." *C R C Critical Reviews in Food Science and Nutrition* 18, no. 3 (January 1, 1983): 231–312. https://doi.org/10.1080/10408398309527364.

Masic, Una, and Martin R. Yeomans. "Umami Flavor Enhances Appetite but Also Increases Satiety." *American Journal of Clinical Nutrition* 100, no. 2 (August 1, 2014): 532–38. https://doi.org/10.3945/ajcn.113.080929.

Mouritsen, Ole G., Klavs Styrbæk, Mariela Johansen, and Jonas Drotner Mouritsen. *Umami: Unlocking the Secrets of the Fifth Taste*. New York: Columbia University Press, 2014.

Nakayama, Tokiko, and Haruko Kimura. "Umami (*Xian-wei*) in Chinese Food." *Food Reviews International* 14, no. 2–3 (May 1998): 257–67. https://doi.org/10.1080/87559129809541160.

Ninomiya, Kumiko. "Natural Occurrence." *Food Reviews International* 14, no. 2–3 (May 1998): 177–211. https://doi.org/10.1080/87559129809541157.

Otsuka, Shigeru. "Umami in Japan, Korea, and Southeast Asia." *Food Reviews International* 14, no. 2–3 (May 1998): 247–56. https://doi.org/10.1080/87559129809541159.

Smriga, Miro, Toshimi Mizukoshi, Daigo Iwahata, Sachise Eto, Hiroshi Miyano, Takeshi Kimura, and Robert I. Curtis. "Amino Acids and Minerals in Ancient Remnants of Fish Sauce (Garum) Sampled in the 'Garum Shop' of Pompeii, Italy."

Journal of Food Composition and Analysis 23, no. 5 (August 1, 2010): 442–46. https://doi.org/10.1016/j.jfca.2010.03.005.

Wahlstedt, Amanda, Elizabeth Bradley, Juan Castillo, and Kate Gardner Burt. "MSG Is A-OK: Exploring the Xenophobic History of and Best Practices for Consuming Monosodium Glutamate." *Journal of the Academy of Nutrition and Dietetics* 122, no. 1 (January 2022): 25–29. https://doi.org/10.1016/j.jand.2021.01.020.

Yamaguchi, Shizuko. "The Synergistic Taste Effect of Monosodium Glutamate and Disodium 5′-Inosinate." *Journal of Food Science* 32, no. 4 (1967): 473–78. https://doi.org/10.1111/j.1365-2621.1967.tb09715.x.

Zanfirescu, Anca, Anca Ungurianu, Aristides M. Tsatsakis, George M. Nițulescu, Demetrios Kouretas, Aris Veskoukis, Dimitrios Tsoukalas, Ayse B. Engin, Michael Aschner, and Denisa Margină. "A Review of the Alleged Health Hazards of Monosodium Glutamate." *Comprehensive Reviews in Food Science and Food Safety* 18, no. 4 (July 2019): 1111–34. https://doi.org/10.1111/1541-4337.12448.

Bitter

Barratt-Fornell, Anne, and Adam Drewnowski. "The Taste of Health: Nature's Bitter Gifts." *Nutrition Today* 37, no. 4 (August 2002): 144.

Di Pizio, Antonella, and Masha Y. Niv. "Promiscuity and Selectivity of Bitter Molecules and Their Receptors." *Bioorganic and Medicinal Chemistry* 23, no. 14 (July 15, 2015): 4082–91. https://doi.org/10.1016/j.bmc.2015.04.025.

Drewnowski, Adam. "The Science and Complexity of Bitter Taste." *Nutrition Reviews* 59, no. 6 (June 1, 2001): 163–69. https://doi.org/10.1111/j.1753-4887.2001.tb07007.x.

Drewnowski, Adam, and Carmen Gomez-Carneros. "Bitter Taste, Phytonutrients, and the Consumer: A Review." *American Journal of Clinical Nutrition* 72, no. 6 (December 1, 2000): 1424–35. https://doi.org/10.1093/ajcn/72.6.1424.

Gutiérrez-Rosales, F., J. J. Ríos, and Ma. L. Gómez-Rey. "Main Polyphenols in the Bitter Taste of Virgin Olive Oil: Structural Confirmation by On-Line High-Performance Liquid Chromatography Electrospray Ionization Mass Spectrometry." *Journal of Agricultural and Food Chemistry* 51, no. 20 (September 1, 2003): 6021–25. https://doi.org/10.1021/jf021199x.

Higgins, Molly J., and John E. Hayes. "Discrimination of Isointense Bitter Stimuli in a Beer Model System." *Nutrients* 12, no. 6 (June 2020): 1560. https://doi.org/10.3390/nu12061560.

Ley, Jakob P. "Masking Bitter Taste by Molecules." *Chemosensory Perception* 1, no. 1 (March 1, 2008): 58–77. https://doi.org/10.1007/s12078-008-9008-2.

Pickenhagen, Wilhelm, Paul Dietrich, Borivoij Keil, Judith Polonsky, Françoise Nouaille, and Edgar Lederer. "Identification of the Bitter Principle of Cocoa." *Helvetica Chimica Acta* 58, no. 4 (1975): 1078–86. https://doi.org/10.1002/hlca.19750580411.

Soler-Rivas, Cristina, Juan Carlos Espín, and Harry J. Wichers. "Oleuropein and Related Compounds." *Journal of the Science of Food and Agriculture* 80, no. 7 (2000): 1013–23. https://doi.org/10.1002/(SICI)1097-0010(20000515)80:7<1013::AID-JSFA571>3.0.CO;2-C.

Subratty, A. H., A. Gurib-Fakim, and F. Mahomoodally. "Bitter Melon: An Exotic Vegetable with Medicinal Values." *Nutrition and Food Science* 35, no. 3 (June 1, 2005): 143–47. https://doi.org/10.1108/00346650510594886.

Sur, Subhayan, and Ratna B. Ray. "Bitter Melon (Momordica Charantia), a Nutraceutical Approach for Cancer Prevention and Therapy." *Cancers* 12, no. 8 (July 27, 2020): 2064. https://doi.org/10.3390/cancers12082064.

Spicy

Boonen, Brett, Justyna B. Startek, and Karel Talavera. "Chemical Activation of Sensory TRP Channels." In *Taste and Smell*, edited by Dietmar Krautwurst, 73–113. Cham: Springer International Publishing, 2016. https://doi.org/10.1007/7355_2015_98.

Gahungu, Arthur, Eric Ruganintwali, Eric Karangwa, Xiaoming Zhang, and Daniel Mukunzi. "Volatile Compounds and Capsaicinoid Content of Fresh Hot Peppers (Capsicum Chinense) Scotch Bonnet Variety at Red Stage." *Advance Journal of Food Science and Technology* 3, no. 3 (June 06, 2011): 211–218.

Govindarajan, V. S. "Pungency: The Stimuli and Their Evaluation." In *Food Taste Chemistry*, edited by James C. Boudreau, 53–92. Washington, D.C.: American Chemical Society, 1979. https://doi.org/10.1021/bk-1979-0115.

Juliani, H. Rodolfo, Cara Welch, Juliana Asante-Dartey, Dan Acquaye, Mingfu Wang, and James E. Simon. "Chemistry, Quality, and Functional Properties of Grains of Paradise (Aframomum Melegueta), a Rediscovered Spice." In *Dietary Supplements*, 100–113. Washington, D.C.: American Chemical Society, 2008. https://doi.org/10.1021/bk-2008-0987.ch006.

Lennertz, Richard C., Makoto Tsunozaki, Diana M. Bautista, and Cheryl L. Stucky. "Physiological

Basis of Tingling Paresthesia Evoked by Hydroxy-α-Sanshool." *Journal of Neuroscience* 30, no. 12 (March 24, 2010): 4353–61. https://doi.org/10.1523/JNEUROSCI.4666-09.2010.

Murakami, Yusuke, Hisakatsu Iwabuchi, Yukie Ohba, and Harukazu Fukami. "Analysis of Volatile Compounds from Chili Peppers and Characterization of Habanero (*Capsicum Chinense*) Volatiles." *Journal of Oleo Science* 68, no. 12 (2019): 1251–60.

Nalli, Marianna, Giorgio Ortar, Aniello Schiano Moriello, Vincenzo Di Marzo, and Luciano De Petrocellis. "Effects of Curcumin and Curcumin Analogues on TRP Channels." *Fitoterapia* 122 (October 2017): 126–31. https://doi.org/10.1016/j.fitote.2017.09.007.

Pedersen, Stine Falsig, Grzegorz Owsianik, and Bernd Nilius. "TRP Channels: An Overview." *Cell Calcium* 38, no. 3 (September 1, 2005): 233–52. https://doi.org/10.1016/j.ceca.2005.06.028.

Rhyu, Mee-Ra, Yiseul Kim, and Vijay Lyall. "Interactions Between Chemesthesis and Taste: Role of TRPA1 and TRPV1." *International Journal of Molecular Sciences* 22, no. 7 (January 2021): 3360. https://doi.org/10.3390/ijms22073360.

Rodríguez-Burruezo, Adrián, Hubert Kollmannsberger, Jaime Prohens, Siegfried Nitz, and Ana Fita. "Comparative Analysis of Pungency and Pungency Active Compounds in Chile Peppers (Capsicum Spp.)." *Bulletin of University of Agricultural Sciences and Veterinary Medicine Cluj-Napoca. Horticulture* 67, no. 1 (September 29, 2010): 270–73. https://doi.org/10.15835/buasvmcn-hort:4972.

Srinivasan, K. "Black Pepper and Its Pungent Principle-Piperine: A Review of Diverse Physiological Effects." *Critical Reviews in Food Science and Nutrition* 47, no. 8 (October 25, 2007): 735–48. https://doi.org/10.1080/10408390601062054.

Sugai, Etsuko, Yasujiro Morimitsu, Yusaku Iwasaki, Akihito Morita, Tatsuo Watanabe, and Kikue Kubota. "Pungent Qualities of Sanshool-Related Compounds Evaluated by a Sensory Test and Activation of Rat TRPV1." *Bioscience, Biotechnology, and Biochemistry* 69, no. 10 (January 2005): 1951–57. https://doi.org/10.1271/bbb.69.1951.

Tsunozaki, Makoto, Richard C. Lennertz, Samata Katta, Cheryl L. Stucky, and Diana M. Bautistaa. "The Plant-Derived Alkylamide, Hydroxy-Alpha-Sanshool, Induces Analgesia Through Inhibition of Voltage-Gated Sodium Channels." *Biophysical Journal* 102, no. 3 (January 2012): 323a. https://doi.org/10.1016/j.bpj.2011.11.1771.

Viana, Félix. "Chemosensory Properties of the Trigeminal System." *ACS Chemical Neuroscience* 2, no. 1 (January 19, 2011): 38–50. https://doi.org/10.1021/cn100102c.

Yang, Xiaogen. "Aroma Constituents and Alkyl-amides of Red and Green Huajiao (Zanthoxylum Bungeanum and Zanthoxylum Schinifolium)." *Journal of Agricultural and Food Chemistry* 56, no. 5 (March 1, 2008): 1689–96. https://doi.org/10.1021/jf0728101.

Zhang, Lu-Lu, Lei Zhao, Hou-Yin Wang, Bo-Lin Shi, Long-Yun Liu, and Zhong-Xiu Chen. "The Relationship Between Alkylamide Compound Content and Pungency Intensity of Zanthoxylum Bungeanum Based on Sensory Evaluation and Ultra-Performance Liquid Chromatography-Mass Spectrometry/Mass Spectrometry (UPLC-MS/MS) Analysis." *Journal of the Science of Food and Agriculture* 99, no. 4 (2019): 1475–83. https://doi.org/10.1002/jsfa.9319.

Fruity

Berenstein, Nadia. (2017). "Flavor Added: The Sciences of Flavor and the Industrialization of Taste in America" Ph.D. diss., University of Pennsylvania, 2017.

———. "The History of Banana Flavoring." *Lucky Peach*, August 2016. http://luckypeach.com/the-history-of-banana-flavoring/.

Boelens, Mans H., and Leo J. van Gemert. "Volatile Character-Impact Sulfur Compounds and Their Sensory Properties," *Perfumer and Flavorist* 18 (May–June 1993): 11.

Cannon, Robert J., and Chi-Tang Ho. "Volatile Sulfur Compounds in Tropical Fruits." *Journal of Food and Drug Analysis* 26, no. 2 (April 1, 2018): 445–68. https://doi.org/10.1016/j.jfda.2018.01.014.

Couture, R., and R. Rouseff. "Debittering and Deacidifying Sour Orange (Citrus Aurantium) Juice Using Neutral and Anion Exchange Resins." *Journal of Food Science* 57, no. 2 (March 1992): 380–84. https://doi.org/10.1111/j.1365-2621.1992.tb05499.x.

Du, Xiaofen, and Michael Qian. "Flavor Chemistry of Small Fruits: Blackberry, Raspberry, and Blueberry." In *Flavor and Health Benefits of Small Fruits*, edited by Michael C. Qian and Agnes M. Rimando, 27–43. Washington, D.C.: American Chemical Society, 2010. https://doi.org/10.1021/bk-2010-1035.ch003.

Gang, David R., ed. *The Biological Activity of Phytochemicals*. New York: Springer, 2011. https://doi.org/10.1007/978-1-4419-7299-6.

Gonda, Itay, Yosef Burger, Arthur A. Schaffer, Mwafaq Ibdah, Ya'akov Tadmor, Nurit Katzir, Aaron Fait, and Efraim Lewinsohn. "Biosynthesis and Perception of Melon Aroma." In *Biotechnology in Flavor Production*, edited by Daphna

Havkin-Frenkel and Nativ Dudai, 281–305. Chichester, Eng.: John Wiley, 2016. https://doi.org/10.1002/9781118354056.ch11.

Jordán, María J., Carlos A. Margaría, Philip E. Shaw, and Kevin L. Goodner. "Volatile Components and Aroma Active Compounds in Aqueous Essence and Fresh Pink Guava Fruit Puree (*Psidium Guajava* L.) by GC-MS and Multidimensional GC/GC-O." *Journal of Agricultural and Food Chemistry* 51, no. 5 (February 2003): 1421–26. https://doi.org/10.1021/jf020765l.

Lan-Phi, Nguyen Thi, Tomoko Shimamura, Hiroyuki Ukeda, and Masayoshi Sawamura. "Chemical and Aroma Profiles of Yuzu (Citrus Junos) Peel Oils of Different Cultivars." *Food Chemistry* 115, no. 3 (August 1, 2009): 1042–47. https://doi.org/10.1016/j.foodchem.2008.12.024.

Mohd-Hanif, Hani, Rosnah Shamsudin, and Noranizan Mohd Adzahan. "UVC Dosage Effects on the Physico-Chemical Properties of Lime (Citrus Aurantifolia) Juice." *Food Science and Biotechnology* 25, no. S1 (March 2016): 63–67. https://doi.org/10.1007/s10068-016-0099-2.

Murakami, Yusuke, Hisakatsu Iwabuchi, Yukie Ohba, and Harukazu Fukami. "Analysis of Volatile Compounds from Chili Peppers and Characterization of Habanero (*Capsicum Chinense*) Volatiles." *Journal of Oleo Science* 68, no. 12 (2019): 1251–60. https://doi.org/10.5650/jos.ess19155.

Pino, Jorge A. "Odour-Active Compounds in Pineapple (*Ananas Comosus* [L.] Merril Cv. Red Spanish)." *International Journal of Food Science and Technology* 48, no. 3 (March 2013): 564–70. https://doi.org/10.1111/j.1365-2621.2012.03222.x.

Takeoka, Gary R., Robert A. Flath, Thomas R. Mon, Roy Teranishi, and Matthias Guentert. "Volatile Constituents of Apricot (Prunus Armeniaca)." *Journal of Agricultural and Food Chemistry* 38, no. 2 (February 1990): 471–77. https://doi.org/10.1021/jf00092a031.

Tsuneya, Tomoyuki, Masakazu Ishihara, and Haruyasu Shiota. "Volatile Components of Quince Fruit (Cydonia Oblonga Mill.)." *Agricultural Biological Chemistry* 47, no. 11 (1983): 2495–2502.

Winterhalter, Peter, and Peter Schreier. "Free and Bound C13 Norisoprenoids in Quince (Cydonia Oblonga, Mill.) Fruit." *Journal of Agricultural and Food Chemistry* 36, no. 6 (November 1, 1988): 1251–56. https://doi.org/10.1021/jf00084a031.

Vegetal

Bunning, Marisa L., Patricia A. Kendall, Martha B. Stone, Frank H. Stonaker, and Cecil Stushnoff. "Effects of Seasonal Variation on Sensory

Properties and Total Phenolic Content of Five Lettuce Cultivars." *Journal of Food Science* 75, no. 3 (2010): S156–61. https://doi.org/10.1111/j.1750-3841.2010.01533.x.

Engelberth, Juergen, Hans T. Alborn, Eric A. Schmelz, and James H. Tumlinson. "Airborne Signals Prime Plants Against Insect Herbivore Attack." *Proceedings of the National Academy of Sciences* 101, no. 6 (February 10, 2004): 1781–85. https://doi.org/10.1073/pnas.0308037100.

Hatanaka, Akikazu. "The Biogeneration of Green Odour by Green Leaves." *Phytochemistry* 34, no. 5 (November 1, 1993): 1201–18. https://doi.org/10.1016/0031-9422(91)80003-J.

Iranshahi, M. "A Review of Volatile Sulfur-Containing Compounds from Terrestrial Plants: Biosynthesis, Distribution and Analytical Methods." *Journal of Essential Oil Research* 24, no. 4 (August 1, 2012): 393–434. https://doi.org/10.1080/10412905.2012.692918.

Maga, Joseph A. "Musty/Earthy Aromas." *Food Reviews International* 3, no. 3 (January 1, 1987): 269–84. https://doi.org/10.1080/87559128709540816.

———. "Potato Flavor." *Food Reviews International* 10, no. 1 (February 1, 1994): 1–48. https://doi.org/10.1080/87559129409540984.

Murray, Keith E., and Frank B. Whitfield. "The Occurrence of 3-Alkyl-2-Methoxypyrazines in Raw Vegetables." *Journal of the Science of Food and Agriculture* 26, no. 7 (1975): 973–86. https://doi.org/10.1002/jsfa.2740260714.

Mutarutwa, Delvana, Luciano Navarini, Valentina Lonzarich, Dario Compagnone, and Paola Pittia. "GC-MS Aroma Characterization of Vegetable Matrices: Focus on 3-Alkyl-2-Methoxypyrazines." *Journal of Mass Spectrometry* 53, no. 9 (2018): 871–81. https://doi.org/10.1002/jms.4271.

Raffo, Antonio, Maurizio Masci, Elisabetta Moneta, Stefano Nicoli, José Sánchez del Pulgar, and Flavio Paoletti. "Characterization of Volatiles and Identification of Odor-Active Compounds of Rocket Leaves." *Food Chemistry* 240 (February 1, 2018): 1161–70. https://doi.org/10.1016/j.foodchem.2017.08.009.

Takeoka, Gary. "Flavor Chemistry of Vegetables." In *Flavor Chemistry: Thirty Years of Progress*, edited by Roy Teranishi, Emily L. Wick, and Irwin Hornstein, 287–304. Boston: Springer US, 1999. https://doi.org/10.1007/978-1-4615-4693-1_25.

Spiced and Herbal

Blank, Imre, Alina Sen, and Werner Grosch. "Sensory Study on the Character-Impact Flavour Compounds of Dill Herb (Anethum Graveolens L.)."

Food Chemistry 43, no. 5 (January 1992): 337–43. https://doi.org/10.1016/0308-8146(92)90305-L.

Calín-Sánchez, Ángel, Krzysztof Lech, Antoni Szumny, Adam Figiel, and Ángel A. Carbonell-Barrachina. "Volatile Composition of Sweet Basil Essential Oil (Ocimum Basilicum L.) as Affected by Drying Method." *Food Research International* 48, no. 1 (August 1, 2012): 217–25. https://doi.org/10.1016/j.foodres.2012.03.015.

Eisenman, Sasha W., H. Rodolfo Juliani, Lena Struwe, and James E. Simon. "Essential Oil Diversity in North American Wild Tarragon (Artemisia Dracunculus L.) with Comparisons to French and Kyrgyz Tarragon." *Industrial Crops and Products* 49 (August 1, 2013): 220–32. https://doi.org/10.1016/j.indcrop.2013.04.037.

Eyres, Graham, Jean-Pierre Dufour, Gabrielle Hallifax, Subramaniam Sotheeswaran, and Philip J. Marriott. "Identification of Character-Impact Odorants in Coriander and Wild Coriander Leaves Using Gas Chromatography-Olfactometry (GCO) and Comprehensive Two-Dimensional Gas Chromatography–Time-of-Flight Mass Spectrometry (GC×GC–TOFMS)." *Journal of Separation Science* 28, no. 9–10 (2005): 1061–74. https://doi.org/10.1002/jssc.200500012.

Güntert, Matthias, Gerhard Krammer, Stefan Lambrecht, Horst Sommer, Horst Surburg, and Peter Werkhoff. "Flavor Chemistry of Peppermint Oil (Mentha Piperita L.)." In *Aroma Active Compounds in Foods*, edited by Gary R. Takeoka, Matthias Güntert, and Karl-Heinz Engel, 119–37. Washington, D.C.: American Chemical Society, 2001. https://doi.org/10.1021/bk-2001-0794.ch010.

Kothari, S. K., A. K. Bhattacharya, S. Ramesh, S. N. Garg, and S. P. S. Khanuja. "Volatile Constituents in Oil from Different Plant Parts of Methyl Eugenol-Rich Ocimum Tenuiflorum L.f. (Syn. O. Sanctum L.) Grown in South India." *Journal of Essential Oil Research* 17, no. 6 (November 1, 2005): 656–58. https://doi.org/10.1080/10412905.2005.9699025.

Lundgren, Lennart, and Gunnar Stenhagen. "Leaf Volatiles from Thymus Vulgaris, T. Serpyllum, T. Praecox, T. Pulegioides and T. x Citriodorus (Labiatae)." *Nordic Journal of Botany* 2, no. 5 (1982): 445–52. https://doi.org/10.1111/j.1756-1051.1982.tb01207.x.

Masanetz, Charly, and Werner Grosch. "Key Odorants of Parsley Leaves (Petroselinum Crispum [Mill.] Nym. Ssp. Crispum) by Odour–Activity Values." *Flavour and Fragrance Journal* 13, no. 2 (1998): 115–24. https://doi.org/10.1002/(SICI)1099-1026(199803/04)13:2<115::AID-FFJ706>3.0.CO;2-6.

Parthasarathy, V. A., B. Chempakam, and T. John Zachariah, eds. *Chemistry of Spices*. Wallingford, Eng.: CABI, 2008.

Ravi, Ramasamy, Maya Prakash, and Kodangala Keshava Bhat. "Characterization of Aroma Active Compounds of Cumin (Cuminum Cyminum L.) by GC-MS, E-Nose, and Sensory Techniques." *International Journal of Food Properties* 16, no. 5 (July 4, 2013): 1048–58. https://doi.org/10.1080/10942912.2011.576356.

Seo, Won Ho, and Hyung Hee Baek. "Characteristic Aroma-Active Compounds of Korean Perilla (Perilla Frutescens Britton) Leaf." *Journal of Agricultural and Food Chemistry* 57, no. 24 (November 16, 2009): 11537–42. https://doi.org/10.1021/jf902669d.

Singh, Gurdip, Om Prakash Singh, M. P. De Lampasona, and César A. N. Catalán. "Studies on Essential Oils. Part 35: Chemical and Biocidal Investigations on Tagetes Erecta Leaf Volatile Oil." *Flavour and Fragrance Journal* 18, no. 1 (2003): 62–65. https://doi.org/10.1002/ffj.1158.

Sonmezdag, Ahmet Salih, Hasim Kelebek, and Serkan Selli. "Characterization of Aroma-Active and Phenolic Profiles of Wild Thyme (Thymus Serpyllum) by GC-MS-Olfactometry and LC-ESI-MS/MS." *Journal of Food Science and Technology* 53, no. 4 (April 2016): 1957–65. https://doi.org/10.1007/s13197-015-2144-1.

Tanaka, Fukuyo, Toshio Miyazawa, and Yumi Ujiie. "Effect of Cultivation Conditions on Odor Character and Chemical Profile of Shiso (Perilla Frutescens) Flavor." In *Proceedings of the 19th World Congress of Soil Science, Soil Solutions for a Changing World*, 1–6. Brisbane, Australia, 2010.

Tohge, Takayuki, Mutsumi Watanabe, Rainer Hoefgen, and Alisdair R. Fernie. "The Evolution of Phenylpropanoid Metabolism in the Green Lineage." *Critical Reviews in Biochemistry and Molecular Biology* 48, no. 2 (March 2013): 123–52. https://doi.org/10.3109/10409238.2012.758083.

Weng, Jing-Ke, and Clint Chapple. "The Origin and Evolution of Lignin Biosynthesis." *New Phytologist* 187, no. 2 (July 2010): 273–85. https://doi.org/10.1111/j.1469-8137.2010.03327.x.

Meaty

Batzer, O. F., A. T. Santoro, M. C. Tan, W. A. Landmann, and B. S. Schweigert. "Precursors of Beef Flavor." *Journal of Agricultural and Food Chemistry* 8, no. 6 (1960).

Calkins, C. R., and J. M. Hodgen. "A Fresh Look at Meat Flavor." *Meat Science* 77, no. 1 (September 1, 2007): 63–80. https://doi.org/10.1016/j.meatsci.2007.04.016.

Giogios, I., N. Kalogeropoulos, and K. Grigorakis. "Volatile Compounds of Some Popular Mediterranean Seafood Species." *Mediterranean Marine Science* 14, no. 2 (June 11, 2013): 343. https://doi.org/10.12681/mms.342.

Kramlich, W. E., and A. M. Pearson. "Separation and Identification of Cooked Beef Flavor Components." *Journal of Food Science* 25, no. 6 (1960): 712–19. https://doi.org/10.1111/j.1365-2621.1960.tb00018.x.

Melton, Sharon L. "Effects of Feeds on Flavor of Red Meat: A Review." *Journal of Animal Science* 68, no. 12 (December 1, 1990): 4421–35. https://doi.org/10.2527/1990.68124421x.

Mottram, D. S., and R. A. Edwards. "The Role of Triglycerides and Phospholipids in the Aroma of Cooked Beef." *Journal of the Science of Food and Agriculture* 34, no. 5 (1983): 517–22. https://doi.org/10.1002/jsfa.2740340513.

Wasserman, Aaron E., and Natalie Gray. "Meat Flavor. I. Fractionation of Water-Soluble Flavor Precursors of Beef." *Journal of Food Science* 30, no. 5 (September 1965): 801–7. https://doi.org/10.1111/j.1365-2621.1965.tb01844.x.

Part 3

Adams, An, Zanda Kruma, Roland Verhé, Norbert De Kimpe, and Viesturs Kreicbergs. "Volatile Profiles of Rapeseed Oil Flavored with Basil, Oregano, and Thyme as a Function of Flavoring Conditions." *Journal of the American Oil Chemists' Society* 88, no. 2 (February 1, 2011): 201–12. https://doi.org/10.1007/s11746-010-1661-3.

Ahmad, Imtiyaz, Shariq Shamsi, and Roohi Zaman. "Sharbat: An Important Dosage Form of Unani System of Medicine." *Medical Journal of Islamic World Academy of Sciences* 24, no. 3 (2016): 83–88. https://doi.org/10.5505/ias.2016.91129.

Christoph, Norbert, and Claudia Bauer-Christoph. "Flavour of Spirit Drinks: Raw Materials, Fermentation, Distillation, and Ageing." In *Flavours and Fragrances: Chemistry, Bioprocessing and Sustainability*, edited by Ralf Günter Berger, 219–39. Berlin: Springer, 2007. https://doi.org/10.1007/978-3-540-49339-6_10.

Janado, Masanobu, and Toshiro Nishida. "Effect of Sugars on the Solubility of Hydrophobic Solutes in Water." *Journal of Solution Chemistry* 10, no. 7 (July 1, 1981): 489–500. https://doi.org/10.1007/BF00652083.

Janado, Masanobu, and Yuki Yano. "The Nature of the Cosolvent Effects of Sugars on the Aqueous Solubilities of Hydrocarbons." *Bulletin of the Chemical Society of Japan* 58, no. 7 (July 1, 1985): 1913–17. https://doi.org/10.1246/bcsj.58.1913.

Li, An, and Samuel H. Yalkowsky. "Solubility of Organic Solutes in Ethanol/Water Mixtures." *Journal of Pharmaceutical Sciences* 83, no. 12 (December 1, 1994): 1735–40. https://doi.org/10.1002/jps.2600831217.

Li, J., E. M. Perdue, S. G. Pavlostathis, and R. Araujo. "Physicochemical Properties of Selected Monoterpenes." *Environment International* 24, no. 3 (April 1, 1998): 353–58. https://doi.org/10.1016/S0160-4120(98)00013-0.

Rajeswara Rao, B. R., P. N. Kaul, K. V. Syamasundar, and S. Ramesh. "Water Soluble Fractions of Rose-Scented Geranium (Pelargonium Species) Essential Oil." *Bioresource Technology* 84, no. 3 (September 1, 2002): 243–46. https://doi.org/10.1016/S0960-8524(02)00057-3.

Shurtleff, William, and Akiko Aoyagi. *History of Soymilk and Other Non-Dairy Milks (1226 to 2013): Extensively Annotated Bibliography and Sourcebook.* Lafayette, Calif.: Soyinfo Center, 2013. http://books.google.com/books?id=MyJPwd72zhgC&printsec=frontcover&source=gbs_ge_summary_r&cad=0#v=onepage&q&f=false.

Yara-Varón, Edinson, Ying Li, Mercè Balcells, Ramon Canela-Garayoa, Anne-Sylvie Fabiano-Tixier, and Farid Chemat. "Vegetable Oils as Alternative Solvents for Green Oleo-Extraction, Purification and Formulation of Food and Natural Products." *Molecules* 22, no. 9 (September 2017): 1474. https://doi.org/10.3390/molecules22091474.

Yilmazer, Mustafa, Sermin Göksu Karagöz, Gulcan Ozkan, and Erkan Karacabey. "Aroma Transition from Rosemary Leaves During Aromatization of Olive Oil." *Journal of Food and Drug Analysis* 24, no. 2 (April 1, 2016): 299–304. https://doi.org/10.1016/j.jfda.2015.11.002.

Part 4

Heat

Adams, An, and Norbert De Kimpe. "Chemistry of 2-Acetyl-1-Pyrroline, 6-Acetyl-1,2,3,4-Tetrahydropyridine, 2-Acetyl-2-Thiazoline, and 5-Acetyl-2,3-Dihydro-4H-Thiazine: Extraordinary Maillard Flavor Compounds." *Chemical Reviews* 106, no. 6 (June 1, 2006): 2299–2319. https://doi.org/10.1021/cr040097y.

Ames, Jennifer M., Robin C. E. Guy, and Gary J. Kipping. "Effect of pH, Temperature, and Moisture on the Formation of Volatile Compounds in Glycine/Glucose Model Systems." *Journal of Agricultural and Food Chemistry* 49, no. 9 (September 1, 2001): 4315–23. https://doi.org/10.1021/jf010198m.

"Analysis of Aroma Compounds in Trapping Solvents of Smoke from Tian Op, a Traditional, Thai, Scented Candle." Accessed November 2, 2019. http://kasetsartjournal.ku.ac.th/kuj_files/2009/a0912011425314687.pdf.

Blank, Imre. "The Role of pH in Maillard-Type Reactions." Presented at the J. de Clerck Symposium XI, Louvain, Belgium, September 8, 2004. http://www.imreblank.ch/11th_De_Clerck_08.09.2004_short.pdf.

———. "Recent Developments in the Maillard Reaction." Presented at the 31st FNK Europe Conference on Potato Processing Leiden, The Netherlands, November 19, 1997. http://www.imreblank.ch/31st_FNK_Leiden_part2.pdf. Accessed October 31, 2019.

Buttery, Ron G., and Louisa C. Ling. "Volatile Flavor Components of Corn Tortillas and Related Products." *Journal of Agricultural and Food Chemistry* 43, no. 7 (July 1, 1995): 1878–82. https://doi.org/10.1021/jf00055a023.

Chitsamphandvej, Winyu, Patcha Saichunyoon, and Suvalak Asavasanti. "Volatile Organic Compounds of Charcoal Combustion Smoke." *Proceedings: The Pure and Applied Chemistry International Conference 2017 (PACCON2017)* 2017, no. 1 (2017).

Chua, XinLing, Elizabeth Uwiduhaye, Petroula Tsitlakidou, Stella Lignou, Huw D. Griffiths, David A. Baines, and Jane K. Parker. "Changes in Aroma and Sensory Profile of Food Ingredients Smoked in the Presence of a Zeolite Filter." In *Sex, Smoke, and Spirits: The Role of Chemistry*, edited by Brian Guthrie et al., 67–79. Washington, D.C.: American Chemical Society, 2019. https://doi.org/10.1021/bk-2019-1321.ch006.

Cutzach, Isabelle, Pascal Chatonnet, Robert Henry, and Denis Dubourdieu. "Identification of Volatile Compounds with a 'Toasty' Aroma in Heated Oak Used in Barrelmaking." *Journal of Agricultural and Food Chemistry* 45, no. 6 (June 1997): 2217–24. https://doi.org/10.1021/jf960947d.

Evtyugina, Margarita, Célia Alves, Ana Calvo, Teresa Nunes, Luís Tarelho, Márcio Duarte, Sónia O. Prozil, Dmitry V. Evtuguin, and Casimiro Pio. "VOC Emissions from Residential Combustion of Southern and Mid-European Woods." *Atmospheric Environment* 83 (February 1, 2014): 90–98. https://doi.org/10.1016/j.atmosenv.2013.10.050.

Fang, M. X., D. K. Shen, Y. X. Li, C. J. Yu, Z. Y. Luo, and K. F. Cen. "Kinetic Study on Pyrolysis and Combustion of Wood Under Different Oxygen Concentrations by Using TG-FTIR Analysis." *Journal of Analytical and Applied Pyrolysis* 77, no. 1 (August 1, 2006): 22–27. https://doi.org/10.1016/j.jaap.2005.12.010.

Ghadiriasli, R., K. Lorber, M. Wagenstaller, and A. Buettner. "Smoky, Vanilla, or Clove-Like?" In *Sex, Smoke, and Spirits: The Role of Chemistry*, edited by Brian Guthrie et al., 43–54. Washington, D.C.: American Chemical Society, 2019. https://doi.org/10.1021/bk-2019-1321.ch004.

Jiang, Bin, Yeting Liu, Bhesh Bhandari, and Weibiao Zhou. "Impact of Caramelization on the Glass Transition Temperature of Several Caramelized Sugars. Part I: Chemical Analyses." *Journal of Agricultural and Food Chemistry* 56, no. 13 (July 1, 2008): 5138–47. https://doi.org/10.1021/jf703791e.

Kung, Hsiang-Cheng. "A Mathematical Model of Wood Pyrolysis." *Combustion and Flame* 18, no. 2 (April 1, 1972): 185–95. https://doi.org/10.1016/S0010-2180(72)80134-2.

Lee, Joo Won, Leonard C. Thomas, John Jerrell, Hao Feng, Keith R. Cadwallader, and Shelly J. Schmidt. "Investigation of Thermal Decomposition as the Kinetic Process That Causes the Loss of Crystalline Structure in Sucrose Using a Chemical Analysis Approach (Part II)." *Journal of Agricultural and Food Chemistry* 59, no. 2 (January 26, 2011): 702–12. https://doi.org/10.1021/jf104235d.

Lee, Joo Won, Leonard C. Thomas, and Shelly J. Schmidt. "Investigation of the Heating Rate Dependency Associated with the Loss of Crystalline Structure in Sucrose, Glucose, and Fructose Using a Thermal Analysis Approach (Part I)." *Journal of Agricultural and Food Chemistry* 59, no. 2 (January 26, 2011): 684–701. https://doi.org/10.1021/jf1042344.

Mall, V., and P. Schieberle. "On the Importance of Phenol Derivatives for the Peaty Aroma Attribute of Scotch Whiskies from Islay." In *Sex, Smoke, and Spirits: The Role of Chemistry*, edited by Brian Guthrie et al., 107–16. Washington, D.C.: American Chemical Society, 2019. https://doi.org/10.1021/bk-2019-1321.ch009.

McGee, Harold. "Caramelization: New Science, New Possibilities." *Curious Cook*, September 13, 2012. https://www.curiouscook.com/site/.

Pons, Isabelle, Christian Garrault, Jean-Noel Jaubert, Jean Morel, and Jean-Claude Fenyo. "Analysis of Aromatic Caramel." *Food Chemistry* 39, no. 3 (January 1, 1991): 311–20. https://doi.org/10.1016/0308-8146(91)90148-H.

Prins, Mark J., Krzysztof J. Ptasinski, and Frans J. J. G. Janssen. "Torrefaction of Wood: Part 2. Analysis of Products." *Journal of Analytical and Applied Pyrolysis* 77, no. 1 (August 1, 2006): 35–40. https://doi.org/10.1016/j.jaap.2006.01.001.

Shafizadeh, Fred. "The Chemistry of Pyrolysis and Combustion." In *The Chemistry of Solid Wood*, edited by Roger Rowell, 489–529. Washington,

D.C.: American Chemical Society, 1984. https://doi.org/10.1021/ba-1984-0207.ch013.

Watcharananun, Wanwarang, Keith R. Cadwallader, Kittiphong Huangrak, Hun Kim, and Yaowapa Lorjaroenphon. "Identification of Predominant Odorants in Thai Desserts Flavored by Smoking with 'Tian Op,' a Traditional Thai Scented Candle." *Journal of Agricultural and Food Chemistry* 57, no. 3 (February 11, 2009): 996–1005. https://doi.org/10.1021/jf802674c.

Fermentation

Corbo, Maria Rosaria, Angela Racioppo, Noemi Monacis, and Barbara Speranza. "Commercial Starters or Autochtonous Strains? That Is the Question." In *Starter Cultures in Food Production*, edited by Barbara Speranza, 174–98. Chichester, Eng.: John Wiley, 2017. https://doi.org/10.1002/9781118933794.ch10.

Demarigny, Yann. "Fermented Food Products Made with Vegetable Materials from Tropical and Warm Countries: Microbial and Technological Considerations." *International Journal of Food Science and Technology* 47, no. 12 (2012): 2469–76. https://doi.org/10.1111/j.1365-2621.2012.03087.x.

Escamilla-Hurtado, M. L., A. Tomasini-Campocosio, S. Valdés-Martínez, and J. Soriano-Santos. "Diacetyl Formation by Lactic Bacteria." *Revista Latinoamericana de Microbiologia* 38, no. 2 (April 1, 1996): 129–37.

Han, Thazin, and Kyaw Nyein Aye. "The Legend of Laphet: A Myanmar Fermented Tea Leaf." *Journal of Ethnic Foods* 2, no. 4 (December 1, 2015): 173–78. https://doi.org/10.1016/j.jef.2015.11.003.

Huang, Hsing Tsung, and Joseph Needham. "Part V: Fermentations and Food Science." In *Biology and Biological Technology*. Vol. 6 of *Science and Civilisation in China*. Cambridge: Cambridge University Press, 2000.

Jeong, Sang Hyeon, Hyo Jung Lee, Ji Young Jung, Se Hee Lee, Hye-Young Seo, Wan-Soo Park, and Che Ok Jeon. "Effects of Red Pepper Powder on Microbial Communities and Metabolites During Kimchi Fermentation." *International Journal of Food Microbiology* 160, no. 3 (January 1, 2013): 252–59. https://doi.org/10.1016/j.ijfoodmicro.2012.10.015.

Kim, Kyung Min, Jaeho Lim, Jae Jung Lee, Byung-Serk Hurh, and Inhyung Lee. "Characterization of Aspergillus Sojae Isolated from Meju, Korean Traditional Fermented Soybean Brick." *Journal of Microbiology and Biotechnology* 27, no. 2 (February 28, 2017): 251–61. https://doi.org/10.4014/jmb.1610.10013.

Kwon, Dae Young. "Scientific Knowledge in Traditional Fermented Foods." *Journal of Ethnic Foods* 5, no. 3 (September 1, 2018): 153–54. https://doi.org/10.1016/j.jef.2018.09.001.

Kwon, Dae Young, Dai-Ja Jang, Hye Jeong Yang, and Kyung Rhan Chung. "History of Korean Gochu, Gochujang, and Kimchi." *Journal of Ethnic Foods* 1, no. 1 (December 1, 2014): 3–7. https://doi.org/10.1016/j.jef.2014.11.003.

Kyung, Kyu Hang, Eduardo Medina Pradas, Song Gun Kim, Yong Jae Lee, Kyong Ho Kim, Jin Joo Choi, Joo Hyong Cho, Chang Ho Chung, Rodolphe Barrangou, and Frederick Breidt. "Microbial Ecology of Watery Kimchi: Watery Kimchi (*Nabak* and *Dongchimi*) Ecology. . . ." *Journal of Food Science* 80, no. 5 (May 2015): M1031–38. https://doi.org/10.1111/1750-3841.12848.

Lee, Sang-Sun. "Meju Fermentation for a Raw Material of Korean Traditional Soy Products." *Korean Journal of Mycology* 23, no. 2 (1995): 161–75.

Machida, Masayuki, Osamu Yamada, and Katsuya Gomi. "Genomics of Aspergillus Oryzae: Learning from the History of Koji Mold and Exploration of Its Future." *DNA Research* 15, no. 4 (August 1, 2008): 173–83. https://doi.org/10.1093/dnares/dsn020.

Majumdar, Ranendra K., Deepayan Roy, Sandeep Bejjanki, and Narayan Bhaskar. "An Overview of Some Ethnic Fermented Fish Products of the Eastern Himalayan Region of India." *Journal of Ethnic Foods* 3, no. 4 (December 1, 2016): 276–83. https://doi.org/10.1016/j.jef.2016.12.003.

Montville, T. J. "Interaction of pH and NaCl on Culture Density of Clostridium Botulinum 62A." *Applied and Environmental Microbiology* 46, no. 4 (October 1, 1983): 961–63.

Narahara, Hideki, Yosuke Koyama, Toshimi Yoshida, Sumalee Pichanigkura, Ryuzo Ueda, and Hisaharu Taguchi. "Growth and Enzyme Production in a Solid-State Culture of Aspergillus Oryzae" *Journal of Fermentation Technology* 60, no. 4 (1982): 311–19.

Pérez-Díaz, I. M., Fred Breidt, R. W. Buescher, F. N. Arroyo-López, R. Jiménez-Díaz, A. Garrido-Fernández, J. Bautista-Gallego, S.-S. Yoon, and S. D. Johanningsmeier. "51. Fermented and Acidified Vegetables." In *Compendium of Methods for the Microbiological Examination of Foods*, 4th ed., 531–32. Washington, D.C.: American Public Health Association, 2013. https://doi.org/10.2105/MBEF.0222.056.

Rhee, Sook Jong, Jang-Eun Lee, and Cherl-Ho Lee. "Importance of Lactic Acid Bacteria in Asian Fermented Foods." *Microbial Cell Factories* 10, no. 1 (August 30, 2011): S5. https://doi.org/10.1186/1475-2859-10-S1-S5.

Shin, Donghwa, and Doyoun Jeong. "Korean Traditional Fermented Soybean Products: Jang."

Journal of Ethnic Foods 2, no. 1 (March 1, 2015): 2–7. https://doi.org/10.1016/j.jef.2015.02.002.

Shurtleff, William, and Akiko Aoyagi. *History of Koji—Grains and/or Soybeans Enrobed with a Mold Culture (300 BCE to 2012): Extensively Annotated Bibliography and Sourcebook*. Lafayette, Calif.: Soyinfo Center, 2012. http://www.soyinfocenter.com/pdf/154/Koji.pdf.

Snyder, Abigail, Fred Breidt Jr., Elizabeth L. Andress, and Barbara H. Ingham. "Manufacture of Traditionally Fermented Vegetable Products: Best Practice for Small Businesses and Retail Food Establishments." *Food Protection Trends* 40, no. 4 (July 1, 2020): 251–63.

Sun, Shu Yang, Wen Guang Jiang, and Yu Ping Zhao. "Profile of Volatile Compounds in 12 Chinese Soy Sauces Produced by a High-Salt-Diluted State Fermentation." *Journal of the Institute of Brewing* 116, no. 3 (2010): 316–28. https://doi.org/10.1002/j.2050-0416.2010.tb00437.x.

Tamang, J. P., and P. K. Sarkar. "Microflora of Murcha: An Amylolytic Fermentation Starter." *Microbios* 81 (1995): 115–22.

Tamang, Jyoti P., Koichi Watanabe, and Wilhelm H. Holzapfel. "Review: Diversity of Microorganisms in Global Fermented Foods and Beverages." *Frontiers in Microbiology* 7 (2016). https://doi.org/10.3389/fmicb.2016.00377.

Xu, Yan, Dong Wang, Wen Lai Fan, Xiao Qing Mu, and Jian Chen. "Traditional Chinese Biotechnology." In *Biotechnology in China II: Chemicals, Energy and Environment*, edited by G. T. Tsao, Pingkai Ouyang, and Jian Chen, 189–233. Berlin: Springer Berlin Heidelberg, 2010. https://doi.org/10.1007/10_2008_36.

Yanfang, Z., and T. Wenyi. "Flavor and Taste Compounds Analysis in Chinese Solid Fermented Soy Sauce." *African Journal of Biotechnology* 8, no. 4 (January 1, 2009). https://www.ajol.info/index.php/ajb/article/view/59911.

Index